Co-ordinated SCIENCE

Biology

second edition

Brian Beckett RoseMarie Gallagher

Oxford University Press

Oxford University Press, Great Clarendon Street, Oxford OX2 6DP

Oxford New York
Athens Auckland Bangkok Bogota Bombay
Buenos Aires Calcutta Cape Town Dar es Salaam Delhi
Florence Hong Kong Istanbul Karachi
Kuala Lumpur Madras Madrid Melbourne
Mexico City Nairobi Paris Singapore
Taipei Tokyo Toronto

and associated companies in
Berlin Ibadan

Oxford is a trade mark of Oxford University Press

First published 1989 (ISBN 0 19 914317 X)
Reprinted 1991, 1992, 1993, 1994

2nd edition published 1996, reprinted 1996, 1997
ISBN 0 19 914653 5 School edition
ISBN 0 19 914674 8 Book Shop edition

Typeset in 11/13pt Palatino by Tradespools Ltd
Printed in Spain by Gráficas Estella

Introduction

Biology is an exciting subject. In it you can discover the mysteries of the living world, find out how living things evolved, and how they survive. Learn about their structures and the way they function.

You will find the book useful if you are studying biology as part of a GCSE Co-ordinated Science course, or as a single GCSE subject.

Everything in this book has been organized to help you find things quickly and easily. It is written in two-page units. Each unit is about a topic that you are likely to study. The units are grouped into sections.

- **Use the contents page**

If you are looking for information on a large topic, look it up in the contents list.

- **Use the index**

If there is something small you want to check on, look up the most likely word in the index. The index gives the page where you'll find more information.

- **Test yourself**

There are questions within units, and at the end of each section. The end-of-section questions test you on areas such as factual recall, understanding, data interpretation, and hypothesis formulation.

We hope that this book will give you a greater understanding of what biology is about, and that you enjoy studying it.

Brian Beckett
RoseMarie Gallagher

January 1996

Contents

6 Feeding and digestion

7 Senses and co-ordination

8 Reproduction

9 Living things and their environment

10 Health and hygiene

1.1 Living things

You are a **living thing**. Grass, whales, and bats are living things too. But stones and rain are non-living things. Living things are different from non-living things in the ways shown below.

Living things move and have senses

Animals walk, or run, or hop, or crawl, or swim, or fly. They find their way using **sense organs**. These are eyes, ears, noses, taste buds, skin, and insect feelers called **antennae**.

Plants move by growing, like these beans growing up bean poles. They don't have sense organs but they can still respond to things. Roots grow down in response to gravity, and to find water. Shoots grow up towards light.

Living things feed

They need food for energy, growth, and repair.

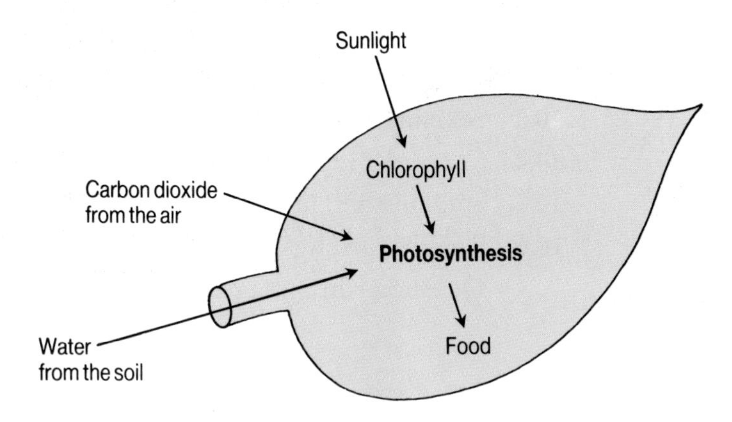

Plants make their own food in their leaves. This is called **photosynthesis**. It needs light, water, carbon dioxide, and a green chemical called **chlorophyll** which is found in leaves.

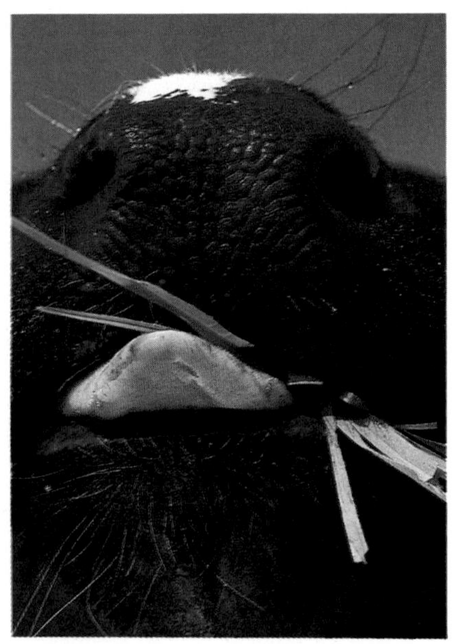

Animals can't make their own food so they eat plants and other animals. Which animal is this? What is it eating?

Living things respire

They get energy from food by a process called **respiration**. This usually needs oxygen.

Food + Oxygen ⇨ ENERGY + Waste: water and carbon dioxide

Living things excrete

All living things produce waste. The removal of waste from their bodies is called **excretion**.

Animals excrete through their lungs and kidneys, and through their skin when they sweat.

Plants store waste in old leaves, which fall in the autumn.

Plants grow all their lives. This giant redwood tree has been growing for over 2000 years!

Living things reproduce and grow

Animals lay eggs, or have babies. Seeds from plants grow into new plants.

Animals stop growing when they reach their adult size.

Questions

1 Name seven ways in which living things are different from non-living things.

2 Name the green stuff, and three other things plants need to make food.

3 Name all your sense organs.

4 What is:
a) respiration?
b) excretion?

1.2 Living things and their needs

Our planet is a huge ball of rock spinning in circles around the Sun. Below is a photograph of it taken from outer space.

This huge ball of rock is home for countless billions of living creatures. They can live on it because it has the six things which living creatures need.

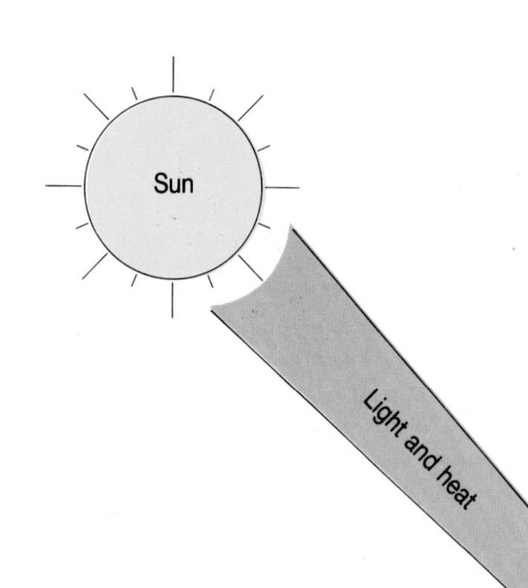

Light and carbon dioxide

The Sun's light provides energy for all living things. Plants need light, and carbon dioxide from the air, to make food by photosynthesis. Animals get their energy by eating plants, or by eating animals which eat plants.

Warmth

If it gets too hot or too cold the chemical changes which are necessary for life will stop. In many parts of the Earth temperatures lie between 25°C and 30°C. Most living things are adapted to live at these temperatures.

Water

Our world has plenty of water. All living things need water because the chemical changes necessary for life take place in it. Our bodies are three-quarters water.

Oxygen

Air is about one-fifth oxygen. Most living things need oxygen to get energy from their food by respiration.

Minerals

The Earth's soils contain minerals. These are essential for health and growth. Plants take in minerals through their roots. Animals get minerals by eating plants or other animals.

The biosphere

The **biosphere** is all those parts of the Earth's surface where living things are found. In fact they are found almost everywhere from about 9000 metres up mountains to at least 5000 metres under the sea.

This diagram shows a slice through the biosphere, from the highest to the lowest places where creatures can live.

Life on land

Springtails and mites can live up to 9000 metres feeding on pollen and seeds carried up by the wind.

Most flowering plants stop at about 6000 metres. Mosses and lichens can live much higher up.

Humans keep animals and grow crops up to 4500 metres.

Conifer forests stop at about 1500 metres. Deciduous forests stop at about 1000 metres.

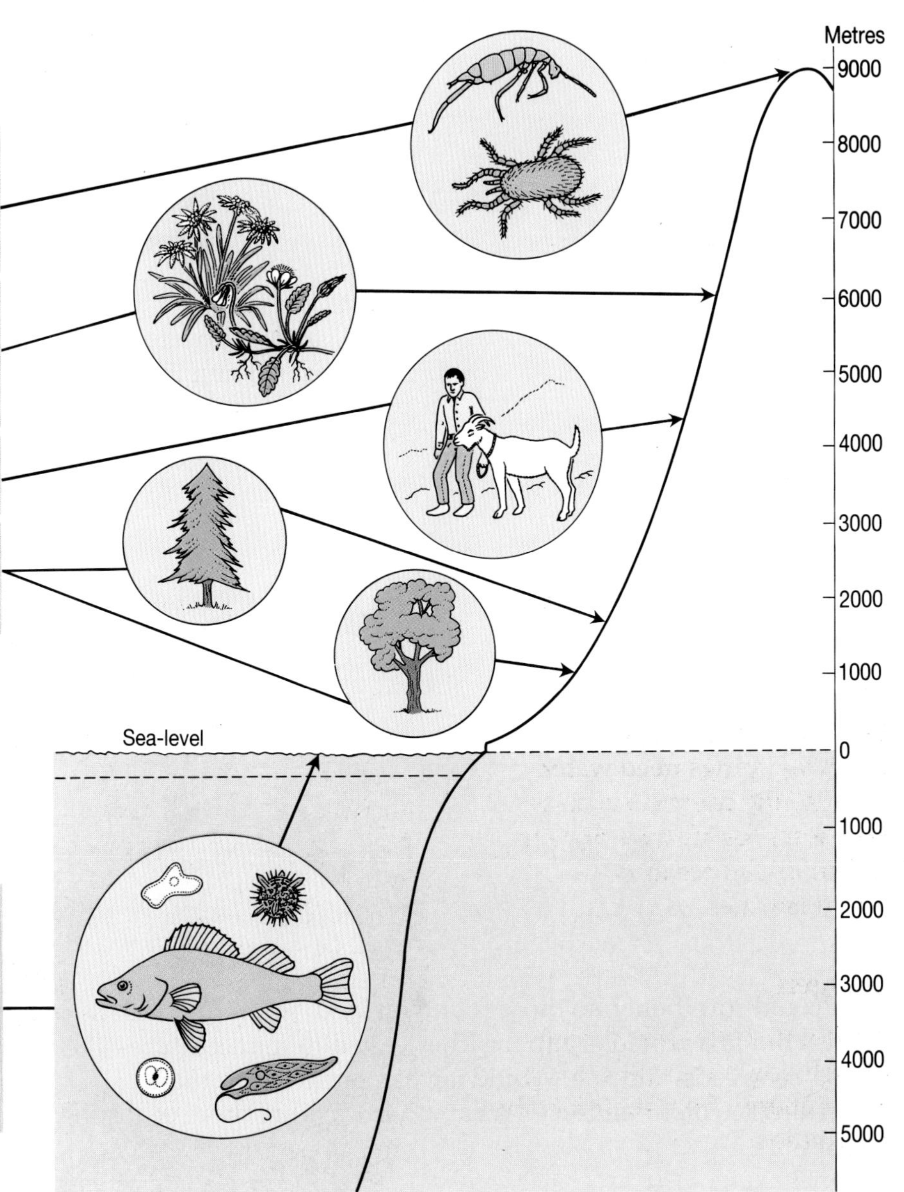

Life in water

Deep water has an upper sunlit zone where photosynthesis is possible, and a dark lower zone.

In clear water the sunlit zone can be up to 100 metres deep. Millions of tiny plant-like creatures called **phytoplankton** live here. They are eaten by tiny animals called **zooplankton**.

Both types of plankton are eaten by larger creatures such as fish.

The pitch dark lower zone is home to creatures which live on dead and living things which fall from the sunlit zone above, as well as cast-off skins, droppings etc.

Question

The Earth is home to billions of creatures but there are no living things on the moon. Write down the six things needed for life. Which of them do you think are found on the moon and which are not?

1.3 Sorting and naming

Spiny anteater.

Jumping spider.

These are just two of the things that live on Earth. Altogether there are over one-and-a-half million different kinds of living thing!

To make it easier to study them biologists sort living things into groups which are alike in some way. Try sorting the six creatures below into groups. Do this now.

You could sort them into those found on land, in the air, and in water. But this would group together very different creatures like crabs, seaweeds, and fish. Would it make sense to sort them according to colour and size? What other ways are there of sorting them?

Biologists first sort living things into five large groups called **kingdoms**. The five kingdoms are:

---

1 Prokaryotes 2 Protoctists 3 Fungi 4 Plants 5 Animals

These huge groups are then split into smaller groups. You can see how this is done in the rest of this chapter.

Naming living things

If you want to find the name of a plant or animal you could look through books until you see a picture of it. Or you could use a **key**.

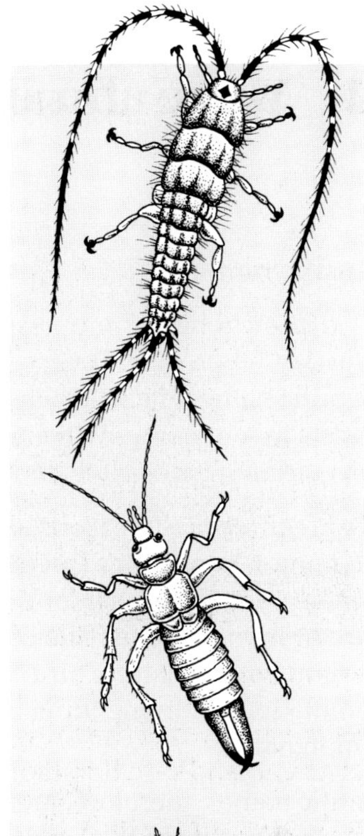

Keys

The simplest keys are made up of short, numbered sentences arranged in pairs. Look at the example below. Read the instructions, then use this key to name the insects drawn on this page.

How to use the key

Read the first pair of descriptions and decide which fit the insect you are trying to name. Opposite the description you choose there is a number. This number tells you which pair of descriptions to read next. Read them, and again decide which describes the insect. Opposite, you will find either the insect's name, or the number of the next pair of descriptions to read. Carry on until you find the insect's name.

An example of a key

1	Wings visible	3
	Wings not visible	2
2	Three-pronged tail	Bristle tail
	Pincers at end of tail	Earwig
3	Two pairs of wings	4
	One pair of wings	5
4	Wings fringed with hairs	Thrip
	Wings not fringed with hairs	6
5	Legs longer than body	Cranefly
	Legs not longer than body	Housefly
6	Wings larger than body	Butterfly
	Wings not larger than body	Wasp

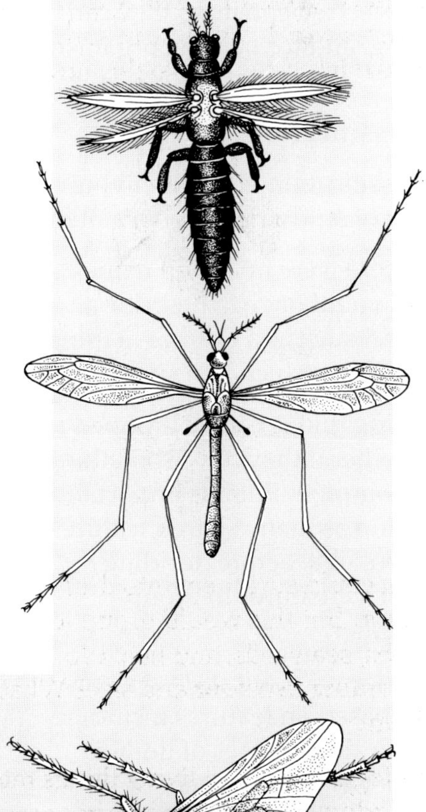

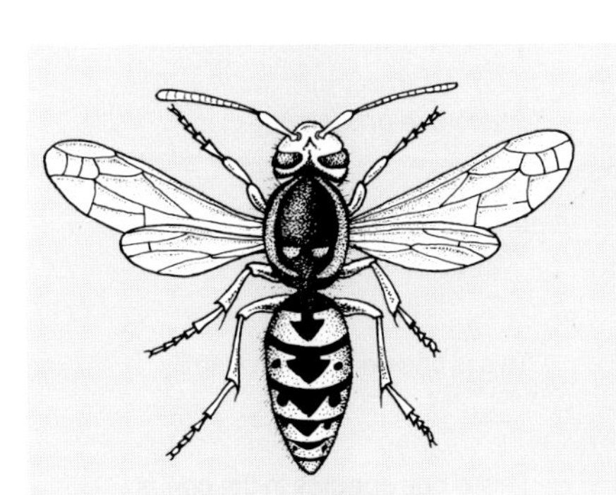

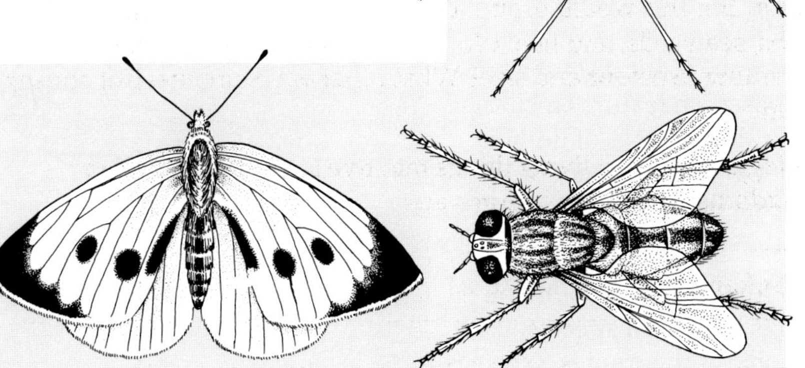

1.4 More about sorting and naming

Scientific names

Do you know what a *Taraxacum officinale* is? It's the scientific name for the dandelion plant. Every living thing known to science is given a scientific name which is then used throughout the world. Why do scientists invent names which are difficult to learn and say, and why aren't common names like dandelion good enough?

There are three main reasons. First, many creatures have more than one common name: they can have several different English names and many names in other languages if found abroad. Second, the same common name may be used for different organisms (see photographs opposite). Third, scientists have found over 1 500 000 different kinds of living thing but only a few have been given a common name.

Scientific names, therefore, are essential for communication between scientists so they can be sure they are discussing the same thing, and for naming organisms which have no common name.

The name 'robin' is given to a type of thrush in North America, and a type of flycatcher in Australia.

An American elk. What we call an elk is called a moose by North Americans, and they use the name elk for what we call a red deer.

Classification

Classification is sorting living things into groups. This makes the task of studying the enormous variety of life far easier.

If there were no group names we would have to describe an object such as a chair as 'a thing with legs which you sit on'.

Living things can be grouped together if they have something in common. But it is important to find as many shared features as possible before deciding which creatures to group together.

Modern classification systems are based on features including body shape, types of limbs, skeleton, arrangement of internal organs and many other features.

The smallest group of living things is called a **species.** Each dot on this diagram is a species.

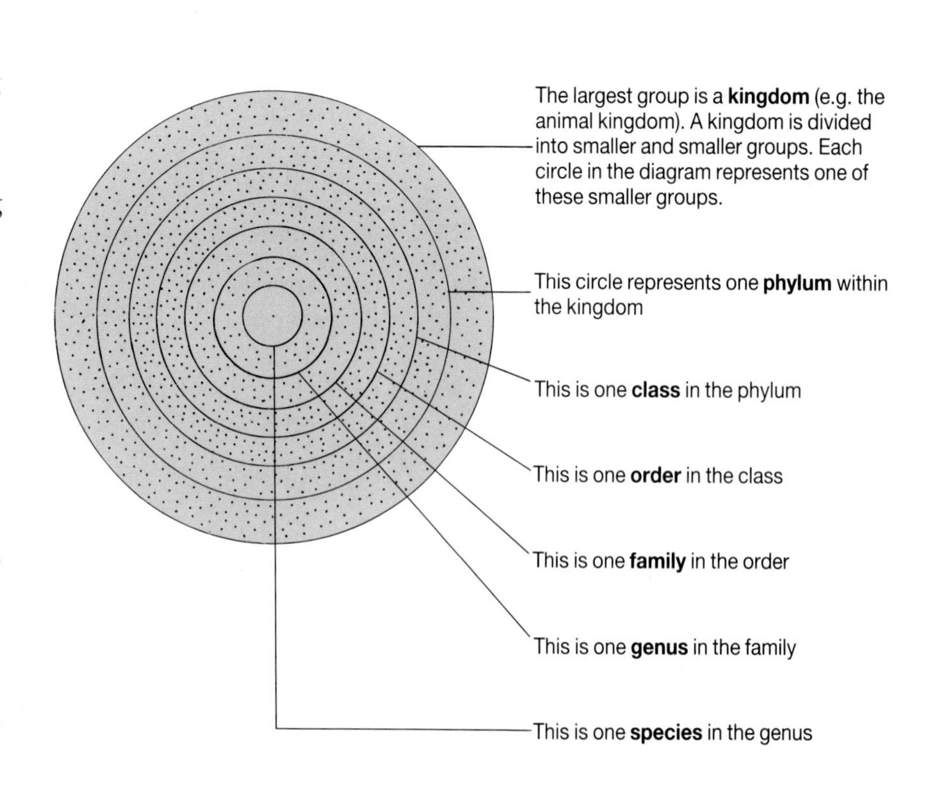

How to classify living things

A kingdom contains a large number of organisms but they have only a few things in common. The **plant kingdom** contains thousands of species, but one of the few features they share is the ability to make food by photosynthesis.

Species

A species is a group of living things which have so many features in common that they can mate together and produce young which are fertile (they too can mate and have young: see the photograph on the right). Humans, dogs, cats and sunflowers are all examples of species.

Every species is given two scientific names: the first is for its genus and the second for its species. *Homo sapiens* is the scientific name for humans.

Humans, monkeys, and apes belong to a group called **primates**. How are humans different from monkeys and apes?

Horses and donkeys are different species but are so alike they can breed. But their young, called mules (above), are infertile.

Question – more about keys

You can learn more about keys by making one for yourself. Study the leaves illustrated below. List those features which make each leaf different from all the others. The table opposite will help you do this. Next, arrange these features together in pairs, like the key in Unit 1.3.

Features	A	B	C	D	E
Leaf made up of small leaflets					
Leaf edge with many teeth					
Leaf edge divided into large teeth					
Leaf veins parallel or branched					

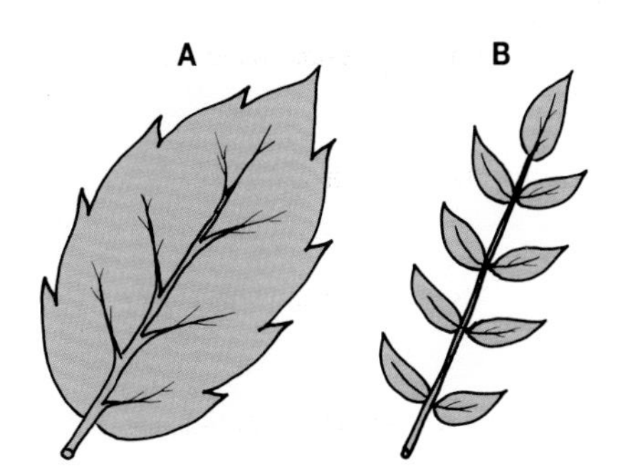

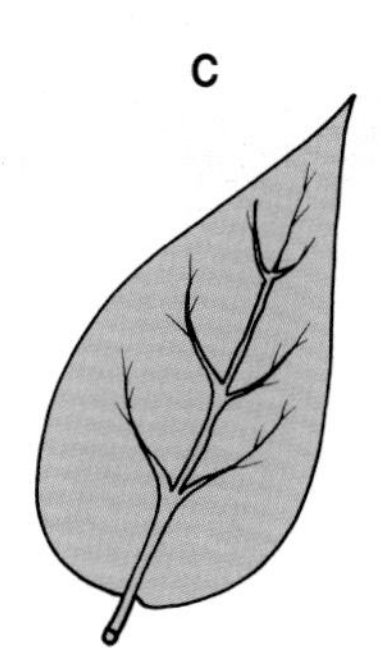

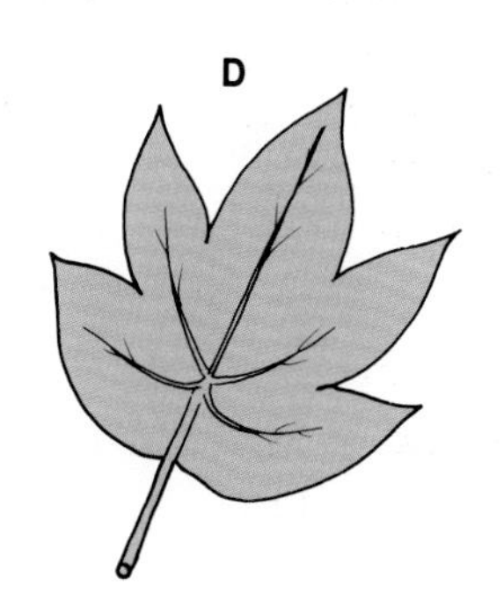

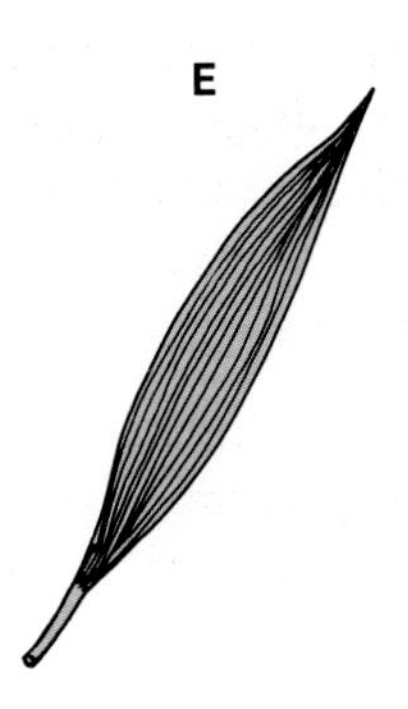

1.5 Groups of living things

This Unit illustrates living things in four of the five kingdoms, and then shows how these can be split into smaller and smaller groups.

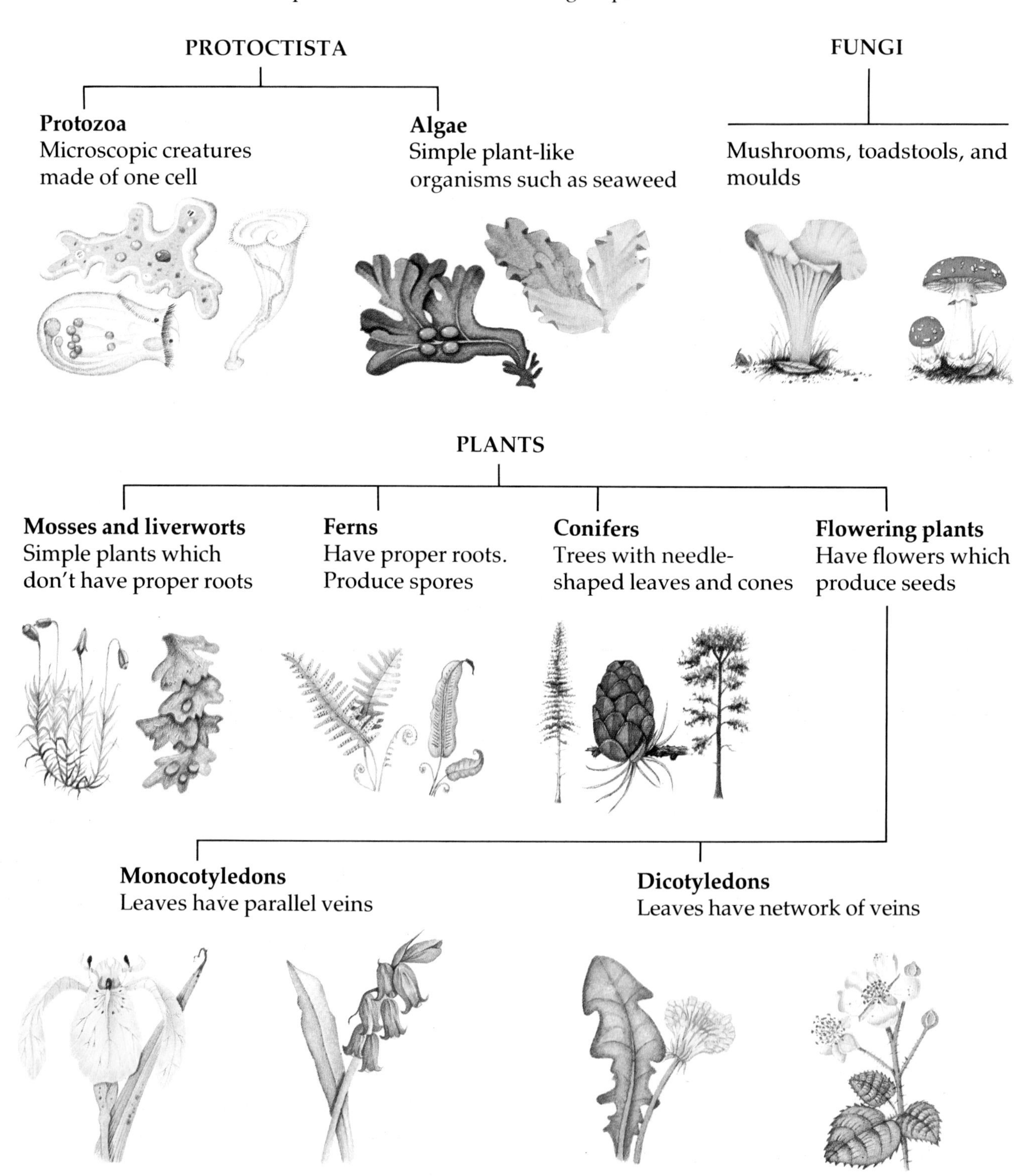

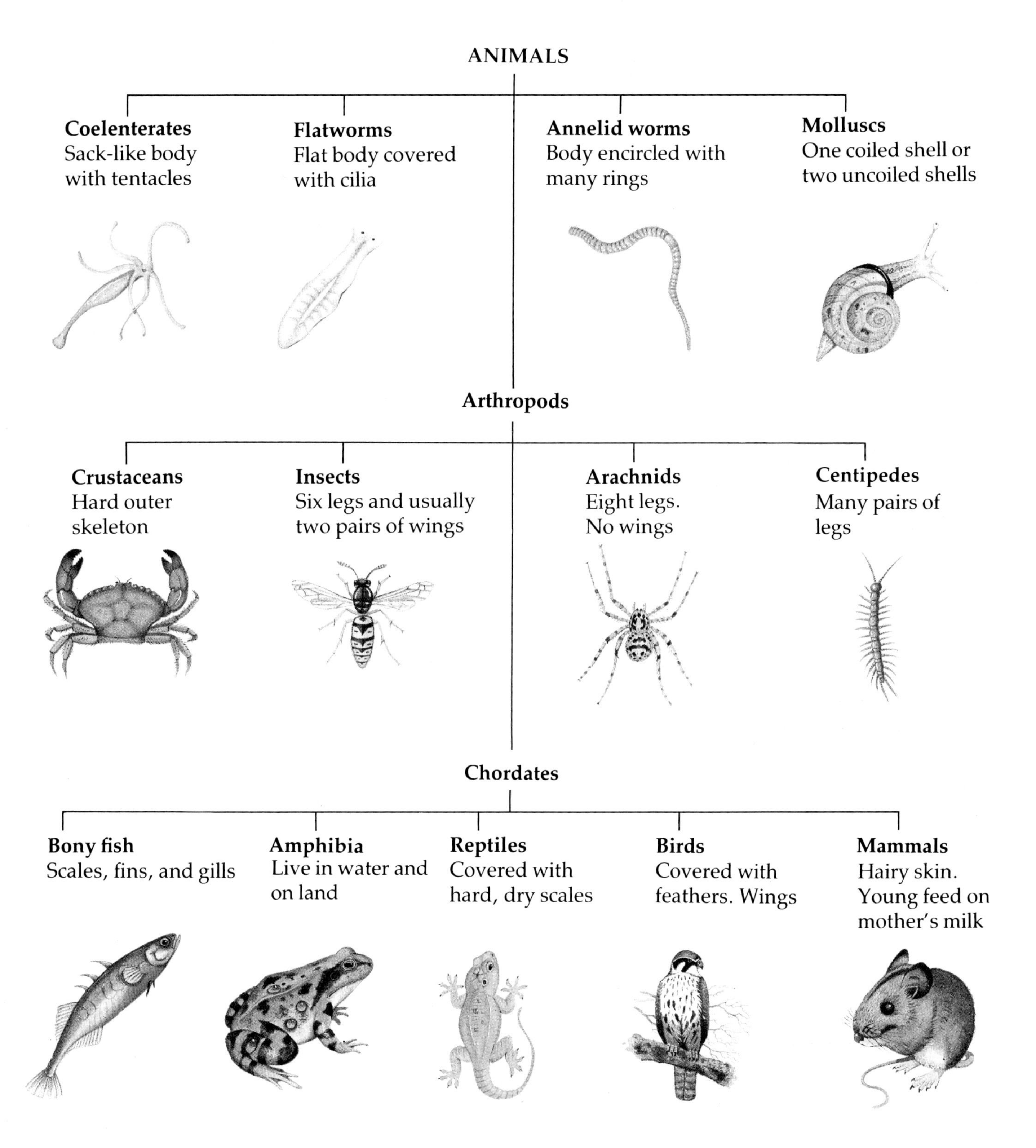

Questions

1 What is the word for:
 a) animals with backbones?
 b) animals without backbones?
2 Do fungi belong to the plant group?
3 Which group do you belong to?
4 Which animals have four pairs of legs?
5 Which plants don't have flowers?
6 What is a dicotyledon?

1.6 Protoctista and fungi

Protoctists

Protoctists are either made up of only one cell or many similar cells. The only feature they share is that their cells contain a nucleus.

Protozoa are microscopic one-celled protoctists which live like animals by eating bacteria and other tiny creatures. *Amoeba* and *Paramecium* are examples. **Euglenoids** are one-celled protoctists which often contain chlorophyll and so can make their own food by photosynthesis like plants do. *Euglena* is an example.

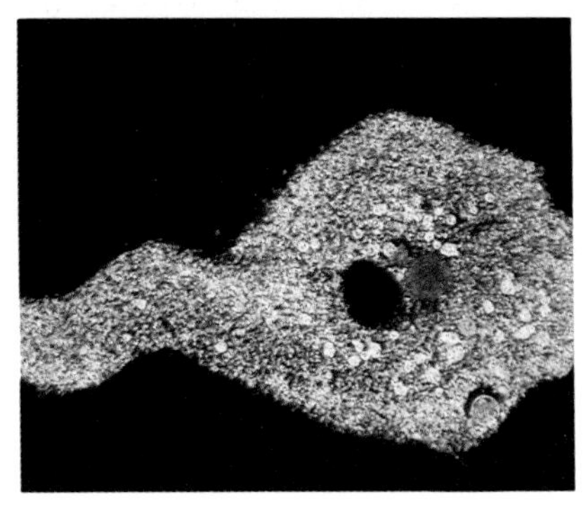

This is an *Amoeba*. It moves by changing its shape. It feeds on bacteria and other tiny creatures.

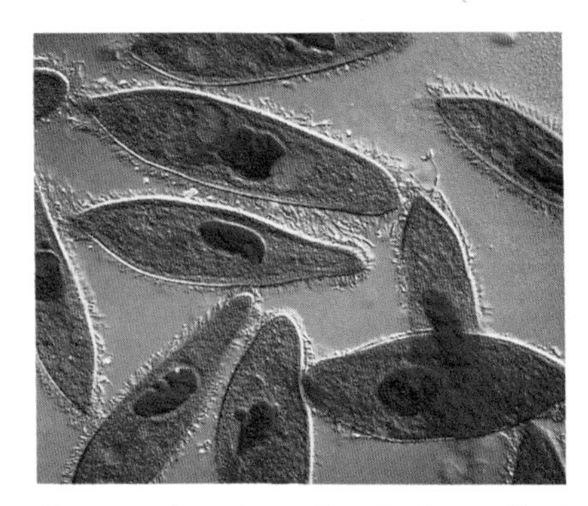

Paramecium has tiny hairs called cilia all over it. These trap its food and help it move.

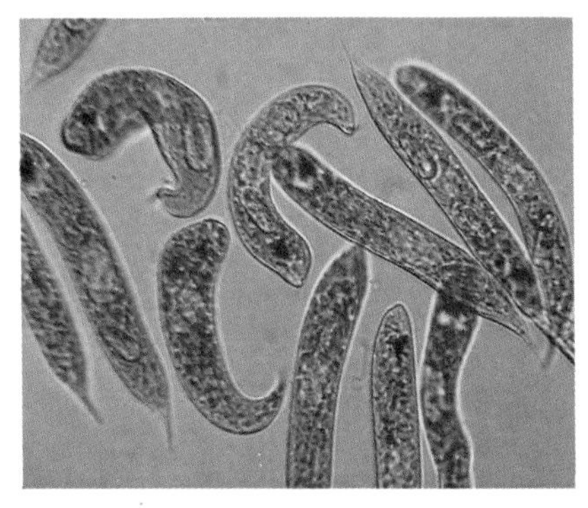

Euglena has chlorophyll and a single tiny hair called a flagellum which it uses like a propeller to help it move.

The red, green, and brown algae are protoctists which can make food by photosynthesis like plants, but are not really plants because they have no proper roots, stems, or leaves.

Fucus, or bladderwrack, is a brown alga which grows on rocky sea-shores and estuaries.

Under a microscope the green alga *Spirogyra* looks like tiny threads. It grows in ponds.

The red alga *Calliblepharis ciliata* is found attached to rocks on the sea-shore.

Fungi

All fungi are made up of fine threads called **hyphae** (pronounced hi-fee). It is easy to see the hyphae in moulds, but in mushrooms and toadstools hyphae are packed tightly together.

Fungi reproduce by forming **spores.** These are tiny cells that get scattered by the wind. When they land in the right place they grow into more fungi.

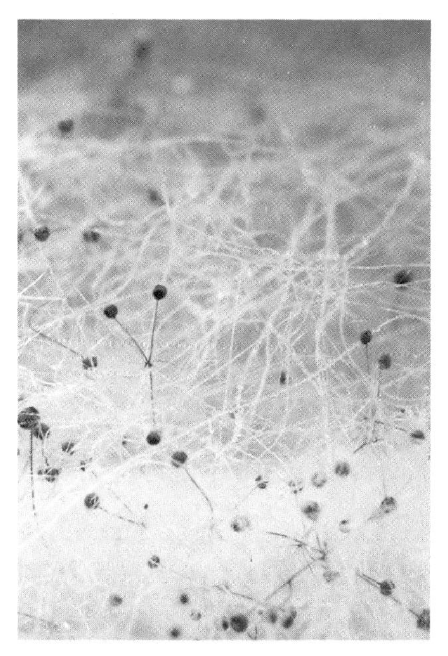

Bread mould growing on a piece of bread. The white threads are the hyphae. Each black dot contains thousands of spores.

These are puff-ball fungi. When a drop of rain hits the puff-ball, a cloud of spores shoots out.

Mushrooms, toadstools, mildews, and moulds are all **fungi**. Some fungi look a bit like plants, but they cannot make their own food. Some fungi feed on dead things, like the remains of plants and animals. They are called **saprophytes**. Others feed on living things, and cause disease. They are called **parasites**. Ringworm in humans and mildew in plants are caused by parasitic fungi.

These mushrooms are fungi you can eat. Mushrooms do not need light, so some people grow them in cellars for sale to shops.

This is a poisonous fungus called a Fly Agaric. It is common in pine and birch woods. If eaten in large amounts it can kill you.

Bracket fungi kill trees, then digest the dead wood. You usually see them on dead trees.

1.7 Animals without backbones I

In this Unit and the next, we look at the enormous variety of animals which have no backbone. These animals are sometimes called **invertebrates**, but this is not the scientific name for a group.

Coelenterates

Hydra, jellyfish, and sea anemones are all **coelenterates**. A coelenterate has a body like a bag. The open end is the mouth. It has tentacles round it to catch small animals for food. The tentacles have sting cells to sting the animals and paralyse them.

Hydra lives in ponds. It moves by turning cartwheels. It grows buds which break off to form new hydras.

This is a sea anemone. It uses sting cells on its tentacles to paralyse small creatures which are then pushed into its mouth.

Jellyfish live in the sea. They move by opening and closing like an umbrella. Some have sting cells which can hurt swimmers.

Flatworms

Flatworms have flat bodies with a mouth at one end. Freshwater flatworms, tapeworms, and flukes all belong to this group.

Freshwater flatworms live in ponds and streams. They move by waving tiny hairs called cilia.

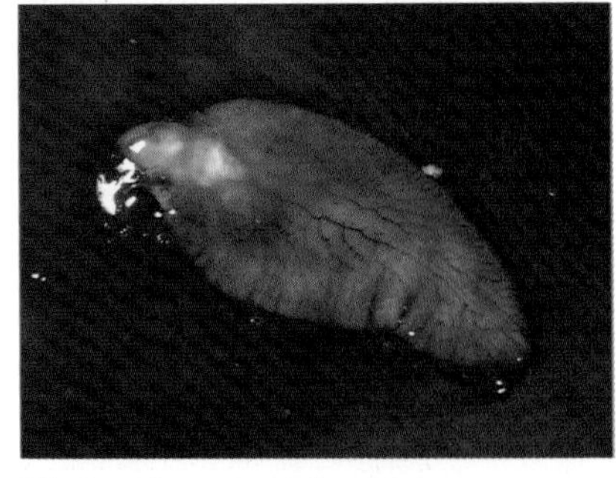

Flukes live inside other animals, and can cause disease. This shows a liver fluke on a sheep's liver. It hangs on by suckers. It can kill the sheep.

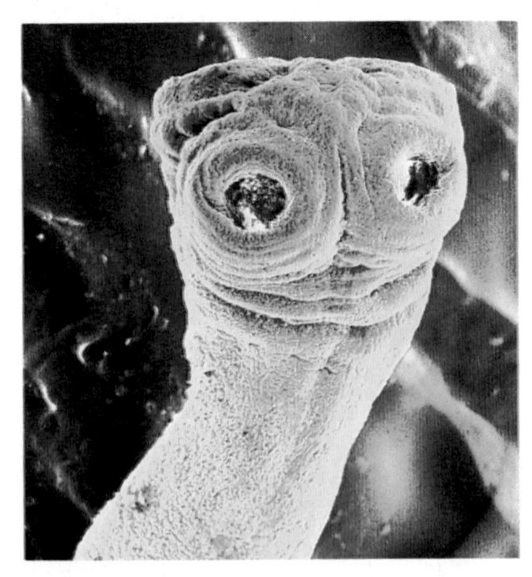

Tapeworms live in the guts of other animals, and eat their digested food. This one lives in humans. It can grow up to eight metres long.

Annelid worms

These have bodies made of rings, or **segments**. They are sometimes called true worms. Examples are earthworms, lugworms, and leeches.

Earthworms 'eat' soil and digest tiny creatures and dead leaves in it. The rest passes through their bodies.

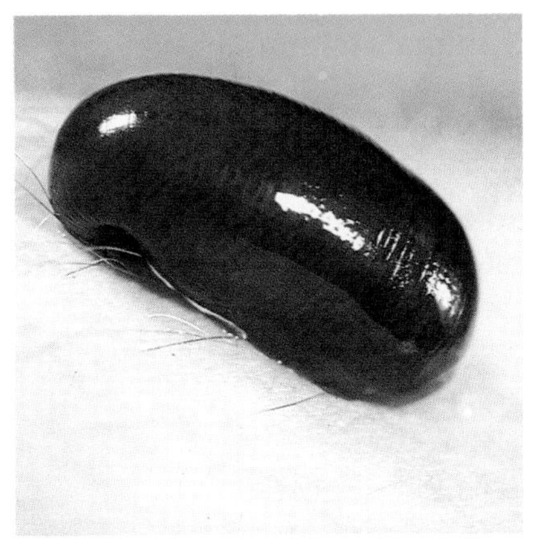

Leeches live on other animals and suck their blood. One type is used by surgeons to clean up infected flesh round wounds.

Lugworms live under sand on the seashore. They suck in sea water and eat any tiny creatures in it.

Molluscs

Snails, mussels, and octopuses are all molluscs. Molluscs have soft bodies. Most have one or two shells.

Snails have coiled shells. They move about on a slimy flat foot. They have long rough tongues for scraping up food.

A mussel has two shells hinged together. It anchors itself to rock by tough threads. It feeds on tiny creatures in sea water.

An octopus has a shell inside its body. It uses tentacles with suckers to move and to catch food.

Questions

1. Which group of animals has a body like a bag?
2. How do coelenterates feed themselves?
3. What do tapeworms eat?
4. What is the difference between flatworms and annelid worms?
5. Give another name for annelid worms.
6. Name three kinds of annelid worm.
7. Write down two features of mollucsc. Name two molluscs.
8. How do snails eat?

1.8 Animals without backbones II

Let's look now at the largest group of animals without backbones – the **arthropods**. Spiders, flies, and shrimps are all arthropods. Arthropods have a tough skin called a **cuticle**. They have jointed legs, and most have feelers or **antennae**. They can have simple eyes with only one lens, or **compound eyes** with thousands of lenses. They are divided into five smaller groups.

Like all flies, the horsefly is an arthropod. Look at its eyes. They are compound eyes, made up of thousands of tiny lenses.

Crustaceans

Crabs, lobsters, shrimps, woodlice, and waterfleas are **crustaceans**. Crustaceans have two pairs of antennae.

Crabs have a thick, hard cuticle and five pairs of legs.

Woodlice have a thin cuticle and live in damp places.

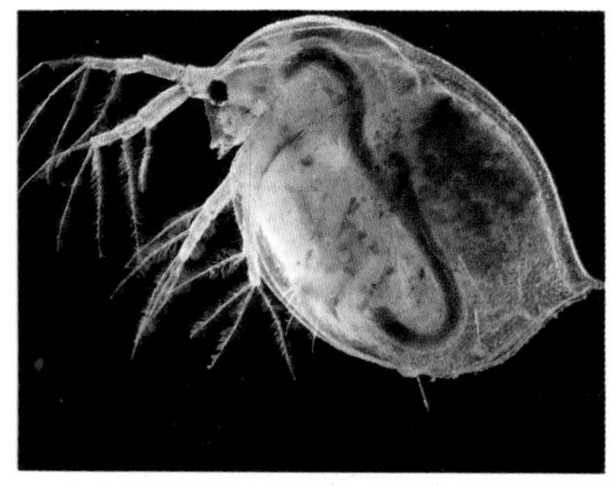

In waterfleas the cuticle is quite thin and soft.

Insects

For every person alive there are about a million insects! About 700 000 different kinds are known. Insects' bodies have three parts: a **head**, a **thorax**, and an **abdomen**.

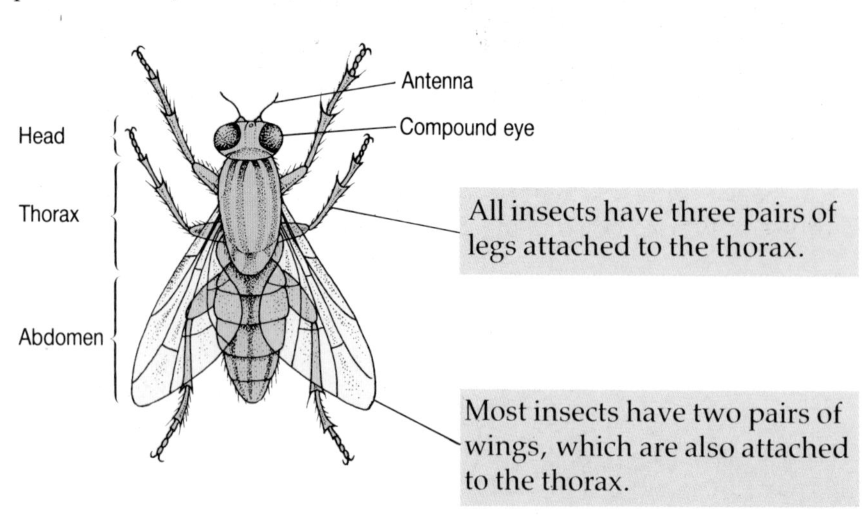

All insects have three pairs of legs attached to the thorax.

Most insects have two pairs of wings, which are also attached to the thorax.

Name as many parts of this horsefly as you can, by comparing it with the diagram on the left.

Many insects cause harm. Some destroy crops. Some spread disease. For example mosquitos spread malaria among humans. Some are a nuisance, like the midges that bite you in summer, and moths that damage clothes.
Many insects undergo big changes during their lives. Every butterfly was once a caterpillar. Every housefly was a maggot!

Colorado beetles are feared by potato growers because they destroy potato crops.

Many butterflies are brightly coloured. Scientists think they might use their colours to signal to each other.

Arachnids

Arachnids are arthropods with four pairs of legs, and no antennae. They use poison fangs to paralyse their prey. Spiders, harvestmen, mites, ticks, and scorpions are all arachnids.

Spiders are the most common arachnids. Many of them spin webs or tripwires of silk, to catch their prey.

A scorpion ready to strike. Its sting is in its tail. The sting from some scorpions can kill humans.

Centipedes and millipedes

These arthropods have long bodies made up of many segments. Each segment has at least one pair of legs.

Centipedes have one pair of legs to each segment. They paralyse prey with poison fangs.

Millipedes have two pairs of legs to each segment. They eat plants.

Questions

1 Which group of arthropods has:
  a) three pairs of legs?
  b) four pairs of legs?
  c) two pairs of antennae?

2 Give two reasons why spiders are not insects.

3 What are the differences between centipedes and millipedes?

4 Name the three parts of an insect's body.

1.9 Animals with backbones I

In this Unit and the next we look at those members of a group (phylum) called **chordates** which possess a backbone. These animals are also known as **vertebrates**, and there are five types: fish, amphibians, reptiles, birds, and mammals.

Fish

There are two types of fish: those with a skeleton of cartilage, which include sharks and dogfish, and those with a skeleton of bone, which include herring, cod, and haddock. Both types have a streamlined shape and are covered with scales. Bony fish are weightless in water: their bodies contain a bladder full of air which buoys them up.

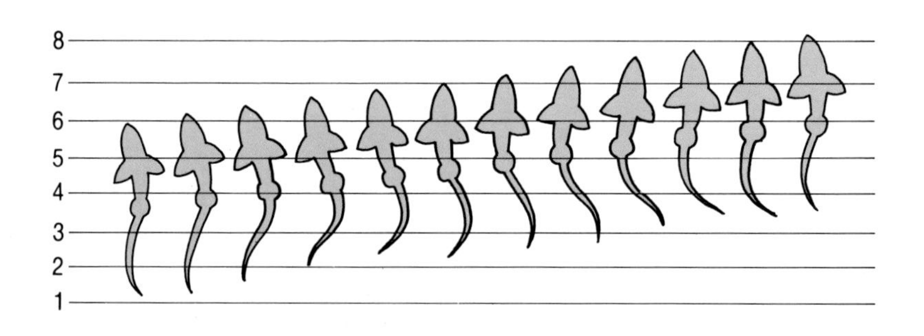

A fish swims by moving its tail from side to side.

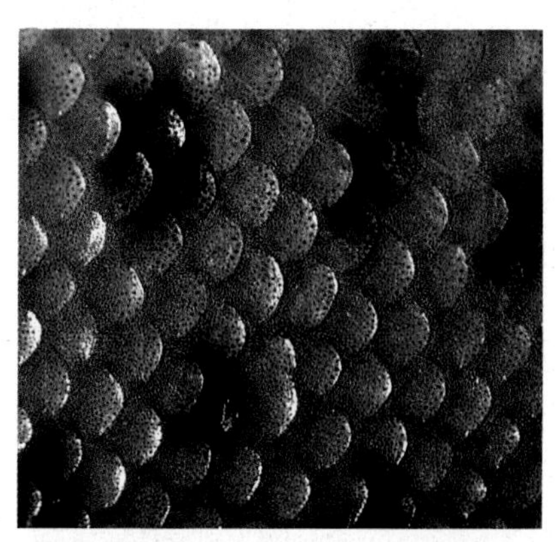

The scales on a trout's body.

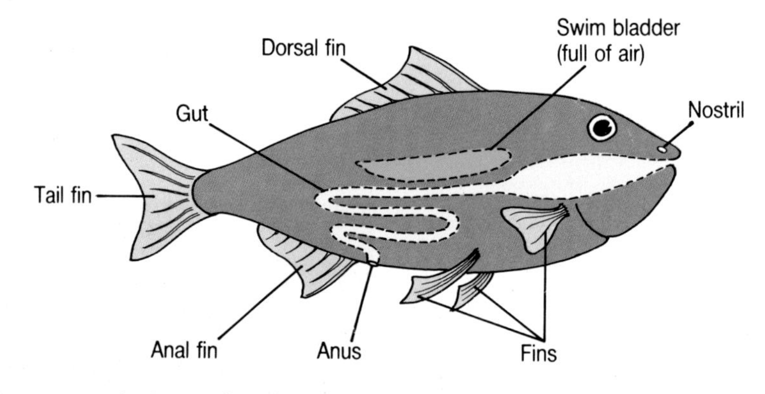

The parts of a bony fish.

Sharks have a cartilage skeleton and visible gill slits.

Many fish, like these trout, reproduce by laying eggs . . .

. . . but these female guppies give birth to live baby fish.

Amphibians

Frogs, toads, salamanders, and newts are all **amphibians**.

Amphibians have four limbs and moist skin. They can live both on land and in water. On land they breathe using lungs. In water they breathe through their skin.

Amphibians lay their eggs in water. These hatch into tadpoles, which swim using tails, and breathe through gills.

This frog has just laid her eggs as frog spawn. In about a week they'll become tadpoles.

Newts live on land for part of the year. But in spring they move into the water to breed.

Reptiles

Lizards, crocodiles, snakes, and tortoises are reptiles. Reptiles have tough, dry, scaly skin. They breathe with lungs. They lay eggs with tough shells like leather.

A snake egg hatches... and out comes the snake's head.

Crocodiles look peaceful and lazy. But don't be fooled...

Some snakes won't harm you. But a viper like this one can kill.

Questions

1 Fish are streamlined. How does this help them?

2 How do fish breathe?

3 How do fish swim?

4 How is a tadpole different from a frog?

5 Explain how frogs can live on land and in water.

6 Do all reptiles have legs?

7 How are reptile eggs different from bird eggs?

8 Name two amphibians and two reptiles.

1.10 Animals with backbones II

Like all birds, penguins have wings – but they cannot fly.

The last two groups of animals with backbones are birds and mammals. Birds and mammals are able to maintain a constant, high body temperature: that is, they are **warm-blooded**. This means they can stay active in both cold and warm weather. In all other animals, body temperature depends on the weather.

Birds

All birds have feathers on their bodies. They also have wings, but not all birds can fly. They have beaks for pecking and tearing food. They lay eggs covered with a hard shell.

Birds have light hollow bones, to help them fly.

Water birds live on fish and other water creatures . . .

. . . but this tawny owl is about to dine on a mouse.

Mammals

Mammals have hair on their bodies. Female mammals have breasts, or **mammary glands**, from which their young ones suck milk. There are three kinds of mammals.

Mammals that lay eggs

Mammals that lay eggs are called **monotremes**. The spiny ant-eater and duck-billed platypus are monotremes.

When monotreme eggs hatch the babies are not fully formed. They are fed on milk for several weeks until their missing parts have grown.

The duck-billed platypus lays its eggs in an underground nest.

Mammals with pouches

Some mammals carry their young in a pouch. They are called **marsupials**. Kangaroos, wallabies, and koalas are marsupials.

Baby marsupials start growing in their mother's womb. Before they are fully formed they crawl out of the womb and into her pouch. The pouch has a teat in it from which the young suck milk.

A kangaroo carrying her baby in her pouch.

Placental mammals

Humans, bats, whales, horses, and sheep are all placental mammals. That means their young develop inside the mother's womb until they are fully formed.

Young are attached to the wall of the womb by a **placenta**. Food and oxygen pass through the placenta from the mother's blood.

Baby rabbits are born below ground, in fur-lined nests.

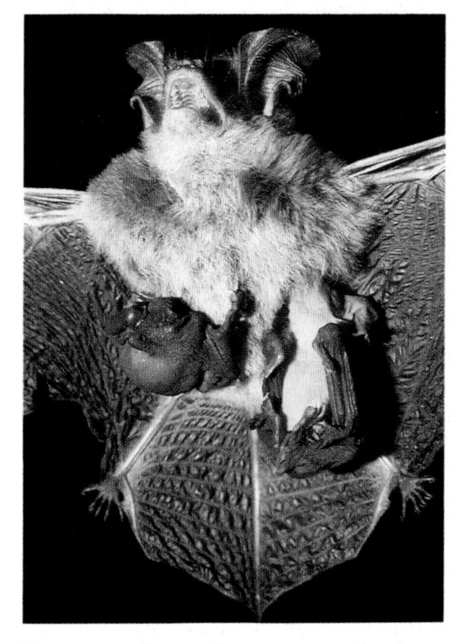

Bats are the only mammals that can fly. A baby bat clings to its mother until it is about two weeks old. Then she hangs it upside down in a safe place while she flies off to find food.

Sea lions are placental mammals too. The pups are suckled until they are about three months old.

Questions

1 Name two things that help birds fly.

2 Name a bird that can't fly.

3 What do human mothers and kangaroo mothers have in common?

4 What are monotremes? Give two examples.

5 Think of two reasons why pouches might be better than eggs, for developing babies.

6 Write down ten examples of placental mammals.

7 Think of three advantages of a mother carrying babies inside her rather than laying eggs.

1.11 Plants

There are two kinds of plants:

1 **Plants without seeds** These are mosses, liverworts, and ferns. They produce **spores** which grow into new plants.

2 **Seed plants** These are conifers and flowering plants. They produce **seeds** which grow into new plants.

Plants without seeds (non-flowering plants)

Mosses and liverworts These usually grow in damp places. Their spores grow in **capsules** on the end of stalks.

Moss plants have stems and tiny leaves. Over 600 different kinds grow in Britain. This one is called bank hair moss.

Moss capsules. When they are ripe their lids drop off, releasing spores which are scattered by the wind.

Unlike mosses, many liverworts have no distinct stems. They look like leaves growing flat on the ground.

Ferns These usually grow in woods and other damp, shady places. Their spores grow in capsules attached to the back of the leaves.

Britain has over 50 different kinds of fern. This one is called hard fern. It is common on moorland and heath.

This kind of fern is called bracken. It can grow nearly two metres tall. It is very common on moorland.

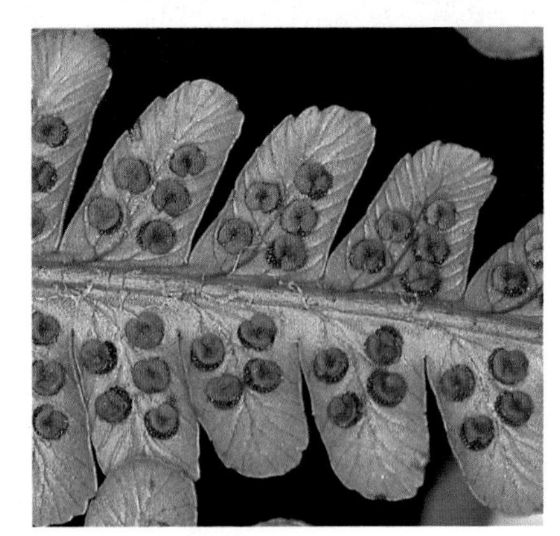

Groups of spore capsules on the back of a fern leaf. Each group is protected by a tiny orange-brown scale leaf.

Seed plants

Conifers Pines, spruces, larches, cedars, and firs are conifers. They produce male and female cones.

The male pine cone produces pollen, which is carried by the wind to a female cone.

The fertilized female cone grows seeds under its flaps, which open when they are ripe.

Flowering plants Cabbages, horsechestnut trees, and grasses are all flowering plants. They have flowers with reproductive organs which produce seeds. There are two kinds of flowering plants, **monocotyledons** and **dicotyledons**.

Monocotyledons, like these daffodils, have long thin leaves with parallel veins.

Dicotyledons, like this water lily, have broad leaves with a network of veins.

A horsechestnut tree in flower. Its hard brown seeds are used in the game of 'conkers'.

Questions

1. What are capsules, cones, and flowers for?
2. Which of these plants have capsules? Which have cones? Which have flowers?
   roses, bog moss, Christmas trees, bracken ferns.
3. Which of these are monocotyledons, and which are dicotyledons?
   buttercups, tulips, grasses, apple trees. Explain your answer.
4. Name the two main groups of plant.

Questions on Section 1

1 *Recall knowledge*

Vocabulary test Write out the numbered phrases then, opposite each, write the word from the list of technical terms which the phrase describes.

respiration, excretion, photosynthesis, chlorophyll, vertebrates, invertebrates, antenna, spores, biosphere, phytoplankton, mammary glands, dicotyledons, marsupials, coelenterates

1. A green chemical in plant leaves
2. Removal of wastes from the body
3. Animals with backbones
4. Insect sense organs
5. Their leaves have a network of veins
6. The parts of our planet which can support life
7. How plants make food
8. They carry their young in a pouch
9. Jellyfish are examples
10. Getting energy from food
11. Mosses and ferns use them to reproduce
12. Animals without backbones
13. They produce milk
14. Microscopic plants

2 *Understanding*

In each of the following groups of living things find the one which is LEAST related to the other four.

Honey bee
Bat
Sparrow
Penguin
Flying fish

Spider
Grasshopper
Housefly
Butterfly
Ant

Lizard
Frog
Alligator
Tortoise
Snake

Haddock
Mackerel
Whale
Cod
Goldfish

Oyster
Octopus
Snail
Slug
Starfish

Iris
Horsechestnut tree
Moss
Stinging nettle
Buttercup

3 *Recall knowledge*

All living things *move, have senses, feed, respire, excrete, reproduce,* and *grow.*

Write out these seven features as headings across a page. Then sort the list below into seven groups under the headings.

fins	seeds
photosynthesis	eggs
kidneys	taste
babies	sunlight
energy	oxygen
muscles	lungs
eyes	mating
root growth	leaf-fall
leaves	seedlings

4 *Recall knowledge*

a) Write down three headings: *Animals only, Plants only* and *Both animals and plants.*

b) Now sort the list below into three groups under these headings.

excretion
photosynthesis
movement from place to place
growth throughout life
respiration
eating other living things
growth towards light
growth stops when adult size is reached

5 *Recall knowledge*

Look at each statement below. Which animal or group of animals does it describe?

a) They have three pairs of legs.

b) They move about on a large slimy foot.

c) They have a thick, hard cuticle, and two pairs of antennae.

d) They have tentacles and a beak like a bird.

e) They carry their babies in a pouch.

f) The body is like a bag, with a mouth and tentacles at one end.

6 *Recall knowledge*

Look at each statement below. Which group of plants does it describe?

a) Spore capsules grow on the backs of their leaves.

b) Grasses, sycamore trees and tulips are examples.

c) Their seeds grow inside cones.

7 *Recall knowledge*

The drawings below show some living things. Which of them:

a) is a vertebrate?
b) is an invertebrate?
c) is a mammal?
d) is an insect?
e) makes food by photosynthesis?
f) lives by making things decay?
g) reproduces with seeds?
h) reproduces with spores?
i) suckles young on milk?
j) has compound eyes?

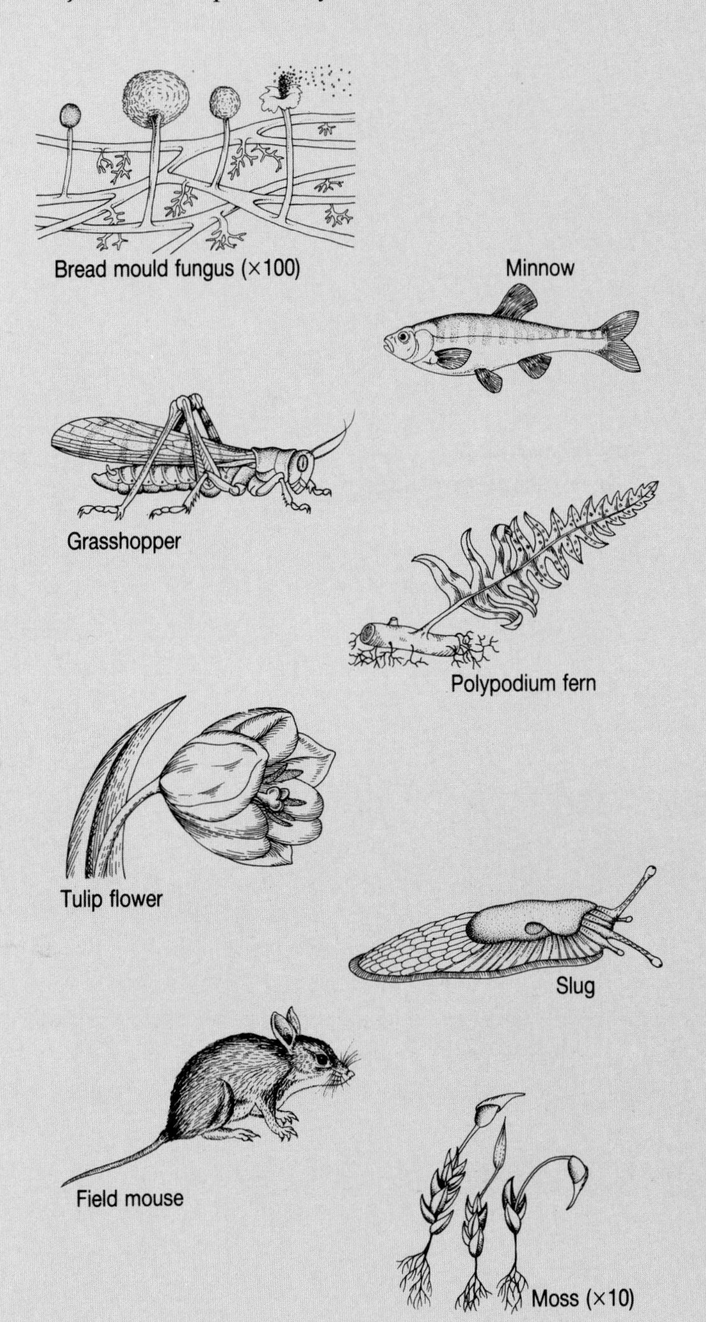

8 *Understanding*

Scientists visited an uninhabited island and discovered some previously unknown insects shown below. Construct a key which would enable a visitor to the island to identify them.

a) Begin by choosing one feature which sorts the insects into two groups.

b) Sort the two groups into smaller groups by choosing other differences, then look for one feature which separates each insect from all the others.

c) Produce a key using the features you have chosen by arranging the features into numbered pairs, like the key on page 11.

d) The first pair of features should separate the insects into two groups, and subsequent pairs should either identify an insect or lead on to another pair of features.

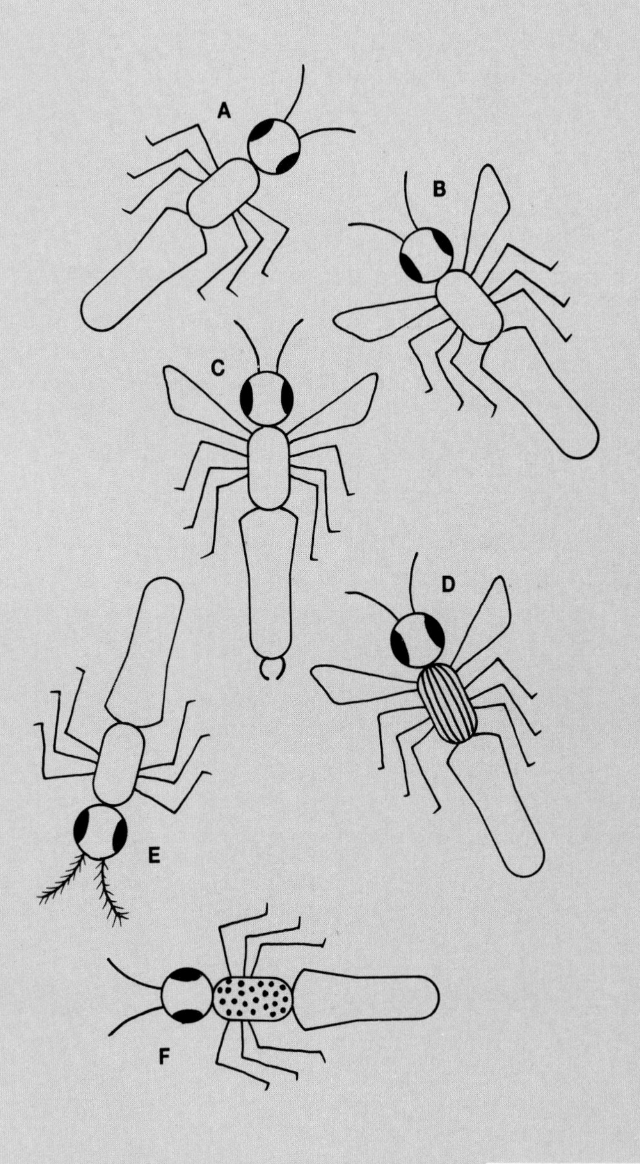

2.1 What are cells?

A house is built of bricks. In the same way, animals and plants are built of **cells**.

Your skin, your bones, your muscles, and your brain are all made of cells. There are around a hundred million million cells in your body. They are very tiny. At least ten cells would fit side by side across this next full-stop.

A cell from the lining of a human cheek, coloured to show up the different parts. What do you think the orange blob is?

Inside an animal cell

Your cells are not very different from the cells of a frog or a cat or a giraffe. In fact all animal cells have these parts:

A cell membrane. This is a thin skin around the cell. It lets some things pass through, but stops others.

Cytoplasm. This is a jelly containing hundreds of chemicals. Lots of chemical reactions go on in it. It fills the cell.

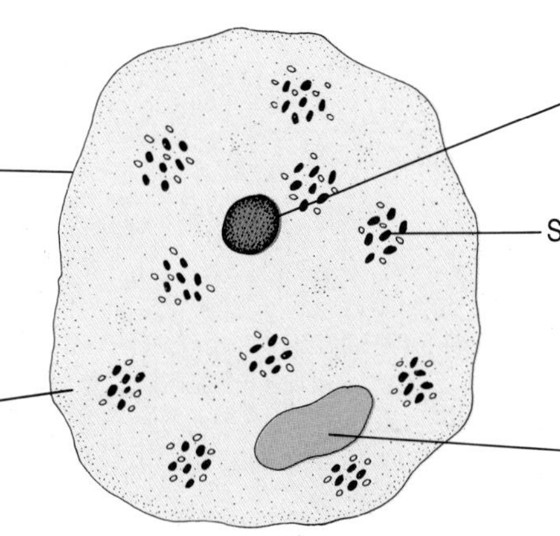

A nucleus. It controls what a cell does, and how it develops.

A vacuole. This is a space within the cell containing air, liquids, or food particles. Animal cells usually have several small vacuoles.

Cells are not all the same shape. There are about twenty different types of cell in your body, all doing different jobs. Below are photographs of three types of cell.

Red blood cells are disc-shaped. Their job is to carry oxygen round the body.

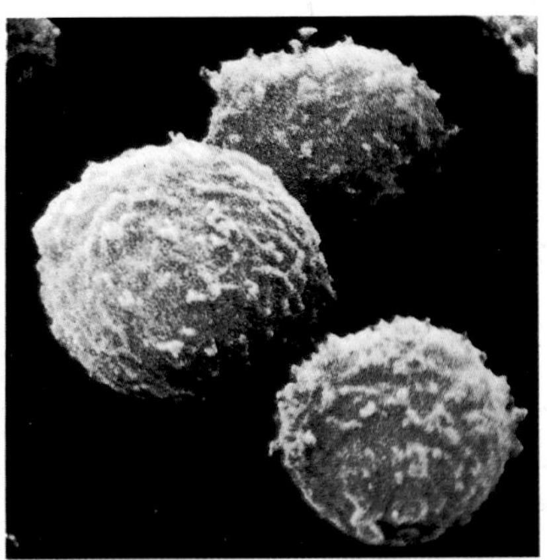

White blood cells, stained to show up clearly. They attack germs, and can change shape.

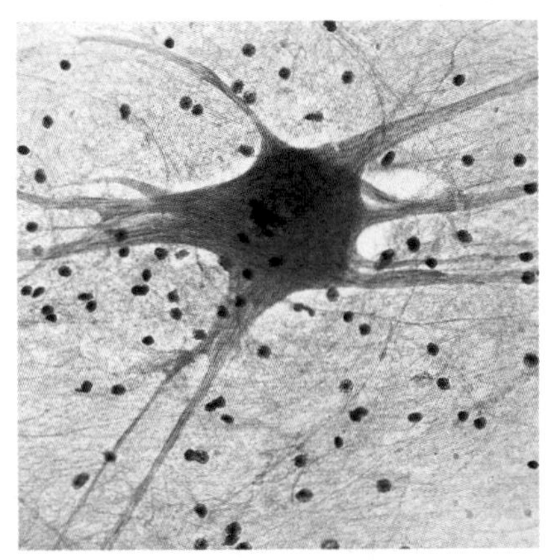

Nerve cells have long thin fibres which carry 'messages' around your body.

Inside a plant cell

All plant cells have these parts:

A cell wall of cellulose. It covers the cell membrane.

Cytoplasm.

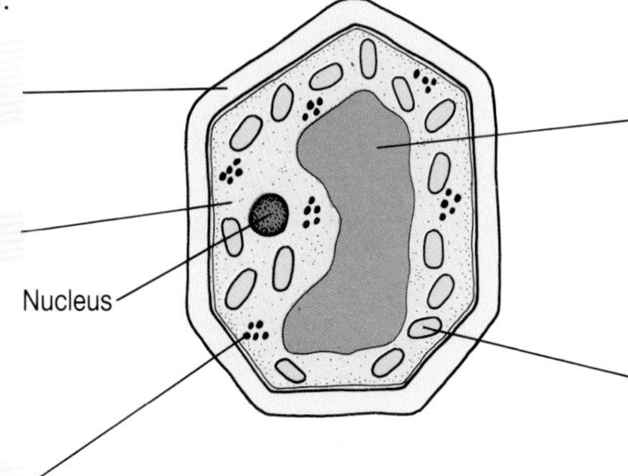

Stored food. Mainly starch.

A vacuole. All plant cells have these. The liquid inside them is called **cell sap**.

Chloroplasts. These are tiny discs full of a green substance called **chlorophyll**. They trap the light energy that plants need for making food by photosynthesis.

Unlike animal cells, there are only a few different shapes of plant cells. This is because there are not so many different jobs for them to do.

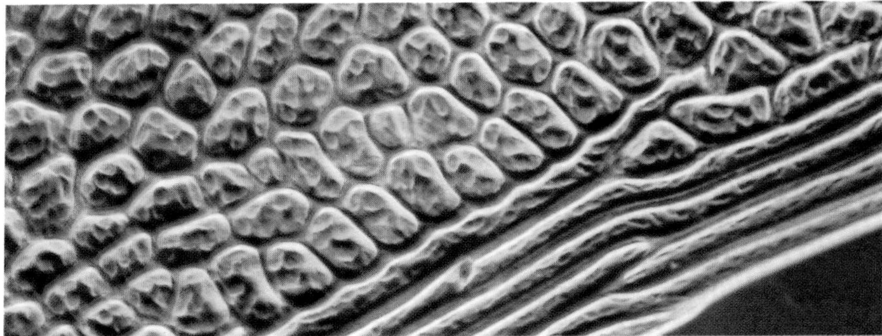

Leaf cells. They look firmer than animal cells because of their cell walls. The tiny green blobs are chloroplasts. There are several in each cell.

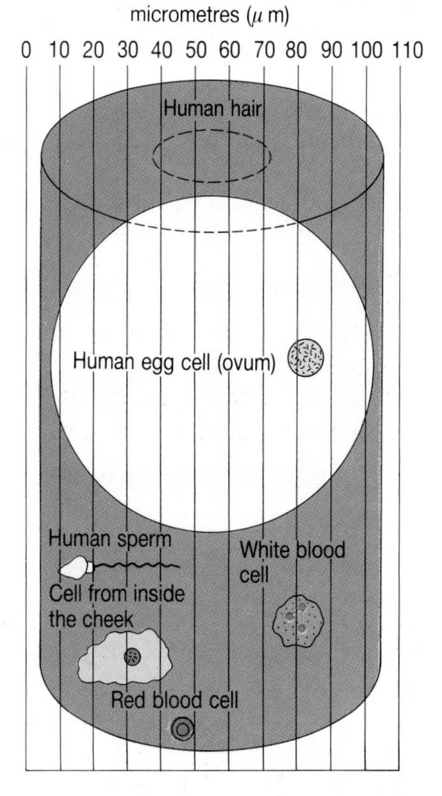

The sizes of microscopic things are measured in micrometres, written **µm**. One µm is 1/1000 millimetre (see question 1).

How plant cells are different from animal cells

	Plant cells		Animal cells
1	Have cellulose cell walls.	1	Do not have cellulose cell walls.
2	Have chloroplasts.	2	Do not have chloroplasts.
3	Always have a vacuole.	3	Sometimes have a vacuole.
4	Have a few different shapes.	4	Have many different shapes.

Questions

1 **a)** How wide is a hair in micrometres (µm)?
b) How wide is a red blood cell in µm?
c) About how many red cells side-by-side equal the diameter of a hair?
d) About how many times would you have to enlarge a cheek cell to make it the same diameter as a tennis ball?

2 Name three different kinds of cell in your body.

3 What are chloroplasts and what do they do?

4 List four things found in *both* plant cells and animal cells.

5 What does the cell nucleus do?

2.2 Cells, tissues, and organs

You started life as a single cell – a fertilized egg cell. You grew because that cell divided to make two cells, these divided to make four, and so on. This is called **cell division**. It is how all living things grow.

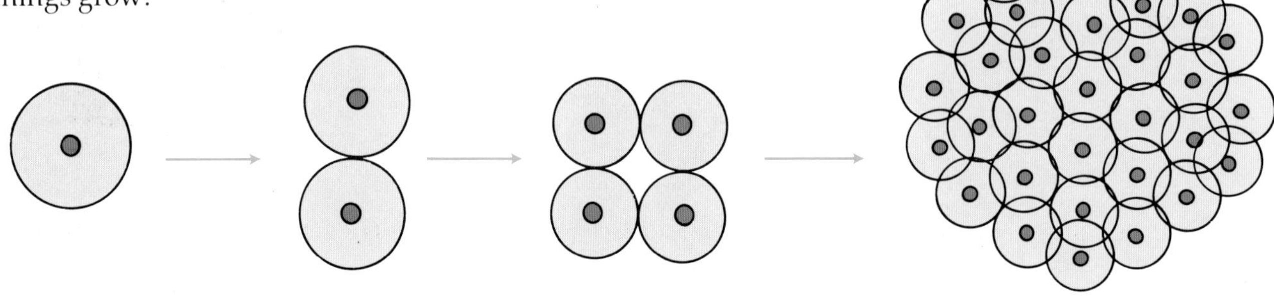

A fertilized egg cell divides to make *two* **daughter** cells, which are identical.

These divide to make *four* identical cells, which divide again and again to make a ball of cells.

Tissues

Some cells in the ball grow and change shape to do a particular job – they become **specialized**. Cells that do the same job group together to form **tissues**.

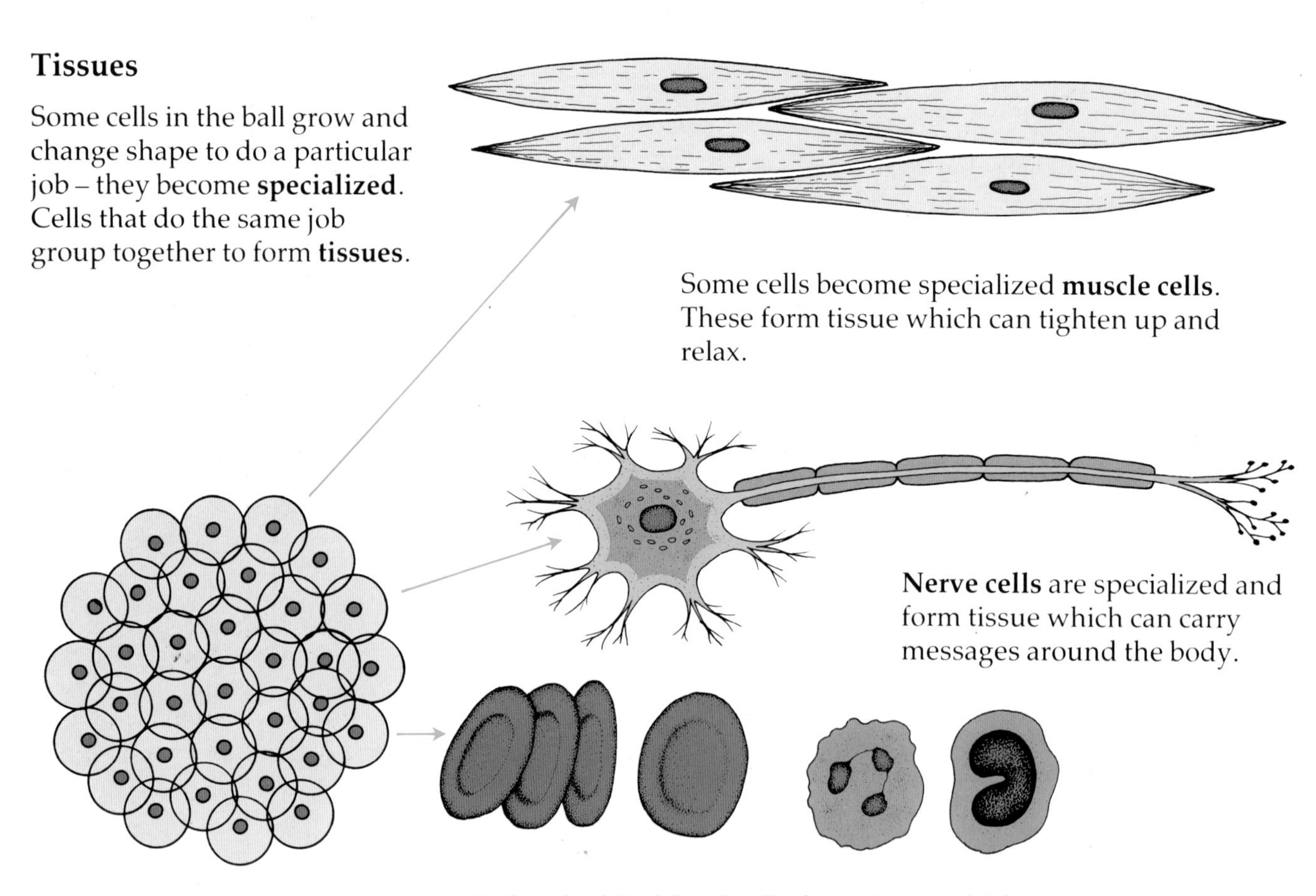

Some cells become specialized **muscle cells**. These form tissue which can tighten up and relax.

Nerve cells are specialized and form tissue which can carry messages around the body.

Red and white blood cells form tissue which can carry oxygen and kill germs. This tissue is called **blood**.

Organs

Different tissues combine to make **organs**.

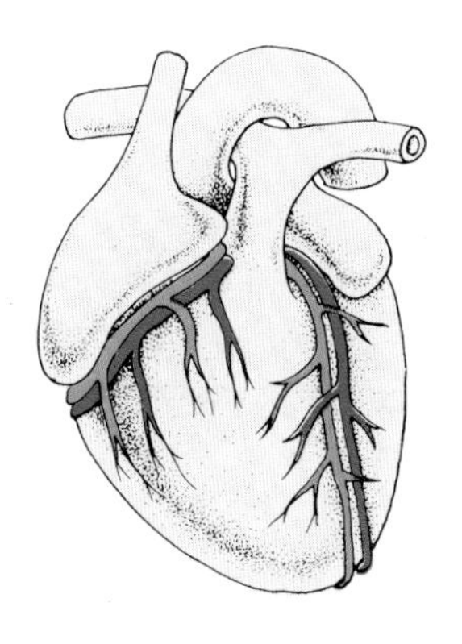

The **heart** is an organ which pumps blood around the body.

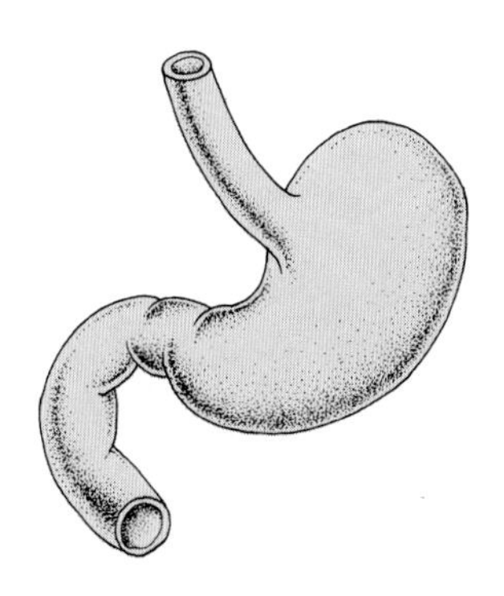

The **stomach** is an organ which digests food.

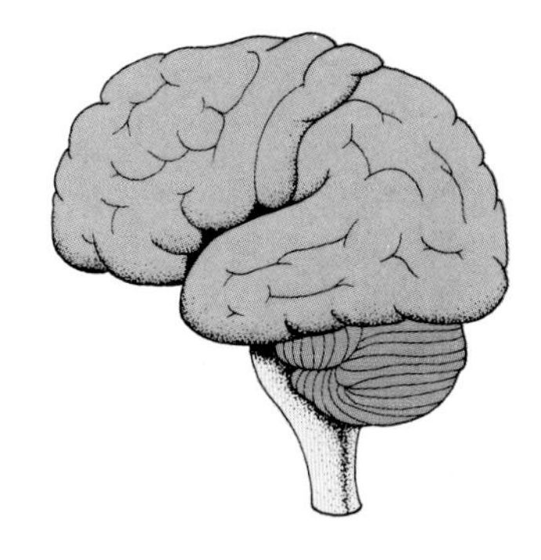

The **brain** is an organ which controls parts of the body.

Organisms and organ systems

You are an **organism**. So is a cat and so is a bird. An organism is made up of many different organs. Some of its organs work together to form **organ systems**.

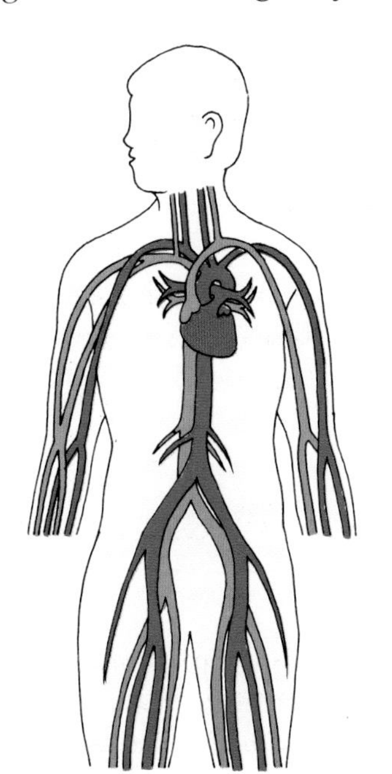

The **circulatory system** is made up of the heart and blood vessels.

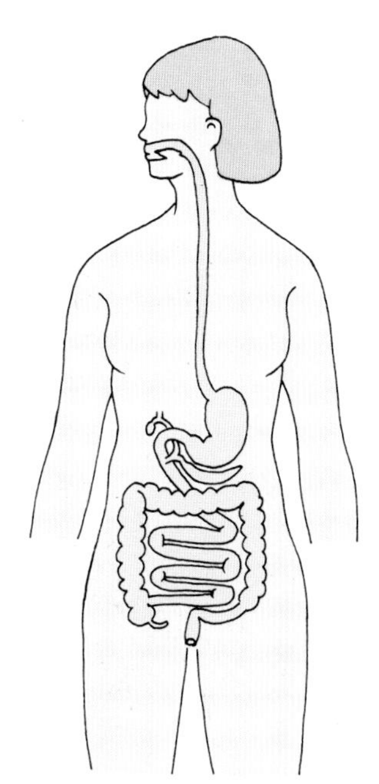

The **digestive system** is made up of the gullet, stomach, and intestine.

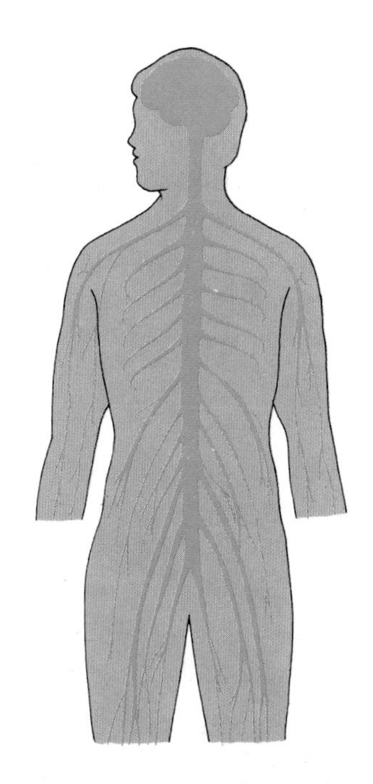

The **nervous system** is made up of the brain, spinal cord, and nerves.

Questions

1 What is a tissue? Give two examples.

2 What is an organ? Give five examples.

3 What is an organ system? Give two examples.

4 Even when you are fully grown some cells carry on dividing. Explain why.

5 Explain what a specialized cell is.

2.3 In and out of cells

Substances pass in and out of cells as tiny particles called **molecules**. The molecules in liquids and gases are never still. They keep moving and bumping into each other all the time.

You can show that molecules move by adding a drop of ink to water. Ink spreads through the water even though it is not stirred. Ink spreads because ink molecules move into the spaces between water molecules, and water molecules move into the spaces between ink molecules.

Movement of molecules so that they mix is called **diffusion**. Molecules diffuse from where they are plentiful to where they are less plentiful. In other words molecules diffuse down a **concentration gradient,** from high to low concentration.

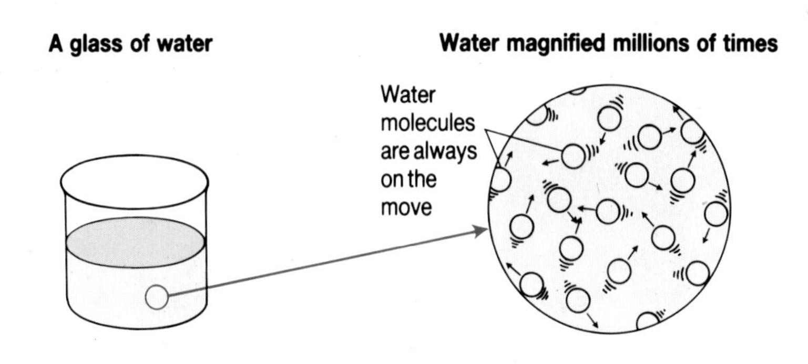

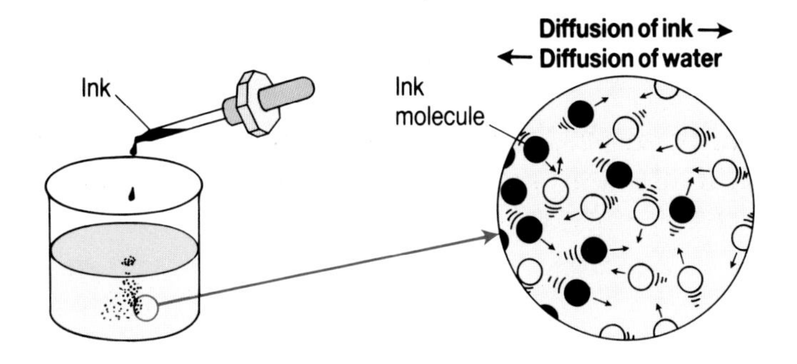

Diffusion in and out of cells

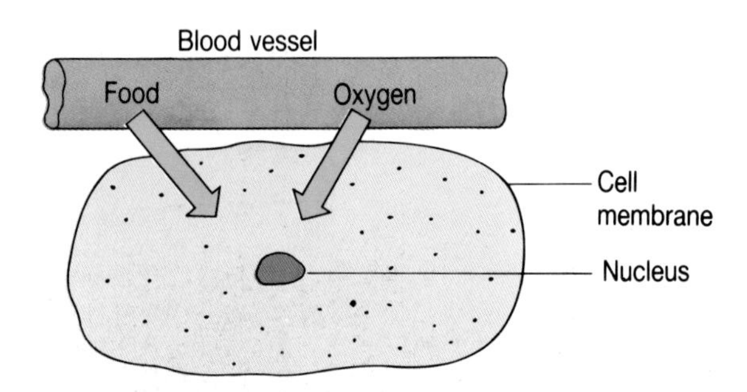

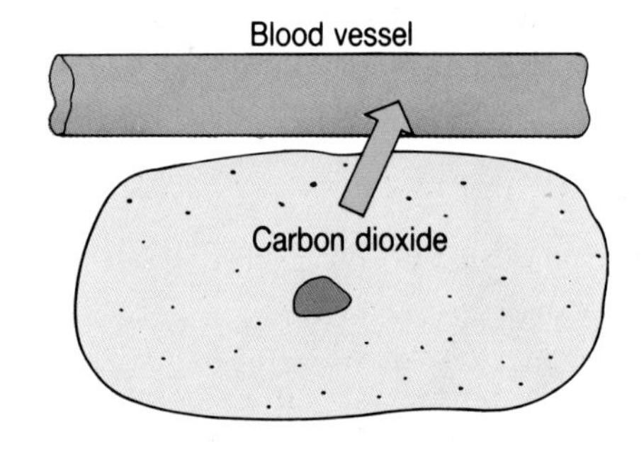

Human body cells need a constant supply of food and oxygen to stay alive and do their jobs. Food and oxygen are carried in the blood, so they diffuse from the blood into each cell.

As cells use food and energy, they produce carbon dioxide waste. This must be removed before it poisons the cells. Carbon dioxide diffuses from cells into the blood which carries it away to the lungs to be breathed out of the body.

Active transport of molecules up concentration gradients

Plants take in minerals from soil *against* concentration gradients (i.e. from low to high concentration) so that their root cells can contain more minerals than the soil water. This is the opposite of diffusion. Plants make it happen by using energy to pull molecules into their cells, which is why it is called **active transport** of molecules.

Osmosis

Osmosis is a special kind of diffusion. It happens when a membrane has tiny holes in it which let water molecules pass through but stop larger molecules, like sugar. A membrane like this is called **semi-permeable**. This experiment shows osmosis:

Sugar molecule
Water molecule

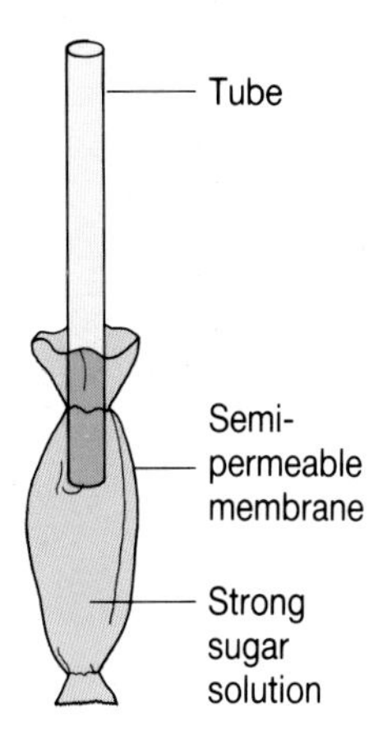

Liquid rises up the tube
Strong solution
Semi-permeable membrane (magnified)
Weak solution

A semi-permeable membrane is tied to a tube. It is then filled with strong sugar solution.

The membrane is stood in a weak sugar solution. Soon liquid starts to rise up the tube.

The liquid rises because water molecules diffuse through the membrane from the weak solution to the strong one. But sugar molecules cannot diffuse like this because they are too big to pass through the membrane.

When a weak solution is separated from a strong one by a semi-permeable membrane, water always flows from the weak solution to the strong one. This diffusion of water is called **osmosis**.

Osmosis in plant cells

Water moves from cell to cell in plants by osmosis. The cell membrane of a plant cell is semi-permeable. So if a cell containing a weak solution is next to a cell with a stronger solution, water moves by osmosis from the weak to the strong solution, as shown in this diagram.

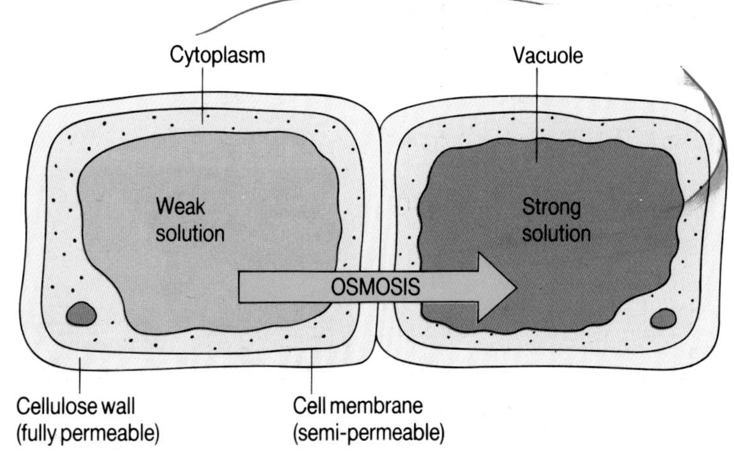

Questions

1 Why does ink move through water even without being stirred?

2 How does oxygen move from blood into body cells?

3 Osmosis is the movement of ______ from a ______ solution to a ______ solution through a ______ membrane.

4 What is a semi-permeable membrane?

2.4 Cell division

You started life as a fertilized egg cell, smaller than a full stop. You grew because this cell divided millions and millions of times to make new cells.

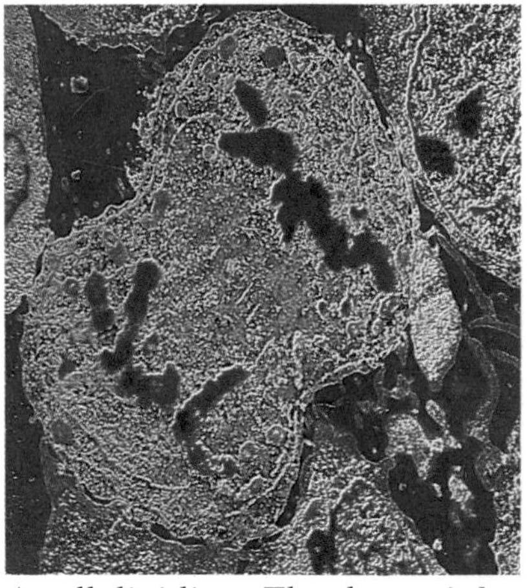

A cell dividing. The deep pink parts are the chromosomes.

Chromosomes

When a cell is ready to divide, long thin threads called **chromosomes** appear in its nucleus.

Chromosomes contain chemicals which control what a cell does. These chemicals also contain all the instructions needed to build a whole new organism from a single fertilized egg cell.

When a cell divides it makes a new set of chromosomes so that these instructions can be passed to the daughter cells.

The chromosomes are always in pairs.

How a cell divides

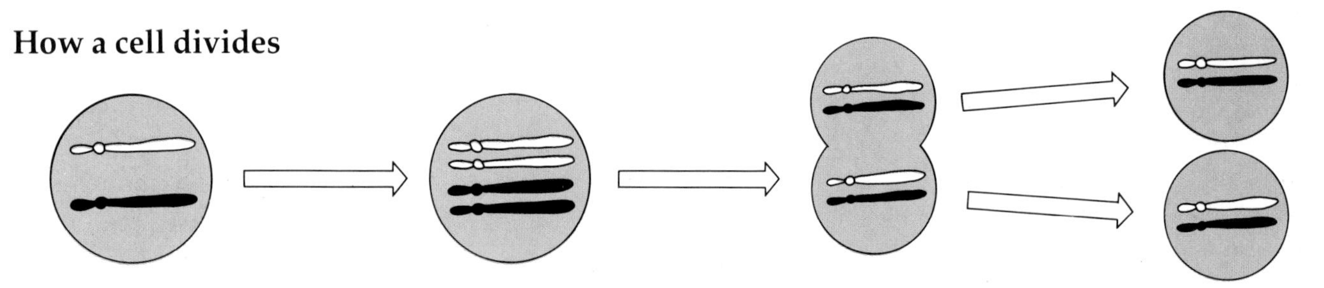

This cell has only two chromosomes. (Human cells have **46** chromosomes altogether.) Each chromosome splits into two, making a second set of chromosomes.

The *cell* divides in two. Each cell gets a *full set* of chromosomes. The two new cells have the same number of chromosomes as the first.

This kind of cell division is called **mitosis**. All the cells in animals and plants, *except* sex cells, are made by mitosis.

Mitosis produces:

1 The cells needed to make an adult organism from a fertilized egg.

2 The cells needed to heal cuts, wounds, and broken bones.

3 The cells that replace dead skin cells, and worn-out red blood cells.

Mitosis produces the cells that let a seed grow into a plant.

Mitosis produces new bone cells to mend a broken arm.

How sex cells are produced

Some of the cells in your body are sex cells, or **gametes**. If you are male they are called **sperms** and are made in your **testes**. If you are female they are called **ova** or eggs, and are made in your **ovaries**. Sex cells are made by a different kind of cell division.

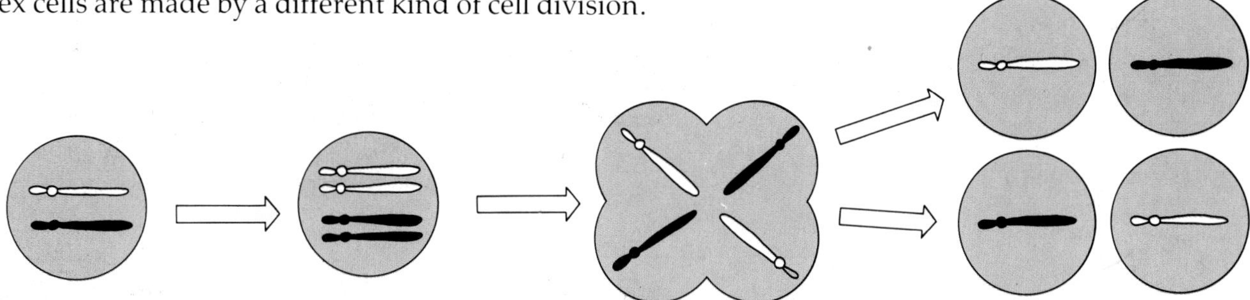

This cell is drawn to show only two chromosomes. Each chromosome splits into two, making a second set of chromosomes.

The cell divides to form *four* sex cells. Each sex cell gets only *half as many* chromosomes as the first cell. (Human sex cells have only **23** chromosomes.)

This kind of cell division is called **meiosis**. Meiosis produces sex cells in humans and other animals, and in plants.

Fertilization

At fertilization a male sex cell joins up with a female sex cell to make a fertilized egg cell called a **zygote**.

This drawing shows fertilization in humans.

In human fertilization a sperm with **23** chromosomes joins an ovum with **23** chromosomes to make a zygote with **46** chromosomes.

Fertilization happens in the same way in other animals, and in flowering plants. In flowering plants, the fertilized sex cell grows into a seed.

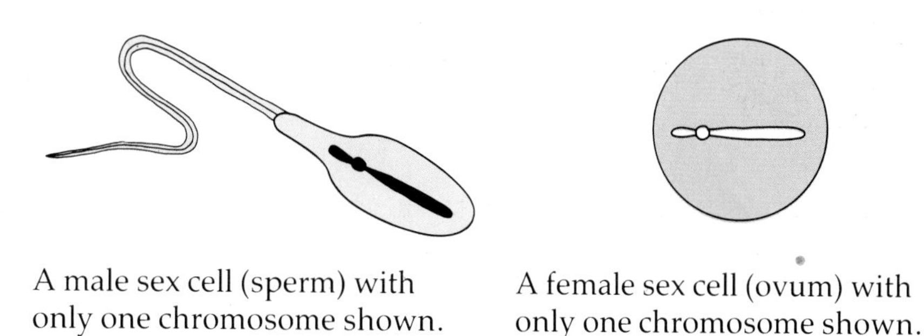

A male sex cell (sperm) with only one chromosome shown.

A female sex cell (ovum) with only one chromosome shown.

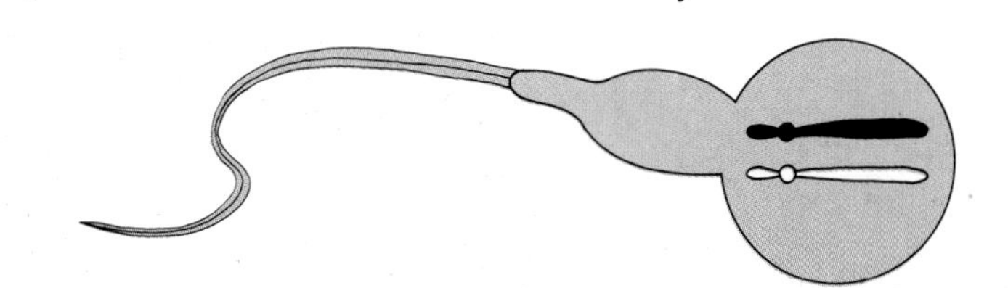

The sperm joins up with the ovum. This produces a zygote with the *full number of chromosomes*. This grows by mitosis into an adult.

Questions

1 Which kind of cell division:
  a) halves the chromosome number?
  b) produces cells to repair wounds?
  c) produces sperms?
  d) produces an adult from a zygote?

2 How are zygotes produced?

2.5 Heredity and variation

Heredity

You started life as a fertilized egg cell with 46 chromosomes. 23 of these chromosomes came from your father, and 23 from your mother. This is why you have some characteristics of your father and some of your mother.

All children inherit certain characteristics from their parents. The study of **inherited characteristics** is called **heredity**.

Inherited characteristics

The photograph above shows some of the characteristics children inherit from their parents. The development of these characteristics is controlled by chromosomes.

The children in the family below inherited some characteristics from each of their parents. Try to sort out which characteristic came from which parent.

Father has red curly hair, brown eyes, a snub nose, and a cleft chin.

Mother has straight black hair, blue eyes, a long thin nose, and a pointed chin.

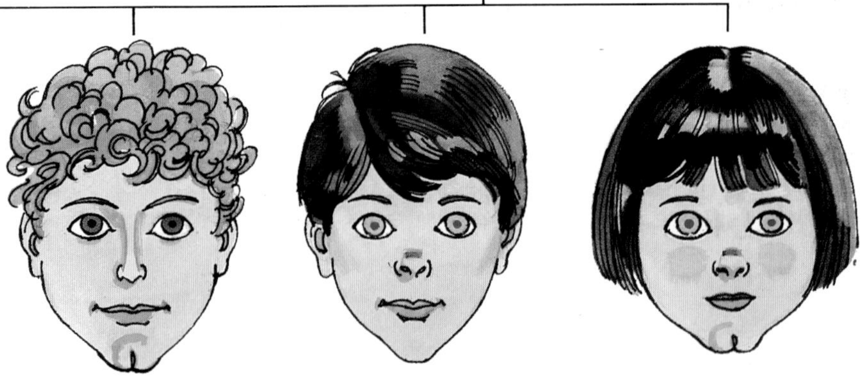

Acquired characteristics

You may know how to swim, or rollerskate, or speak French. Or you may have a scar from a cut. These are **acquired characteristics**. You pick them up (acquire them) as you go through life. You are not born with them, and you cannot pass them on to your children.

Variation

No two people are exactly the same. Even identical twins are different in some ways.

People are different heights and weights. Their hair and eyes are different colours and their faces have different shapes. Their eye colour and hair colour show **variation**.

Continuous variation

These people are arranged in a line from the shortest to the tallest. Their height shows **continuous variation**. It varies from short to tall with many small differences in between.

Intelligence also shows continuous variation. Can you think of other variations which are continuous?

People's height shows continuous variation.

Discontinuous variation

People can either roll their tongues or they cannot. This is an example of **discontinuous variation**. You can either do it or you can't. There is no 'in between' state.

Blood groups show discontinuous variation. You belong to only one group: A, B, AB, or O. Find out what blood group each member of your class belongs to.

Can you think of any other variations that are discontinuous?

Tongue rolling. Either you can. . . or you can't.

Questions

1. Write down eight inherited characteristics.
2. Write down six acquired characteristics.
3. Intelligence shows continuous variation. What does this mean?
4. Colour blindness shows discontinuous variation. What does this mean?
5. Write down four of your own acquired characteristics.

2.6 Chromosomes and genes

You inherit characteristics from your parents through their chromosomes.

A chromosome has small parts called **genes** all the way along it. The genes are made of a chemical called **DNA**. Genes control the development of inherited characteristics. For example, there are genes which control eye colour, hair colour and skin colour.

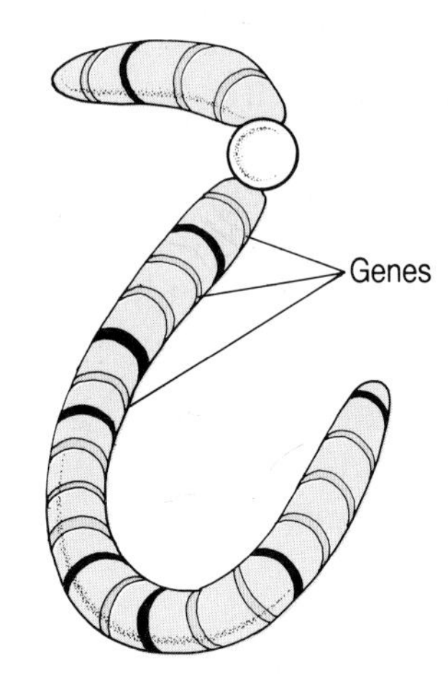

This is what a chromosome would look like if you could see the genes.

What happens to genes during fertilization

A sperm has 23 chromosomes with genes from a man. An ovum (egg) has 23 chromosomes with genes from a woman. During fertilization the sperm and ovum join together.

Each chromosome from the sperm then pairs up with a matching chromosome from the ovum. This brings the two sets of genes together. The genes for hair colour pair up, the genes for skin colour pair up, and so on. *Genes always work in pairs.*

The genes in a pair may be identical. For example, they may both produce black hair. But if one is a gene for black hair and the other a gene for blond hair they are in competition.

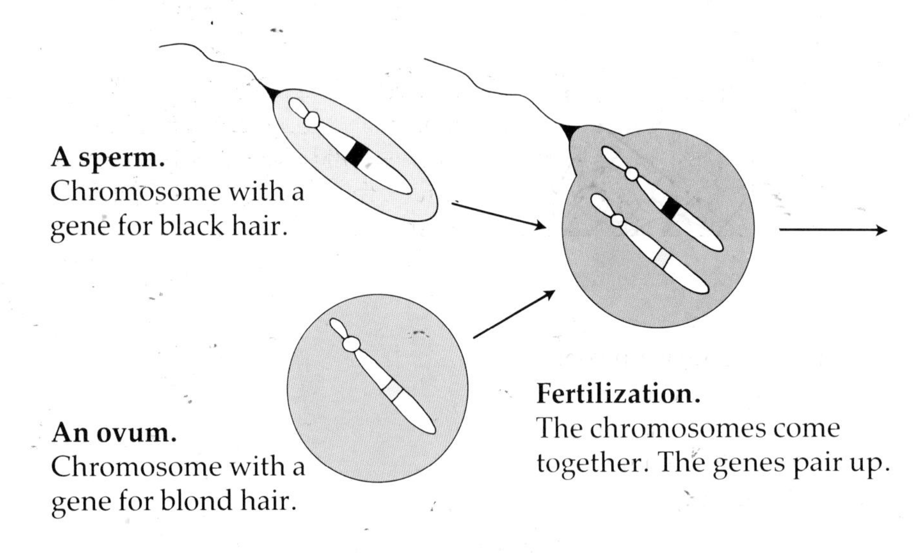

A sperm.
Chromosome with a gene for black hair.

An ovum.
Chromosome with a gene for blond hair.

Fertilization.
The chromosomes come together. The genes pair up.

The child has black hair. Only the gene for black hair has worked.

Dominant and recessive genes

The child has black hair because the gene for black hair is more powerful than the gene for blond hair. It **dominates** the gene for blond hair and produces the final hair colour. Genes which dominate other genes are called **dominant genes.**
Genes which are dominated are called **recessive genes.** When a dominant gene pairs up with a recessive gene, the dominant one produces the final effect.

Genes in action

In diagrams, genes are usually shown by letters. Capital letters are used for dominant genes and small letters for recessive genes. In the diagram below, H is the gene for black hair and h is the gene for blond hair.

Mother's cells contain two genes for black hair (HH) so she has black hair.

Father's cells contain two genes for blond hair (hh) so he has blond hair.

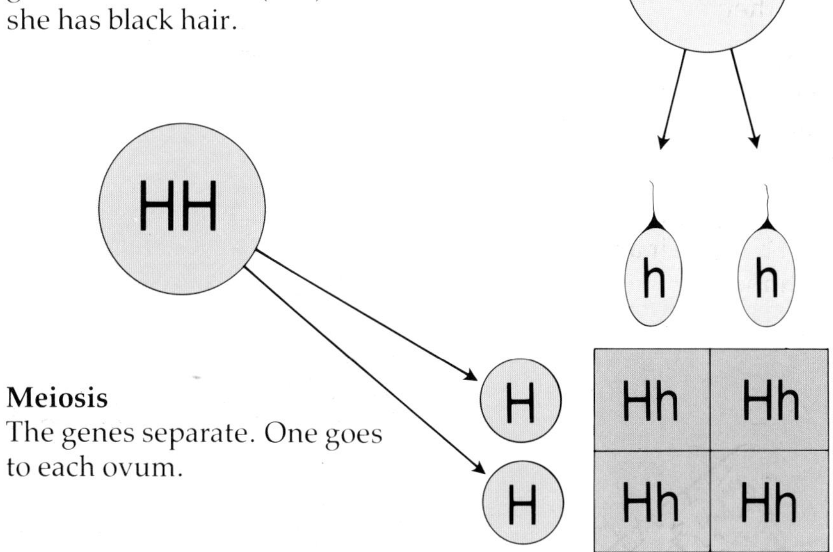

Meiosis
The genes separate. One goes to each sperm.

Meiosis
The genes separate. One goes to each ovum.

Zygotes
All have the genes Hh. So all the children will have black hair, because the gene for black hair is dominant.

Homozygous and heterozygous

A person with two identical genes for a characteristic is pure-bred or **homozygous** for that characteristic. (HH is a pure-bred gene.) A person with two different genes (one dominant and one recessive) for a characteristic is **hybrid** or **heterozygous** for that characteristic. (**Homo** means the same and **hetero** means different.)

Questions

1 Which parts of chromosomes control inherited characteristics?

2 What are genes made of?

3 In the gene pair Hh (for black and blond hair) which gene is dominant?

4 If E is the gene for brown eyes, and e is the gene for blue eyes, what eye colour will the following gene pairs give?
a) EE **b)** Ee **c)** ee

5 Look at question 4 and pick out a gene pair which is heterozygous, and a gene pair which is homozygous.

2.7 More about chromosomes

On the previous page all the children had dark hair, even though their father was blond. This happened because they inherited a dominant gene (H) and a recessive gene (h). They were **hybrids**, or **heterozygous** (Hh) for hair colour.

This does not mean that blond hair has been lost altogether. Look what happens when two hybrids have children.

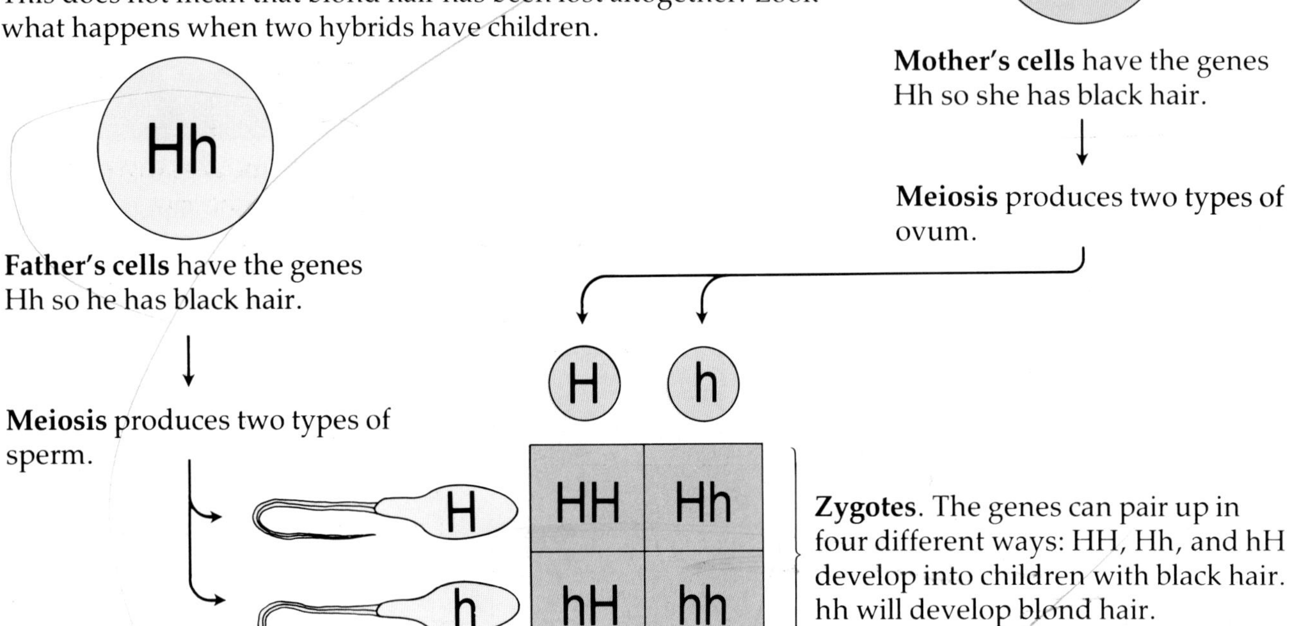

Blond hair appears this time because two recessive genes (hh) can come together. But the children are three times more likely to have black hair than blond hair.

Other dominant characteristics

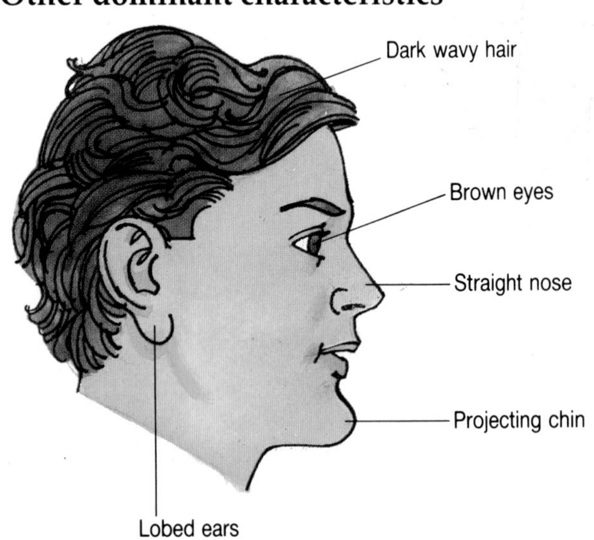

Other recessive characteristics

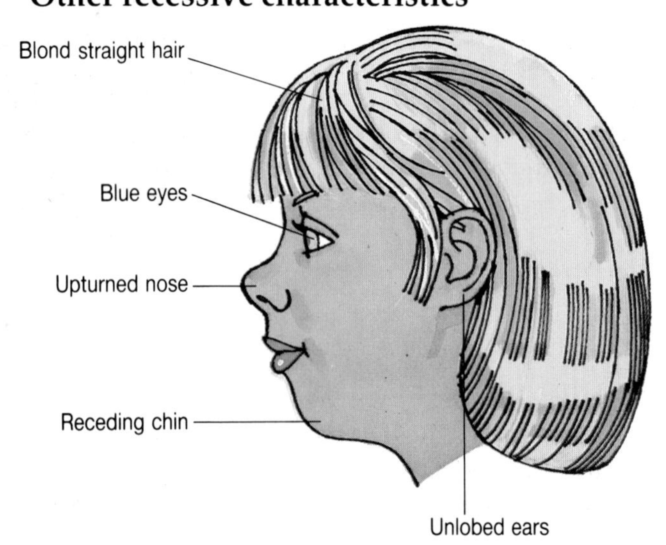

Boy or girl?

Human cells have 46 chromosomes altogether. There are 22 matching pairs. But the last pair does not always match. These two chromosomes are the **sex chromosomes**. They control whether a baby develops into a boy or a girl.

If you are male your cells have an X chromosome, and a smaller Y chromosome. If you are female your cells have two X chromosomes. This diagram shows how sex is inherited.

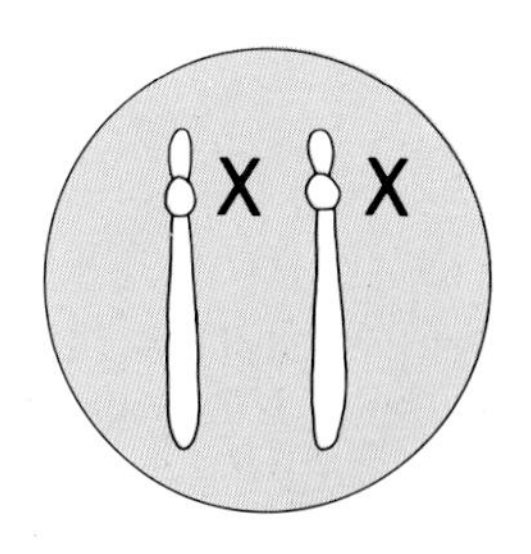

Mother's cells contain two X chromosomes.

Meiosis produces ova with one X chromosome each.

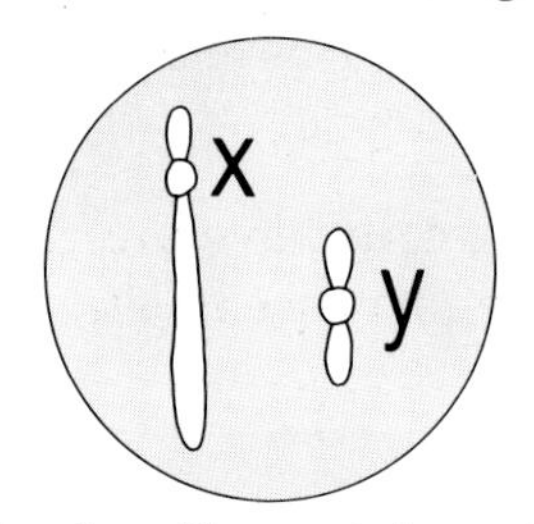

Father's cells contain an X and a Y chromosome.

Meiosis produces equal numbers of X and Y sperms.

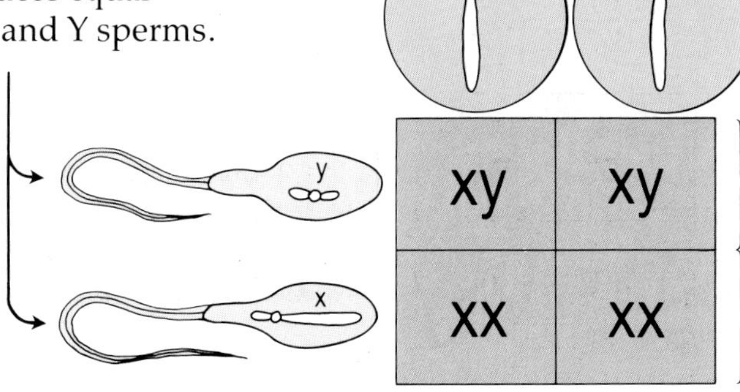

XY zygotes develop into boys.

XX zygotes develop into girls.

Since there are equal numbers of X and Y sperms a child has an equal chance of being a boy or a girl.

This picture shows the 46 chromosomes of a man. They were photographed under a microscope, then cut out and sorted into 22 matching pairs, and a pair of X and Y chromosomes.

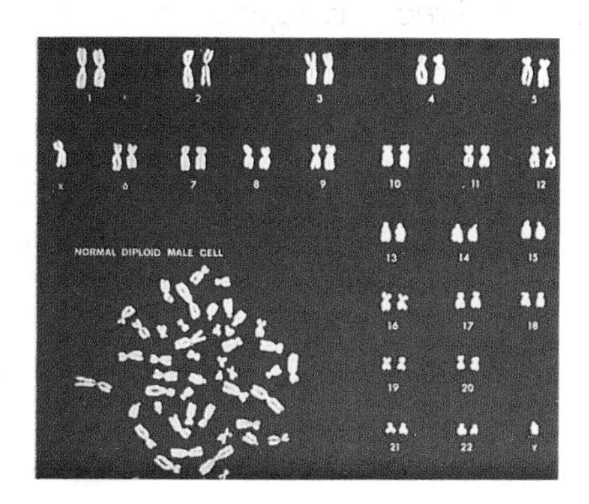

Questions

1. W is the gene for wavy hair, and w is the gene for straight hair. Father has the genes WW. Mother has the genes Ww. Draw a diagram like the one on the previous page to show what type of hair their children could inherit.
2. What do XY zygotes develop into? What do XX zygotes develop into?
3. Why are there roughly equal numbers of boys and girls?

2.8 Patterns of inheritance I

Alleles

People may have brown eyes, green eyes, or blue eyes. In other words, the gene which controls the development of eye colour exists in several different forms. All the different forms of a gene are called **alleles.** So brown, green, and blue are alleles of the gene for eye colour. The alleles of a gene alter the way in which the characteristic controlled by that gene appears in an organism.

Phenotype and genotype

A **phenotype** is a characteristic you can see such as black hair, an upturned nose, or the colour of a flower. A **genotype** is the set of genes which produce a phenotype.

The phenotype depends on the alleles of a gene present in the cells of an organism. Every cell contains two alleles of each gene – one from each parent. The phenotype they produce depends on whether the alleles are dominant or recessive.

Flower colour is a phenotype produced by the genotype CC or Cc. White flowers (flowers with *no* colour) have the genotype cc.

Black hair is a phenotype produced by the genotype HH or Hh. Blond hair is a phenotype produced by the genotype hh.

Genes and the environment

Organisms with the same genotype should look the same. But they may look different if they grow in different surroundings or **environments.** The environment can affect the way genes work.

These maize plants have the same genotype. But they look different because they were grown in different soils.

The ones captioned 'Complete' were grown in soil that had everything they needed. The ones captioned 'No nitrogen' were grown in soil that had everything *except* nitrogen. So the different soils have given different phenotypes from the same genotype.

Complete nutrients.

No nitrogen.

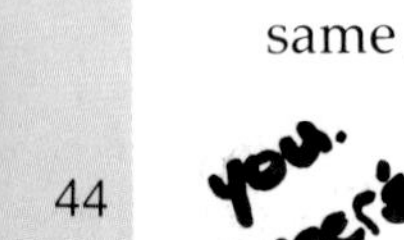

Mutations

A **mutation** is a change in a gene or chromosome which alters the way an organism develops. If a mutation affects gametes (sex cells), it can be inherited. If only body cells are affected, it will not normally be inherited.

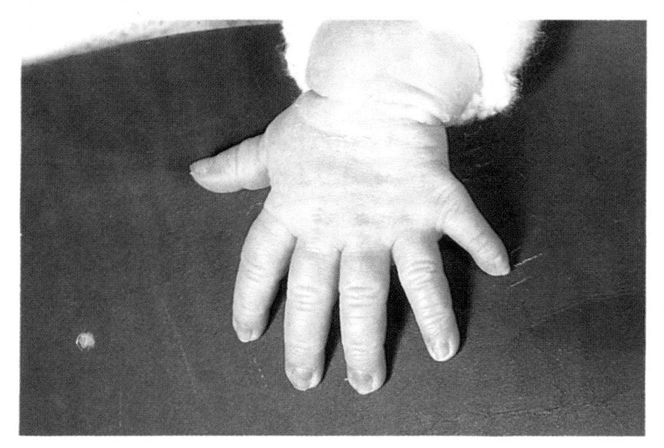

A gene mutation can cause extra fingers and toes to grow, as you can see on this photo of a baby's hand. This does not happen very often.

Sickle cell anaemia is a painful inherited disease, caused by a mutation which makes red blood cells become sickle-shaped in low oxygen concentrations.

What causes mutations?

Mutations can occur naturally. They can also be caused by X-rays, ultraviolet and other radiation, and chemicals such as nitrosamines, found in cigarette smoke, which cause cancer.

The fruitfly on the left was exposed to gamma radiation, which caused a gene mutation. As a result, its offspring (on the right) has crumpled wings.

A natural chromosome mutation can cause a woman to produce an ovum with 24 chromosomes instead of 23. If the ovum is fertilized, the baby has 47 chromosomes instead of 46. If the extra one belongs to chromosome pair 21, the baby develops **Down's syndrome.**

Questions

1 In fruitflies, R gives red eyes and r gives white eyes. What is the phenotype of these genotypes:
a) RR? **b)** Rr? **c)** rr?

2 In mice, B gives dark hair and b gives light hair. What are the genotypes for dark-haired mice?

3 What is a mutation? What causes mutations?

4 Describe a harmful mutation, and a helpful mutation.

2.10 DNA and the genetic code

The instructions to make a human being, or a spider, or a dandelion, or almost any living thing, are contained within molecules of DNA (**deoxyribonucleic acid**) coiled up inside the chromosomes of a cell. These instructions are stored using the **genetic code**.

A gene is a length of DNA containing a small part of this genetic code. There is one DNA molecule per chromosome and this may contain up to 4000 genes. Together, all the chromosomes in human cells contain about 100 000 genes.

By studying the structure of DNA we can begin to understand what these instructions are, and how they form the genetic code which is used to construct a complete living organism out of a single, fertilized egg cell.

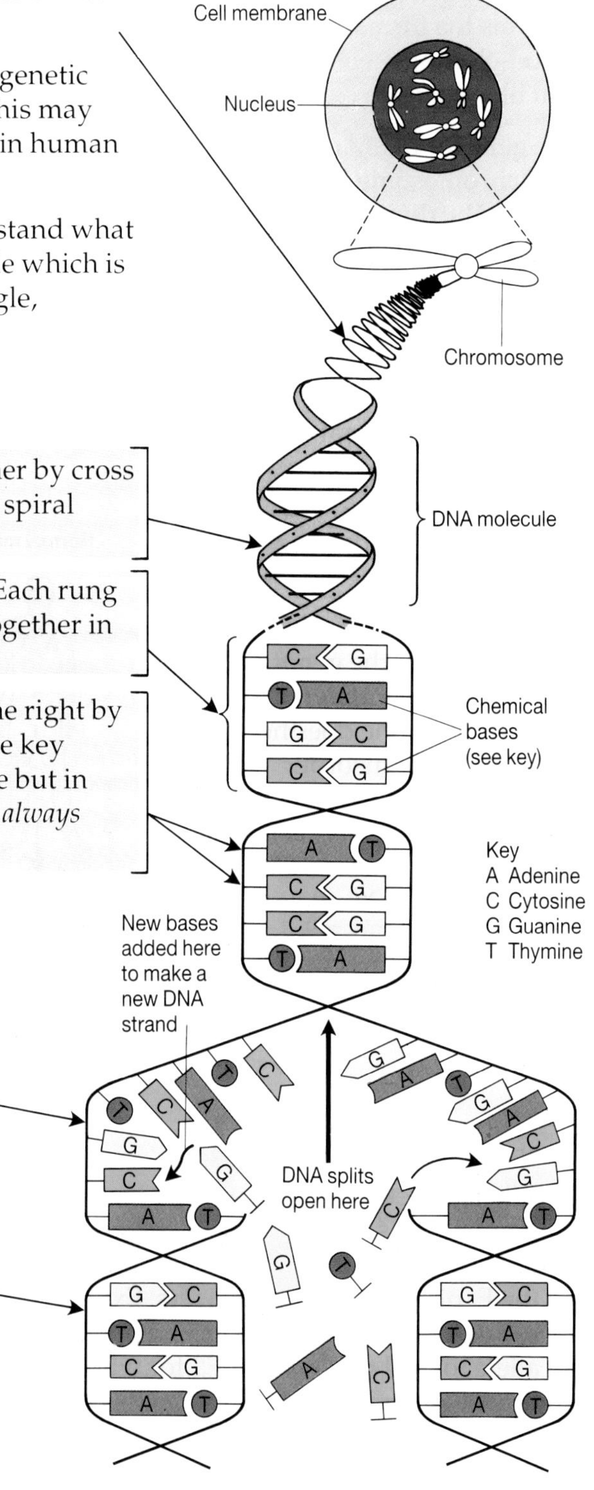

Structure of DNA

A molecule of DNA consists of two strands joined together by cross pieces, rather like a ladder which has been twisted into a spiral shape, called a **double helix**.

The rungs of the ladder are chemical units called **bases**. Each rung of the ladder is made up of two bases which are joined together in the middle.

There are four types of base, shown on the diagram on the right by the letters **A, G, C** and **T** (these letters are explained in the key opposite). There are millions of rungs in a DNA molecule but in every one base A is *always* opposite base T, and base C is *always* opposite base G.

How DNA is copied before cell division

Before a cell can divide, all of its DNA molecules must make exact copies of themselves so that the new cell has a copy of the coded information. To do this the DNA molecule splits into two single strands, rather as though it had been unzipped up the middle.

Each single strand is now built up into a double strand again, using new bases supplied by the cell. One base A always joins with a T, and a base C always joins with a G.

Very rarely, an error occurs as the DNA is copied, which causes a change in the genetic instructions. This change is one way in which a mutation can occur.

We are what is written in our genes

Genes control the development of inherited characteristics such as eye and hair colour. Genes can do this because they contain the instructions (genetic code) for making proteins out of amino acids. Proteins, particularly enzymes, control all the chemical processes of life as well as the structure and functions of cells. Therefore, by controlling which proteins a cell makes, genes control the development, and the structure and functions of a whole organism.

How genes make proteins

A protein is a number of different amino acids linked together. It is the sequence of the bases A, T, C and G along a DNA molecule which tells a cell which amino acids to use, and the order in which they must be linked up to make a particular protein. A gene contains the sequence of bases for one complete protein molecule.

Reading the code

Imagine the letters A, T, C and G are the alphabet in which the words of the coded information are written. The letters CAA (i.e. base sequence cytosine, adenine, adenine) would be a code word for the amino acid valine, and the word AAT (adenine, adenine, thymine) would be a code word for leucine. In other words the coded information is written in words of three letters (i.e. groups of three bases), each of which tells a cell which amino acid to link with another. Let's now look at how this link-up is achieved.

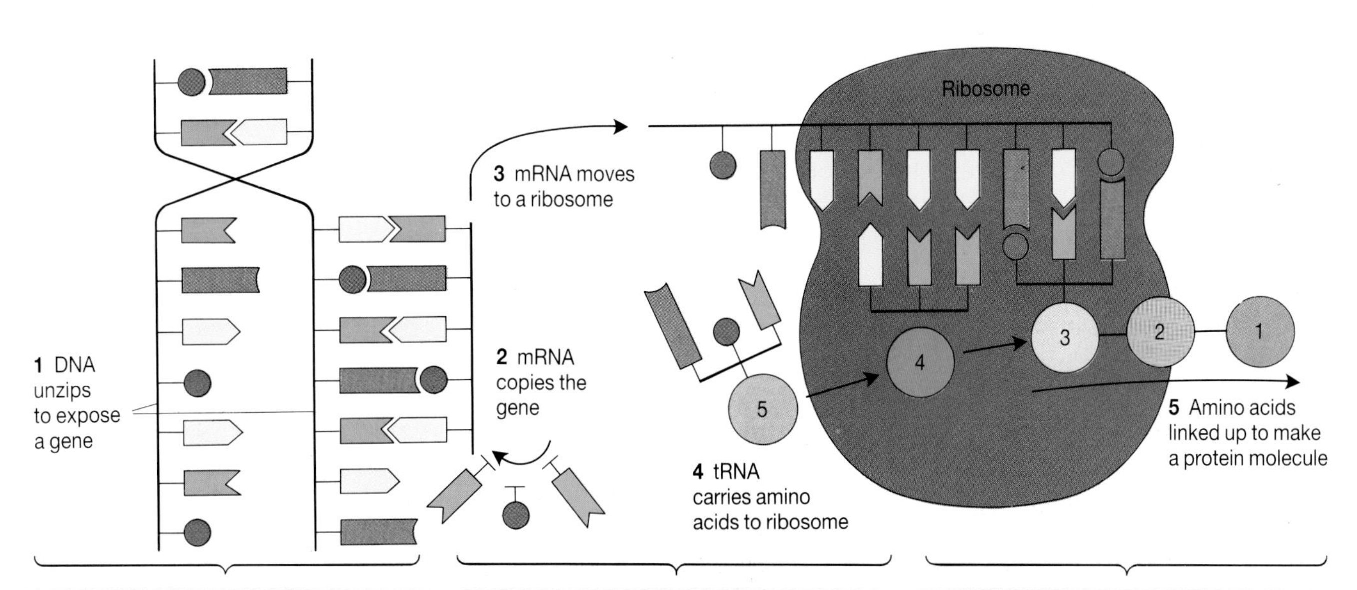

DNA unzips Part of a DNA molecule opens up exposing the gene for a protein **(1)**. The sequence of bases (the code for a protein) in the gene is copied. But this time the new strand, which is an exact copy of the gene, is not made of DNA, but a similar chemical called messenger RNA **(2),** written mRNA for short.

mRNA carries the code for a protein to a ribosome mRNA moves out of the nucleus of the cell and carries the copy of a gene to a structure in the cytoplasm called a **ribosome (3)**. Here it meets a second type of RNA, called transfer RNA (tRNA). tRNA picks up amino acids and transfers them to a ribosome **(4)** where they are linked up to make proteins.

Amino acids are linked together A tRNA molecule has three bases which can only fit against three matching bases on the mRNA. When this happens, the amino acid it carries is linked to a chain of amino acids forming a protein molecule **(5)**. The amino acid sequence, and so the type of protein, depends on the sequence of bases copied by mRNA from the gene.

2.11 Genetic engineering

Modern wheat varieties were produced by crossing selected wild grasses thousands of times.

For centuries, when farmers wanted a cow with more milk or a bigger yield of wheat, they carefully selected only the most productive animals and plants to breed (cross) together. But this process of **selective breeding** is slow, and hit-and-miss.

Farmers did not know it at the time but they were changing animals and plants by moving genes from one living thing to another. Now scientists can do this with far greater reliability, and in a single generation, using the techniques of **genetic engineering.** A single gene can be isolated from millions of others in an organism, then moved to almost any other living thing, where it works as if still in the organism it was taken from. They have transferred genes from fish to tomatoes, from humans to sheep, and even humans to bacteria.

Human genes can be transferred to cows to make them secrete human blood-clotting factor in their milk.

Manipulating the stuff of life

People who are diabetic need supplies of insulin because their bodies do not naturally make it. The gene for insulin can be removed from a chromosome in a human pancreas cell using an enzyme which cuts out the exact piece of DNA containing the gene. The gene can then be transplanted into bacteria where it combines with the microbes' DNA to turn the bacteria into tiny factories for making insulin.

Cutting out a gene An enzyme cuts out one gene from thousands of others in a chromosome taken from a donor cell.

Transplanting a gene A circular strand of bacterial DNA, called a **plasmid,** is cut open with an enzyme, and the donor gene inserted.

Extracting the products Millions of identical engineered bacteria grow in a fermentation tank. These are harvested and the useful products separated.

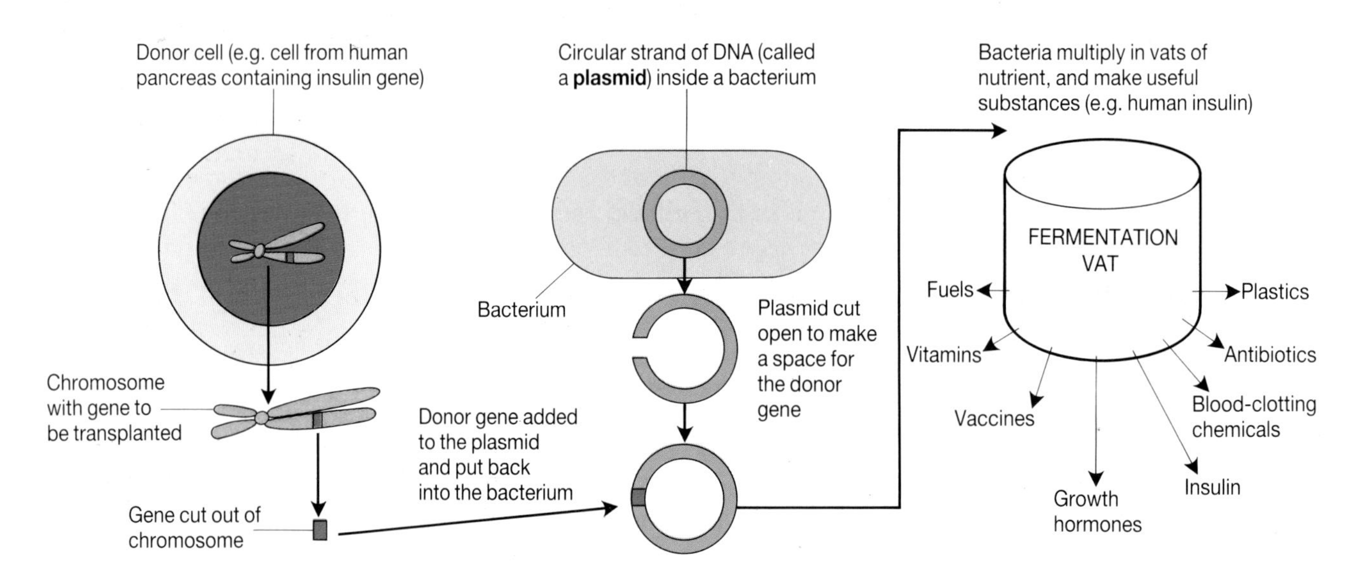

Gene clones Growing millions of identical bacteria, each with the same genes, is an example of **gene cloning.** Plants, and some animals, can be grown from a single cell to which a foreign gene has been added. Unlimited copies of this cell can be grown and used to form clones of whole plants and animals. Theoretically, humans could be cloned in this way. But would this be a good idea?

These tomatoes have been given genes from fish which makes them frost-resistant and fresh for longer. Would you eat them?

Gene therapy – a cure for inherited diseases

One day inherited diseases may be cured by replacing defective genes with healthy ones. Possible candidates are disorders such as haemophilia, caused by one faulty gene. But the problem is how to put healthy genes into the DNA of millions of body cells.

Microbe carriers One way is to use viruses, and certain bacteria, which attack cells by injecting their own genes into the DNA of a host cell. If the microbe's disease-causing genes are removed and replaced with a 'good' gene, the microbe could be used to take the new gene directly into the DNA of another organism's cells.

This mouse was injected with a growth hormone gene. Is this ethically acceptable?

Reaping the genetic harvest

The benefits of genetic engineering in medicine, agriculture and industry seem endless. Within decades we may have:

- Plants with chemicals in their leaves which kill insects, fungi and other pests.
- Plants engineered to live in polluted, arid, or salty soils.
- Plants which produce grains of plastic which can be harvested from their leaves.
- Wheat, potatoes, tomatoes and other crops which can 'fix' nitrogen from the air to make proteins without costly fertilizers.
- Tomatoes and other fruit which remain fresh for much longer.
- Sheep and cows which produce human growth hormones, antibiotics and blood-clotting chemicals in their milk.
- Bacteria which 'eat' oil slicks, and bacterial 'miners' which dissolve everything from metal ore, leaving almost pure metal.
- Farm animals turned into giants by growth hormone genes.

These cotton plants carry a gene which allows them to make their own insecticide.

Questions

1 Mice can be given genes which make them develop cancers, cystic fibrosis and other diseases, for testing possible treatments. Is this cruel? Should such research be stopped?

2 Crop plants can be given genes which make them resistant to chemicals which kill weeds, or able to make chemicals which kill insects. What are the benefits and drawbacks of this?

3 If oranges, bananas, and other tropical fruits could be engineered to live in cool climates, how would this harm the economies of poor countries?

4 Foreign genes put into insect-pollinated crops could quickly spread into related wild plants. How could this harm the environment?

2.12 Biotechnology

Biotechnology means using plants, animals, and microbes (bacteria and fungi), to produce useful substances. The methods go back thousands of years. There is nothing new in using microbes to turn milk into yoghurt and cheese, or for making beer, wine, vinegar, and bread.

The biotechnology industry of today makes use of these ancient techniques, plus new ones including genetic engineering (Unit 2.11), to produce a huge variety of useful things including fuels, foods for humans and animals, antibiotics and other medicines, plastics, and industrial chemicals. It can also dispose of sewage, refuse, and spilled oil. In these ways, biotechnology can help solve the world's food, health, and energy problems.

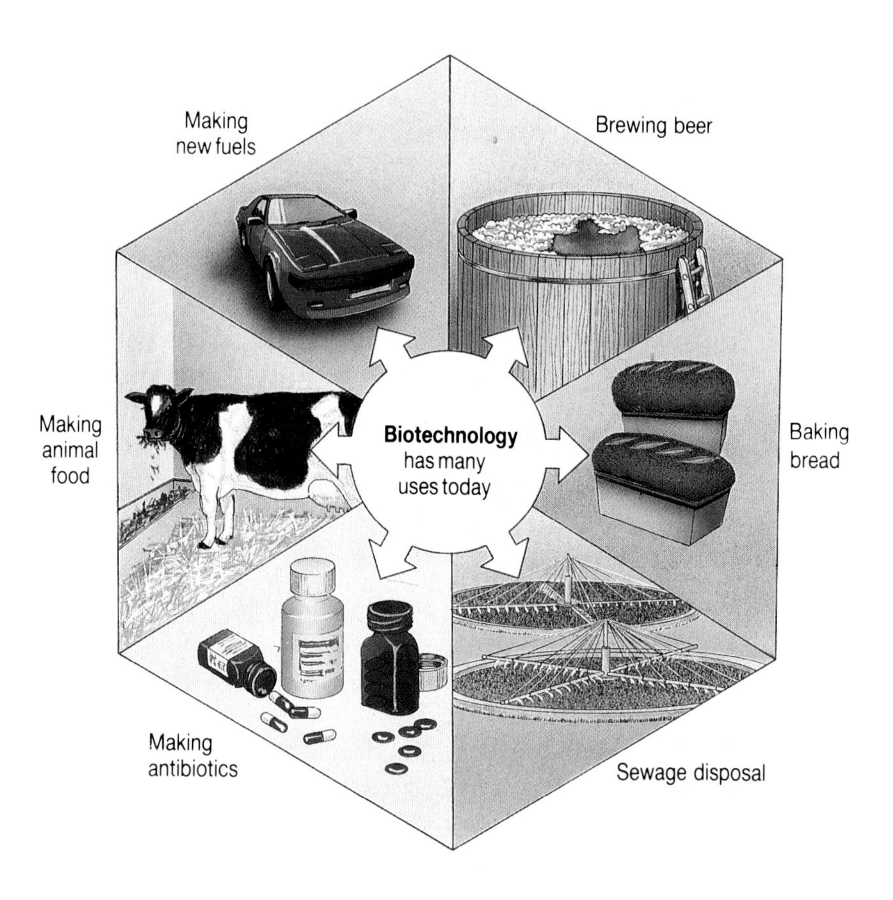

Biotechnology in action

Antibiotics These are medicines, obtained from bacteria and moulds, which kill or slow the growth of other microbes. The antibiotic **penicillin** was discoverd in 1928 by Alexander Fleming. Since then, antibiotics have been isolated from microbes all over the world and over a hundred are now in use.

Enzymes These are substances which speed up chemical reactions inside living cells. Digestive enzymes are put into biological washing powders to dissolve food stains. They are used in the leather industry to help remove hair from skins; also in the production of whisky and other alcoholic drinks, to turn the starch in grain into sugar. This is then changed into alcohol by fermentation.

Fuels from microbes When petrol burns in a car engine, it has gone forever, and the exhaust gases produced pollute the air. But some sources of energy are clean, and can be replaced indefinitely. These include alcohol and methane gas.

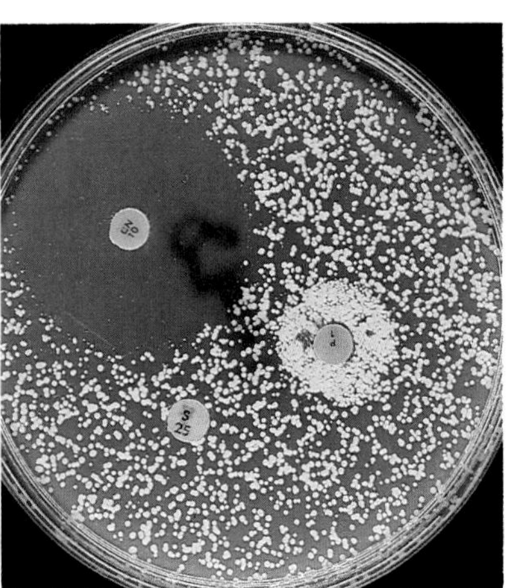

Colonies of bacteria growing on a special jelly. The discs contain antibiotics under test. The clear area around the top left-hand disc shows that it contains an antibiotic which kills this type of bacteria.

Alcohol Instead of putting petrol in their cars, motorists in Brazil can fill up with alcohol made by fermenting sugar from sugar-cane with yeast. Unfortunately sugar and corn starch (another source of alcohol) are more valuable as foods than fuels. One solution would be to make alcohol from other plant materials, including the leaves of crop plants. To do this, genes could be transplanted from moulds to yeast, so that the yeast can digest the cellulose in the leaves.

Methane gas Another use for the leaves of crops and other plants is to place them in containers to rot. This produces methane gas which is used for heating, cooking, or making electricity. What is left makes animal food and fertilizer. Sewage and animal dung can also be used to make methane. They can be used to feed algae which are harvested from time to time and digested by bacteria which produce methane.

Eating microbes This may seem a strange idea, but if you like Marmite, made from processed yeast, you already eat microbes. Algae, bacteria, and fungi can also make foods. They are called single-cell proteins (SCP) and are very rich in protein, fat, vitamins, and minerals.

Edible mould Food with the same nutrients as meat can be made from the mould *Fusarium*. This is found in shops under the brand-name of Quorn.

Edible algae Algae will grow in water containing dissolved minerals, and produce unlimited amounts of proteins, vitamins, and minerals. They can be dried, compressed, and used to feed humans and animals. Algae growing in lakes could yield ten times more food than wheat grown on the same area of land.

Food from gas and oil Bacteria grown in methanol gas, water, ammonia, and mineral salts can produce a nutritious animal food called Pruteen. Also, there are bacteria which live on cheap chemicals found in crude oil.

Sort these foods into those produced by microbes, and those made out of microbes.

Questions

1 What does biotechnology mean?

2 What are the oldest examples of biotechnology?

3 How could biotechnology help solve the world's food and energy problems?

4 a) What is genetic engineering?
  b) What part does genetic engineering play in modern biotechnology?

2.13 Evolution I

Where have the millions of different living things come from? The most likely answer is that they were produced by **evolution.**

What is evolution?

Evolution means change and improvement from simple beginnings. Modern aircraft were evolved by changing and improving the first simple aircraft designs.

Evolution of living things

The first living things appeared about 3500 million years ago. They were no more than bubbles full of chemicals but they could reproduce.

Some of the young were different from their parents. Some of the differences meant they could survive better than their parents. Over billions of years, these changes and improvements led to all the different creatures alive today.

The first powered aircraft evolved to become Concorde.

This diagram shows how vertebrates could have evolved from simple beginnings.

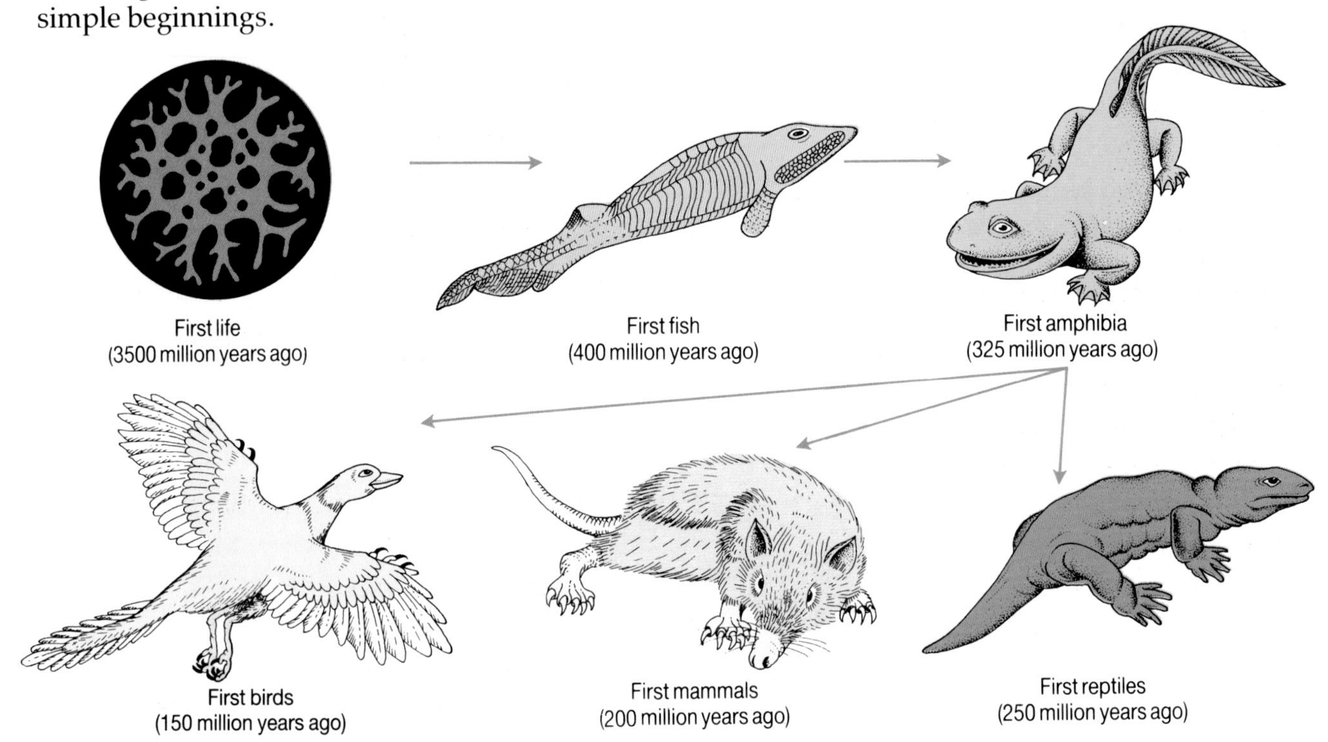

Natural selection

Living things are at risk from disease, predators, and other dangers. If the young are born with differences that make them tougher, or stronger, or better suited to their surroundings, they will be able to live longer and have more babies than their weaker brothers and sisters.

This idea is called **survival of the fittest,** or **natural selection**. Nature *selects* the fittest and strongest for survival by killing off the weaklings. The idea was first put forward by Charles Darwin about a hundred years ago.

Darwin suggested that the young inherit improvements from their parents. In this way a species could keep on improving until it produced a new species.

Can you spot the pale Peppered moth? Here its colouring helps to protect it from enemies.

These photographs show an example of natural selection.

They show dark and light varieties of the Peppered moth. In soot-blackened city areas the light ones are easily seen and eaten by birds, so the dark ones are common. But in clean country areas the dark ones are easily found by birds, so that light ones are more common.

The dark Peppered moth is safer resting on dark surfaces, like this sooty tree trunk. . .

. . . but a pale Peppered moth on a dark surface is easily spotted by birds.

Artificial selection

Animal and plant breeders have proved that selection can change a species. They use the **artificial selection** process.

They select and breed sheep with the longest coats, pigs with the most meat, trees with the juiciest fruit, and plants with the most colourful flowers. So we now have thousands of plants and animals that are very different from their wild ancestors.

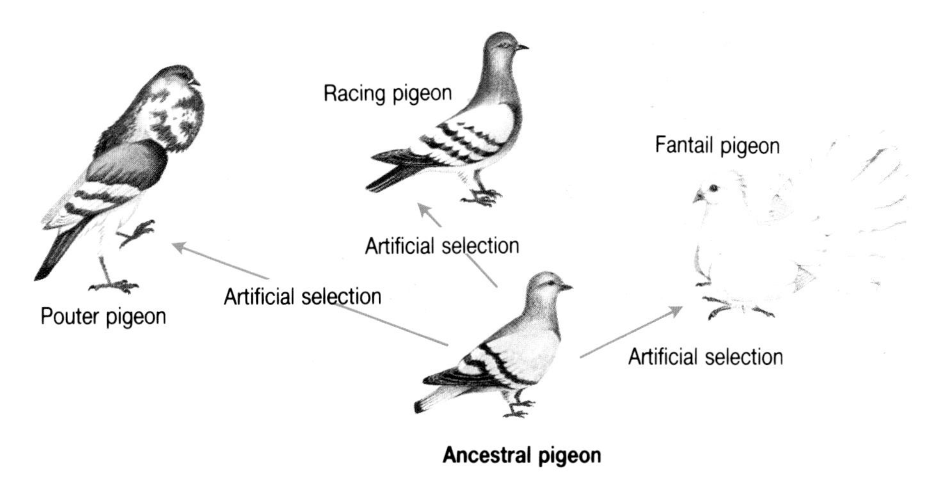

All these different pigeons were produced by artifical selection.

Questions

1. What does evolution mean?
2. What does natural selection mean?
3. How does natural selection make dark Peppered moths more common near factories than light ones?
4. Explain how one species may produce a new species over millions of years.
5. What does artificial selection mean?
6. Give one example of artificial selection.

2.14 Evolution II

Fossils – the record in the rocks

In many parts of the world, if you could dig a tunnel deep into the ground, it would be like taking a journey into the past. As you tunnelled deeper and deeper, the rocks would be older the further you went. If you could collect fossils on your way down, you would end up with a selection from the most recent to the most ancient organism which ever lived.

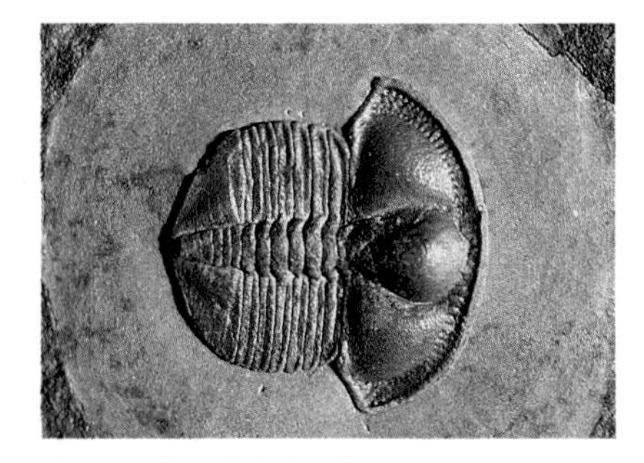

A fossil trilobite from the Ordovician period. Most trilobites lived on the sea-bed.

If life has evolved from simple beginnings, you would expect the oldest rocks to contain fossils of primitive creatures, and that younger rocks would contain fossils of more complex creatures. You would also expect that, if evolution means change, fossilized creatures which lived millions of years ago ought to be different from those alive today, and that some would have become extinct.

In fact, the oldest rocks contain no fossils, since they were formed before life began. Fossils in younger rocks show an increase in complexity with time, and include an enormous variety of creatures which are different from today's plants and animals. Many fossils are of creatures which no longer exist. This is why the **fossil record** is the strongest evidence we have that evolution occurred.

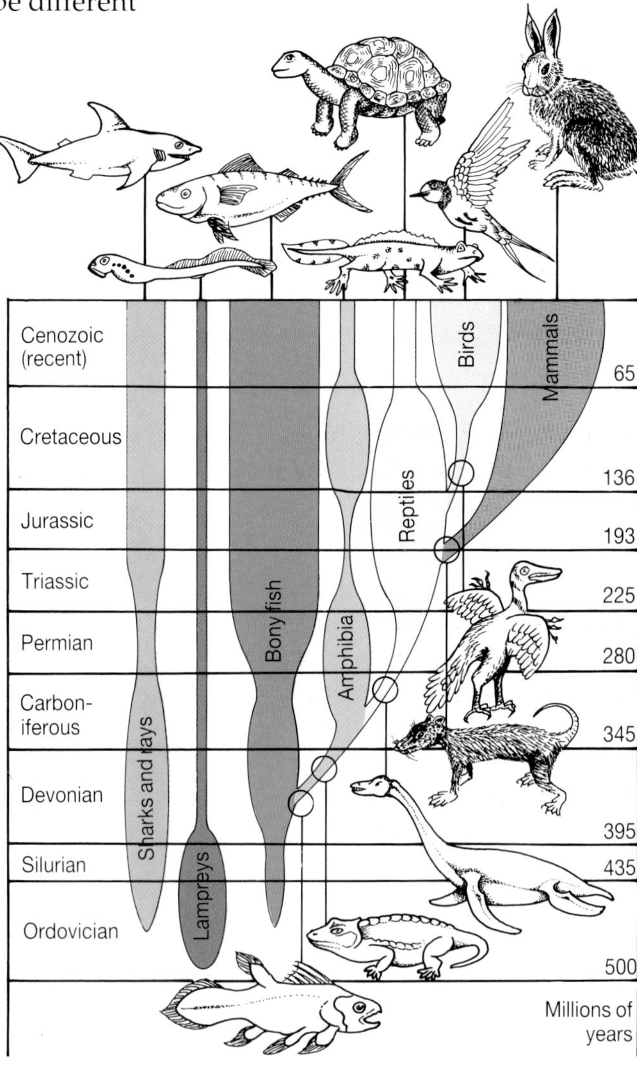

A chart showing the evolution of the main groups of vertebrate.

How fossils are formed

If the body of an animal or plant is washed into a river, or comes to rest at the bottom of a shallow sea, it will eventually be covered with sand and silt which settle on top of it. The soft parts will quickly decay but the hard parts such as animal bones, crustacean exoskeletons, and plant fibres, may survive long enough to absorb minerals from the water. In time these minerals replace the materials which made up the creature's hard parts, literally turning them into stone. A fossil has been formed.

As a deeper and deeper weight of sediment builds up, pressure causes the lowest layers to harden into **sedimentary rock.** Millions of years later, the fossil may be pushed up to the surface by movements of the Earth's crust, and then exposed by cracking, erosion, or the quarrying of the rocks.

How organisms become extinct

There are many reasons why some creatures gradually declined in numbers until they became extinct. Climate can change from burning deserts to freezing ice ages, so that, unless living things can migrate to better conditions, or evolve and adapt to new ones, they will die. Or new creatures may evolve which are in some way better equipped to survive. These take over the living space and food supply of some existing species, depriving them of the necessities of life.

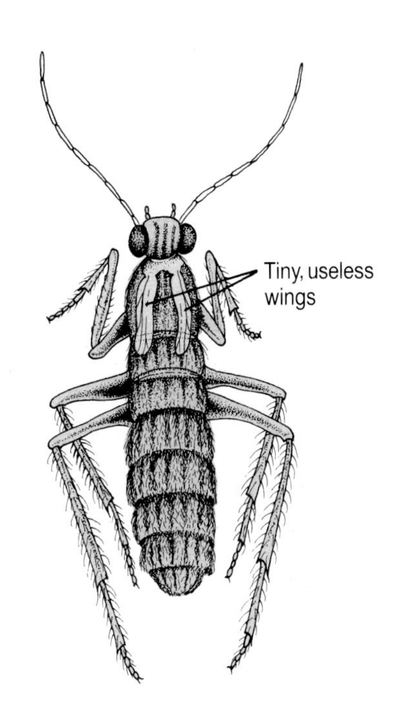

Mutations – the raw material of evolution

Evolution would be impossible without change, and mutation produces that change. When a mutation occurs, the hazards of natural selection decide whether it causes death, or improvements which are passed to future generations. An example can be found on the island of Kerguelen in the Indian Ocean, where mutant moths with tiny, useless wings live. Normally they wouldn't live as long as their winged cousins, but Kerguelen is very windy and winged insects get blown out to sea. Crawling, flightless insects are safe.

How new species form

New species can evolve if creatures become isolated by water, mountains, or other barriers, so can no longer breed with their own kind. This happened 5 million years ago soon after the islands of Hawaii were formed by volcanic eruptions.

Birds called honey creepers, from mainland America, found the islands and, because there were no other birds to compete with, evolved a variety of beak shapes for eating the different types of available food. Eventually they evolved into separate species.

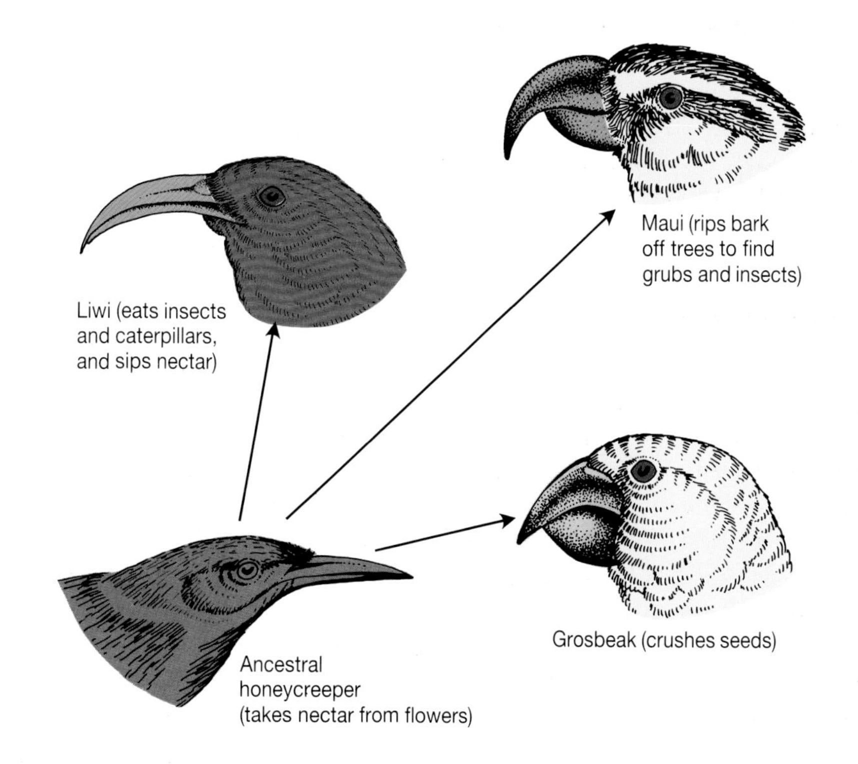

Questions

1. From which group of animals did amphibia, reptiles, birds and mammals evolve?
2. How are fossils formed? How do they provide evidence that evolution has occurred?
3. Mutations are the raw material of evolution. Explain this statement.

Questions on Section 2

1 *Recall knowledge*

Vocabulary test Write out the numbered phrases below then, opposite each, write the word from the list of technical terms which the phrase describes.

nucleus, vacuole, cytoplasm, tissue, organ, osmosis, chromosome, fertilization, gametes, zygote, genes, homozygous, heterozygous, mitosis, meiosis, genotype, phenotype, mutation

1 A sudden change in a chromosome or gene.

2 A gene pair, the members of which are different (i.e. one dominant and one recessive).

3 The visible, as opposed to the genetic, characteristics of an organism.

4 A group of cells specialized to do a particular job.

5 Movement of water through a semi-permeable membrane from a weak to a strong solution.

6 A space in a cell usually filled with liquid.

7 It happens when a sperm and ovum come together.

8 Made up of several different tissues.

9 A gene pair, the members of which are identical.

10 When a cell divides and produces two cells with a full set of chromosomes.

11 All of a cell except the nucleus.

12 The part of a cell which contains chromosomes.

13 They control the development of inherited characteristics (e.g. eye colour).

14 Sperms and ova.

15 Made up of many different genes.

16 A name for a full set of genes possessed by an organism.

17 A fertilized egg cell.

18 When a cell divides and produces two cells with half the normal number of chromosomes.

2 *Interpret data and design*

An uncooked potato was peeled. A piece from the middle was cut into two strips X and Y, each 50 mm long. One strip was placed in a strong sugar solution. Another strip was placed in pure water.

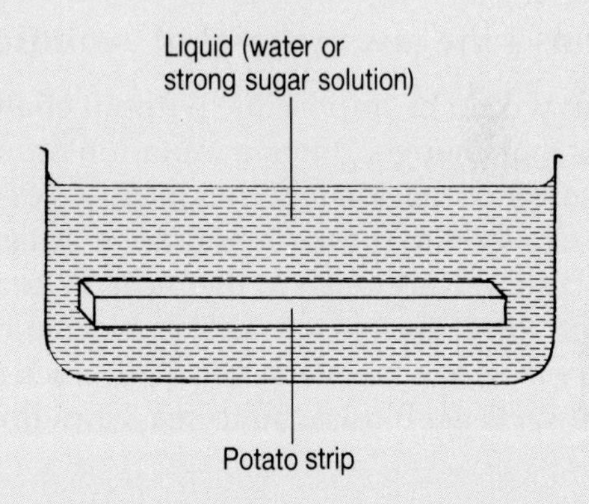

After 30 minutes strip X had grown longer. Strip Y had shrunk.

a) What could have happened to the cells in each strip, to make one strip swell and the other shrink?

b) Which strip had been placed in the sugar solution? Explain your answer.

c) The experiment was repeated using strips of cooked potato. These strips did not change in length. Can you explain why?

d) If you had some sugar solution which was exactly the same osmotic pressure (see page 46) as a potato cell and you placed a potato strip in it how, if at all, would it change?

e) Design an experiment using potato strips to produce a sugar solution with the same osmotic pressure as a potato cell. List your apparatus, and describe your method in full detail.

3 *Recall knowledge*

a) What is a tissue? Name two examples.

b) What is an organ? Name two examples.

c) What is an organ system? Name the organ system which:
carries 'messages' around the body;
changes food into a liquid;
carries food, oxygen and wastes around the body.

4 *Interpret data*
The diagram below shows how wing shape is inheritied in fruitflies. W is the gene for normal wings and w is the gene for curly wings. Flies 1, 2, 7 and 8 are homozygous.

a) What are the genotypes of flies 3, 4, 9 and 10?
b) What do their phenotypes tell you about the effects of the environment on the inheritance of wing shape.

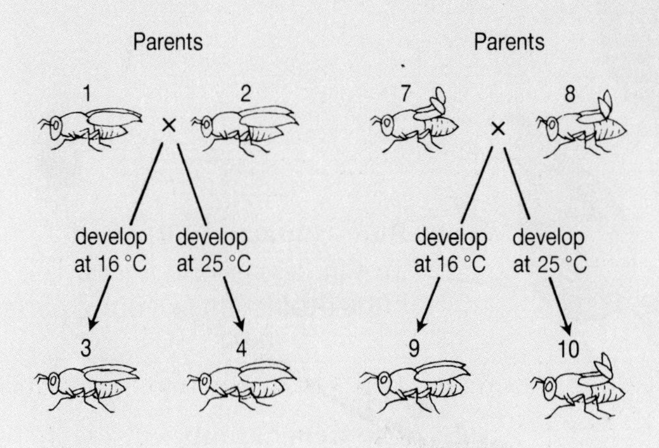

5 *Interpret data*

a) A strong sugar solution was poured into a bag made of semi-permeable membrane.
b) The bag was placed in a beaker of water (diagram A below).
c) After about half an hour it looked like diagram B.
d) Explain why the bag changed shape in this way.
e) What would happen if the bag was placed in a stronger sugar solution than the solution inside it?

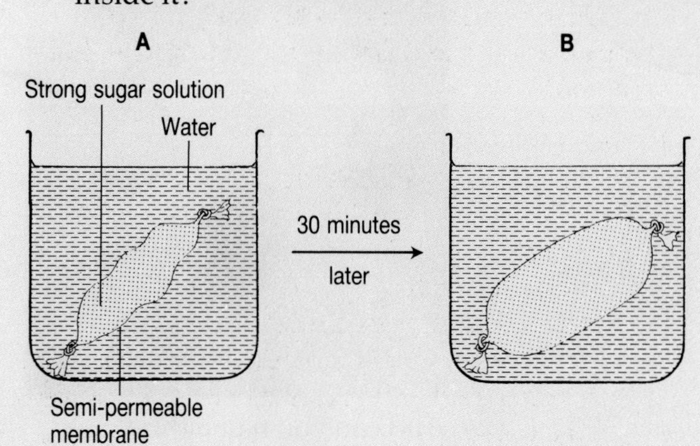

6 *Identify a problem and formulate hypotheses*
A family noticed that their pet cat was infested with fleas. They dusted it once with insecticide powder and almost every flea was killed.

A few months later the cat was again badly infested with fleas. They had to dust it several times, using insecticide from the same tin, in order to kill most of the fleas.

A few weeks later the cat was again badly infested with fleas. This time dusting with powder from the same tin had very little effect on the fleas.

a) Identify a problem which arises from these observations.
b) Suggest as many hypotheses as you can to explain what is happening. At least one hypothesis should be based on the information given on p.43 of Unit 2.8. Other hypotheses need have nothing to do with genetics.

7 *Interpret data*

a) E is the gene for brown eye colour, and e is the gene for blue eye colour. Which gene is dominant?
b) Both father and mother have the genes Ee in their cells. What colour are their eyes?
c) Copy the diagram below. Complete it to show how the genes in each sex cell come together at fertilization.

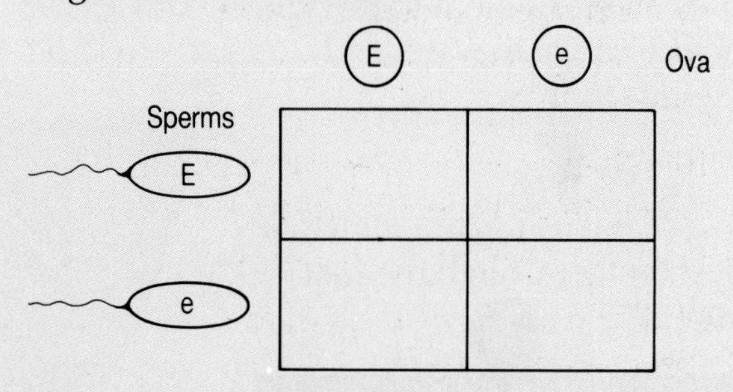

d) Which of the fertilized cells (zygotes) are homozygous for eye colour? Which are heterozygous?

e) Which zygotes will produce children with blue eyes? Which will produce children with brown eyes?

f) Complete the diagram again but this time with a father whose eye-colour genes are EE, and with a mother whose eye-colour genes are Ee. How does this affect the chance that they will have a blue-eyed or a brown-eyed child?

3.1 Flowering plants

There are thousands of plants that flower. Daisies, tomato plants, grasses, and chestnut trees are just some of them. Can you think of others? Flowers have male and female **sex organs**. These make **seeds** from which new plants grow.

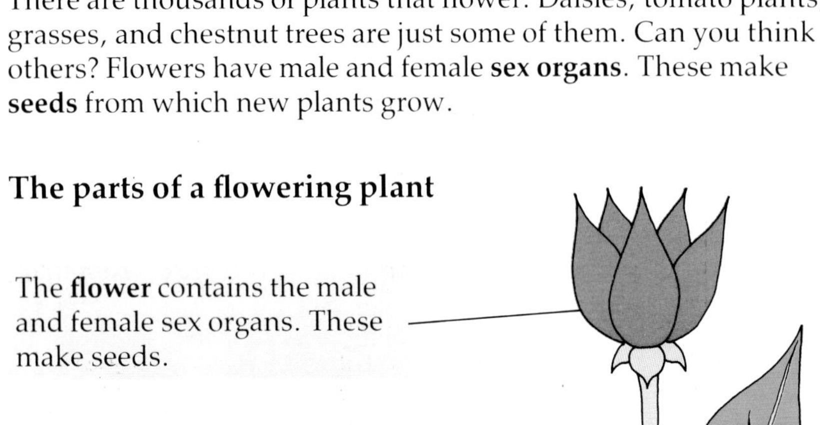

The parts of a flowering plant

The **flower** contains the male and female sex organs. These make seeds.

Buds contain small, partly-grown, leaves or flowers. The buds protect these young parts.

The **leaves** make food for the plant by photosynthesis. For this they need sunlight, carbon dioxide from the air, and water and minerals from the soil.

The **stem** has tubes inside it. Some tubes carry water and minerals from the roots to the leaves. Other tubes carry food up and down the plant.

Roots anchor a plant in the soil. They also take in water and minerals from it.

Soil level

This is a **main root**.

Most of the water and minerals are taken in through the **root hairs**.

This is a **side root**, or **lateral root**.

Which flowering plant is this?

And can you name this one?

How plants make food

Plants make food by **photosynthesis**. Photosynthesis means making things with light. That is what plants do. They use light energy to make food from carbon dioxide and water.

Inside a leaf is a green substance called **chlorophyll**. It can trap energy from sunlight.

Carbon dioxide is taken into the leaf through tiny holes. Water is carried into the leaf from the stem through tubes.

Trapped light energy makes the carbon dioxide and water combine, to make a sugar called glucose, and oxygen. Glucose is used for food. Oxygen is given off into the air and keeps air breathable for all living things.

At night, photosynthesis stops.

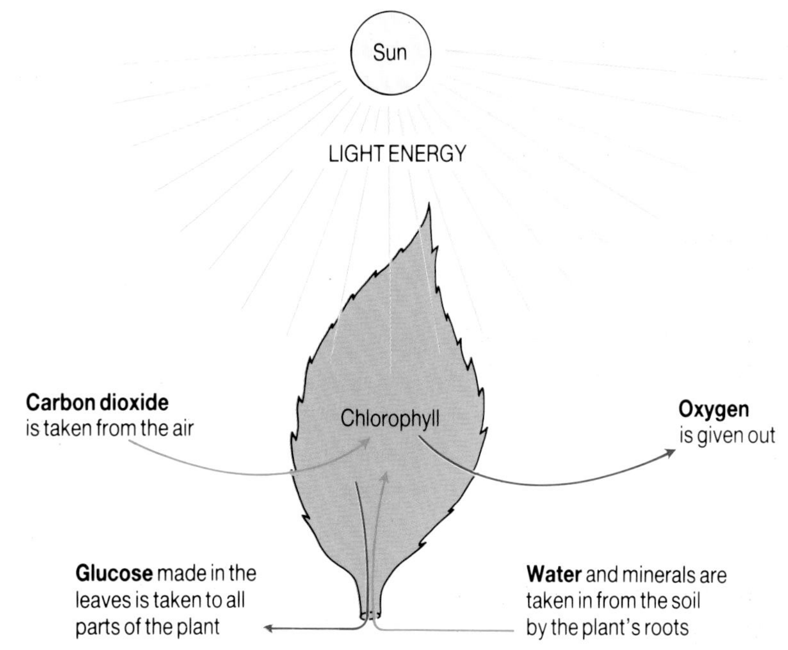

Here is a short way to show what happens during photosynthesis:

$$\text{water} + \text{carbon dioxide} \xrightarrow[\text{chlorophyll}]{\text{sunlight}} \text{glucose} + \text{oxygen}$$

Food factories – busy making food for a horsechestnut tree.

How plants use the glucose from photosynthesis

1 Some is used straight away, to give the plant energy.
2 Some is stored up. It is changed into starch or oil, and stored in stems, roots, seeds, and fruits. When it is needed it is changed back to glucose.
3 Some is used to make cellulose for cell walls.
4 Some is combined with minerals, and used to make proteins and the other things plants need for growth.

Questions

1 Where are the sex organs in a flowering plant?
2 What two things do roots do?
3 What is photosynthesis and where does it occur?
4 What is the green substance in leaves called?
5 What does this green substance do?
6 What does a plant need to make glucose?
7 What does a plant use the glucose for?
8 How is the glucose stored?
9 What else is produced during photosynthesis?

3.2 More about plant nutrition

How plants and animals recycle oxygen

Billions of living things are breathing in oxygen day and night to release energy from food during respiration. Why isn't all the oxygen used up? The answer, given in Unit 3.1, is that plants produce oxygen during daylight as a by-product of photosynthesis.

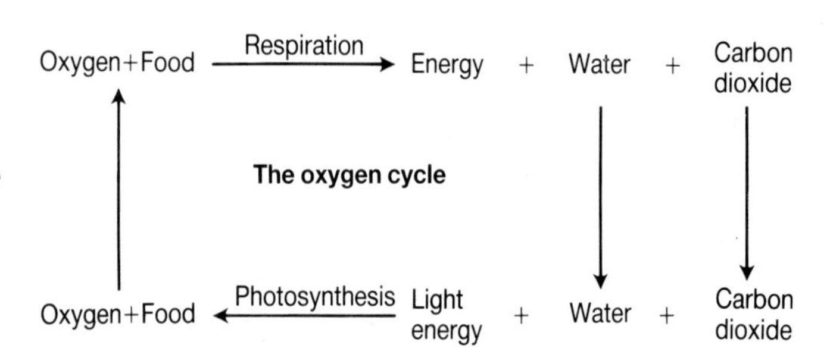

The oxygen released by plants returns to them in two ways.

1 Some returns as they take it in for respiration.

2 The remaining oxygen returns in water used for photosynthesis. The diagram above shows that respiration produces water and some of this will find its way into plants.

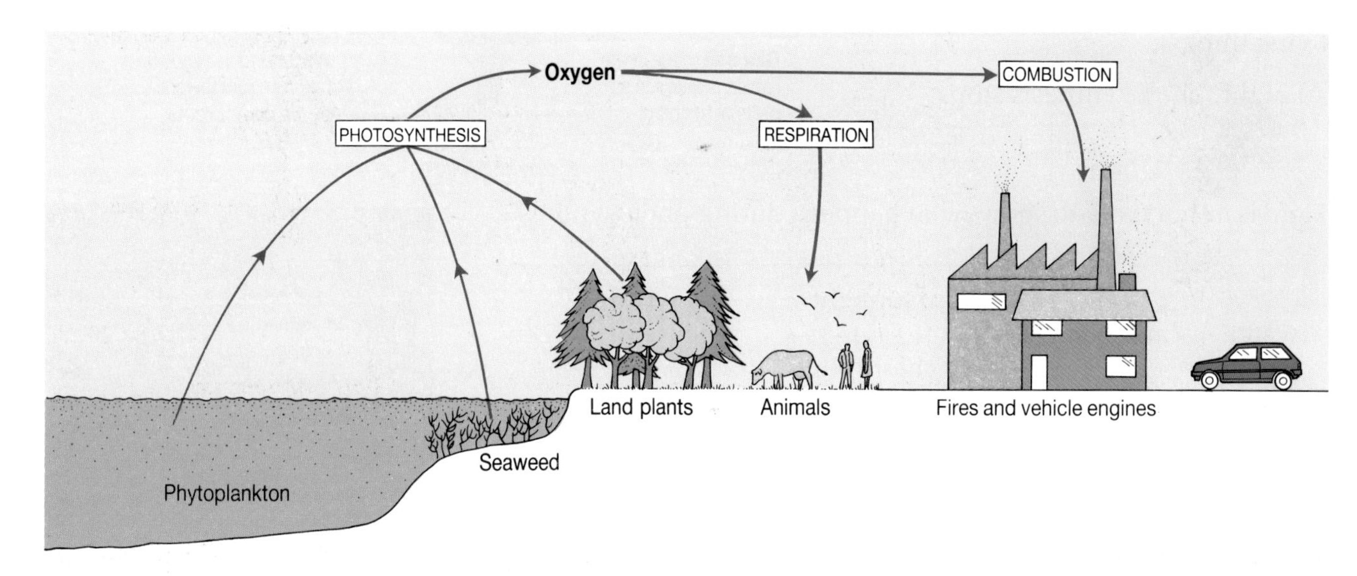

Land plants, especially trees, produce up to 80 per cent of the world's oxygen supplies. The remainder comes from seaweeds, and microscopic plant-like creatures (phytoplankton) in the sea.

While both plants and animals take in oxygen for respiration, during the day plants produce more oxygen than they use for respiration. Oxygen is also used up by burning (combustion).

Factors which limit the rate of photosynthesis

The rate at which a photosynthesis occurs depends on the amount of light and carbon dioxide a plant receives, and on the temperature.

Light and carbon dioxide If light intensity is slowly increased the rate of photosynthesis increases for a time, then stops. If the plant is now given more carbon dioxide, photosynthesis will increase again, so the rate of photosynthesis was limited by carbon dioxide.

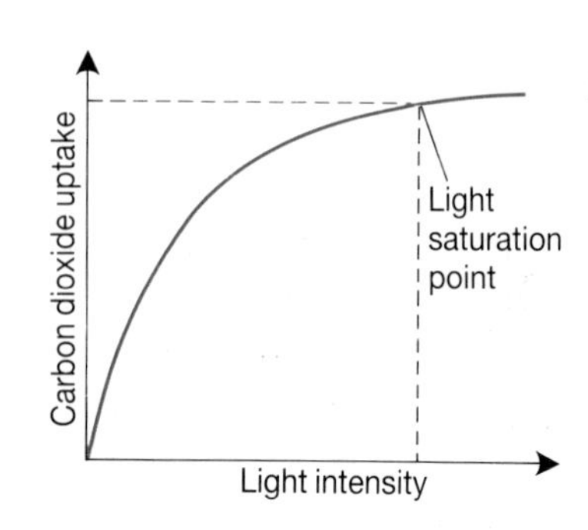

Alternatively, if a plant is in light of a fixed intensity and its carbon dioxide supply increased, photosynthesis will increase until it is limited by the amount of light it is receiving.

Temperature If a plant is kept at a low temperature and the light intensity increased, photosynthesis will increase for a little time then stop. If this is repeated at a higher temperature, the rate of photosynthesis increases much further before it stops. This shows that temperature affects the rate of photosynthesis.

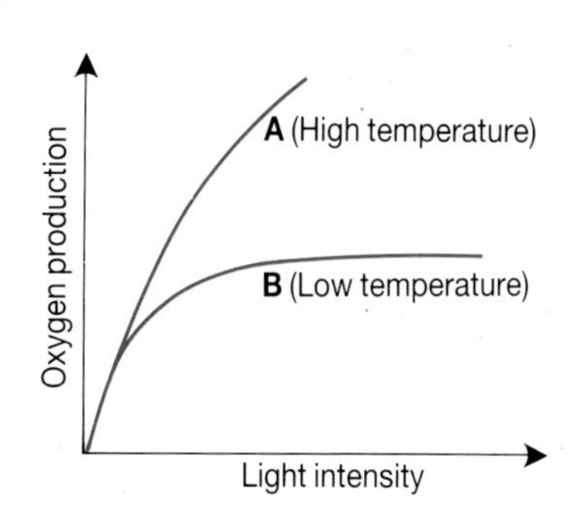

Plants need minerals

Plants need more than carbon dioxide and water to thrive. They need several mineral elements as well. If a plant is deprived of minerals, it develops defects called **mineral deficiency symptoms**, some of which are illustrated below.

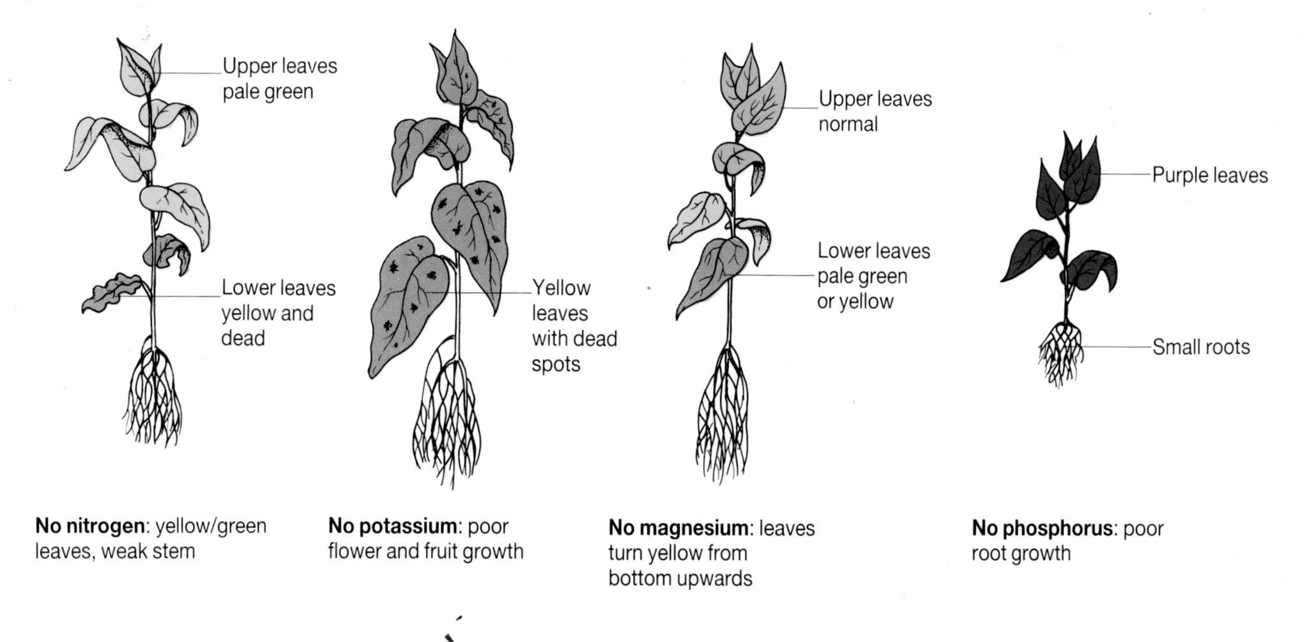

No nitrogen: yellow/green leaves, weak stem

No potassium: poor flower and fruit growth

No magnesium: leaves turn yellow from bottom upwards

No phosphorus: poor root growth

Major elements are the minerals plants need in quite large amounts (several hundred parts per million). They are nitrogen, phosphorus, sulphur, potassium, calcium, and magnesium. Nitrogen is known as the leaf maker, phosphorus the root maker, and potassium the flower and fruit maker. **Trace elements** are needed in very small amounts (as low as one part per million). Examples are manganese, copper, iron, boron, and molybdenum.

Questions

1 Tropical rain forests cover an area **eight** times the size of the British Isles. Give one reason, **connected with this Unit,** why their destruction endangers life.

2 If a plant's carbon dioxide *and* light supply are increased together, photosynthesis increases and then levels out. What is limiting the photosynthetic rate?

3 From the information given in this Unit, decide which minerals should be given regularly to tomatoes, roses, lettuce, grass and fruit bushes.

3.3 Leaves

Most leaves are thin and flat, like this:

These are **veins**.

This is the **mid-rib** of the leaf.

The stalk of a leaf is called a **petiole**.

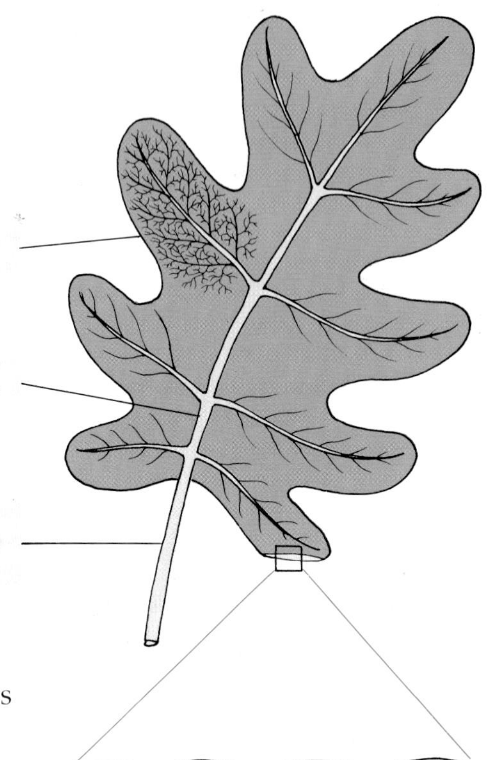

The mid-rib and veins in this leaf are made of tiny tubes. Some of these tubes carry water and minerals into the leaf. Some carry food out to all parts of the plant.

Inside a leaf

This is what a slice of leaf looks like under a microscope.

The **upper skin** of the leaf. It has a layer of wax on it called a **cuticle**. This makes the leaf waterproof.

This thick layer of cells in the middle of a leaf makes glucose by **photosynthesis**.

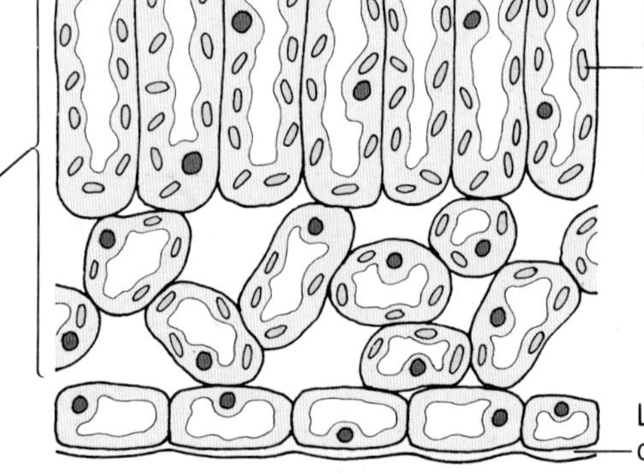

Chloroplasts. These are tiny discs inside cells that contain **chlorophyll**. They trap light energy for photosynthesis.

This is another view of a piece of leaf.

The tiny tubes in a **vein**. Some carry water and minerals into the leaf. Some carry food out.

Upper skin

Chloroplasts

Vein

Tiny hole in a leaf called a **stoma**.

Stomata

Most leaves have pores called **stomata** in their lower surface. One pore, or **stoma**, is a hole between a pair of **guard cells.** Stomata let carbon dioxide into the leaf and oxygen and water vapour pass out.

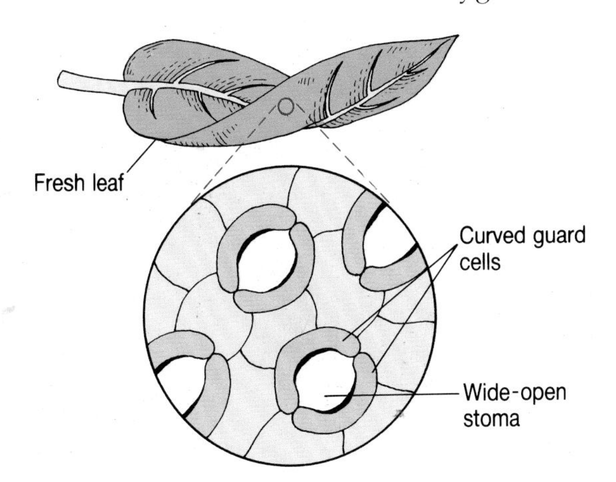

When a plant has plenty of water, the guard cells become curved and the stoma between them opens. This allows water to escape from the leaves of the plant.

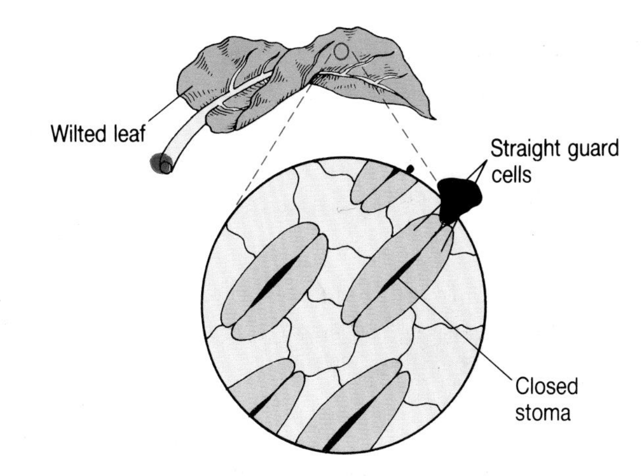

When a plant begins to lose water faster than its roots can take it up, guard cells become less curved. This closes the stomata and slows down the loss of water from the plant.

How leaves are suited to photosynthesis

1 Many leaves are flat, to absorb as much light as they can.
2 They are thin so carbon dioxide can reach inner cells easily.
3 They have plenty of stomata in the lower skin, to let carbon dioxide in, and oxygen and water vapour out.
4 They have plenty of veins to carry water to the photosynthesizing cells, and carry glucose away.

Dicotyledons, like this rhubarb, have broad leaves with a network of veins.

Monocotyledons, like this grass, have long narrow leaves with parallel veins.

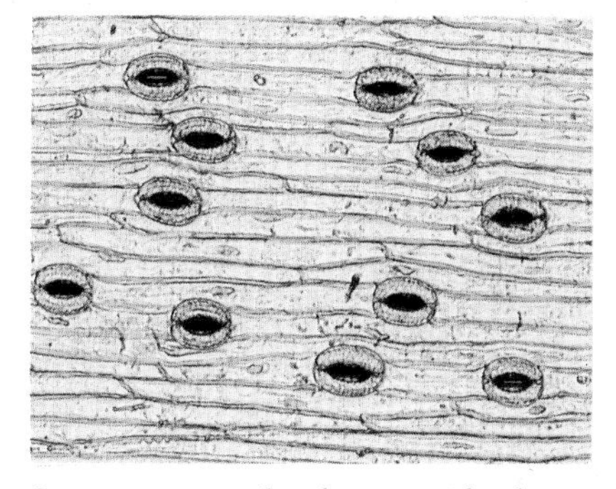

Stomata on a leaf, magnified 8000 times. All the stomata are open.

Questions

1 What makes a leaf waterproof?
2 What are chloroplasts? Where are they found?
3 What is inside a leaf vein?
4 What are stomata? How do they help leaves make food?
5 Explain why leaves are thin and flat, and well supplied with veins.

3.4 Transport and support in plants

A plant's transport system

A plant has lots of thin tubes inside it. They carry liquids up and down the plant. They are the plant's **transport system**.

Some tubes carry glucose solution from the leaves to every part of the plant. These tubes are called **phloem**.

Some tubes carry water and minerals up from the soil. They are called **xylem vessels**.

Xylem and phloem are grouped together in **vascular bundles**.

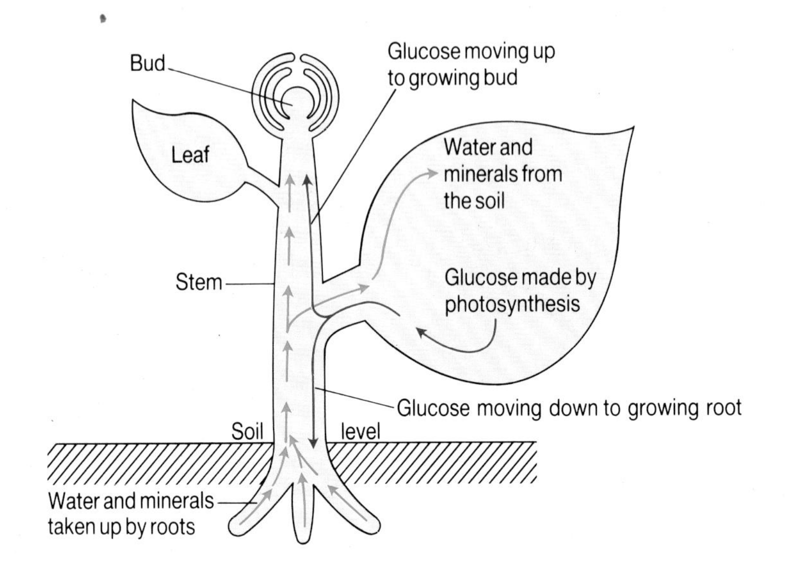

Inside a stem

This is what a thin slice of stem would look like, greatly magnified.

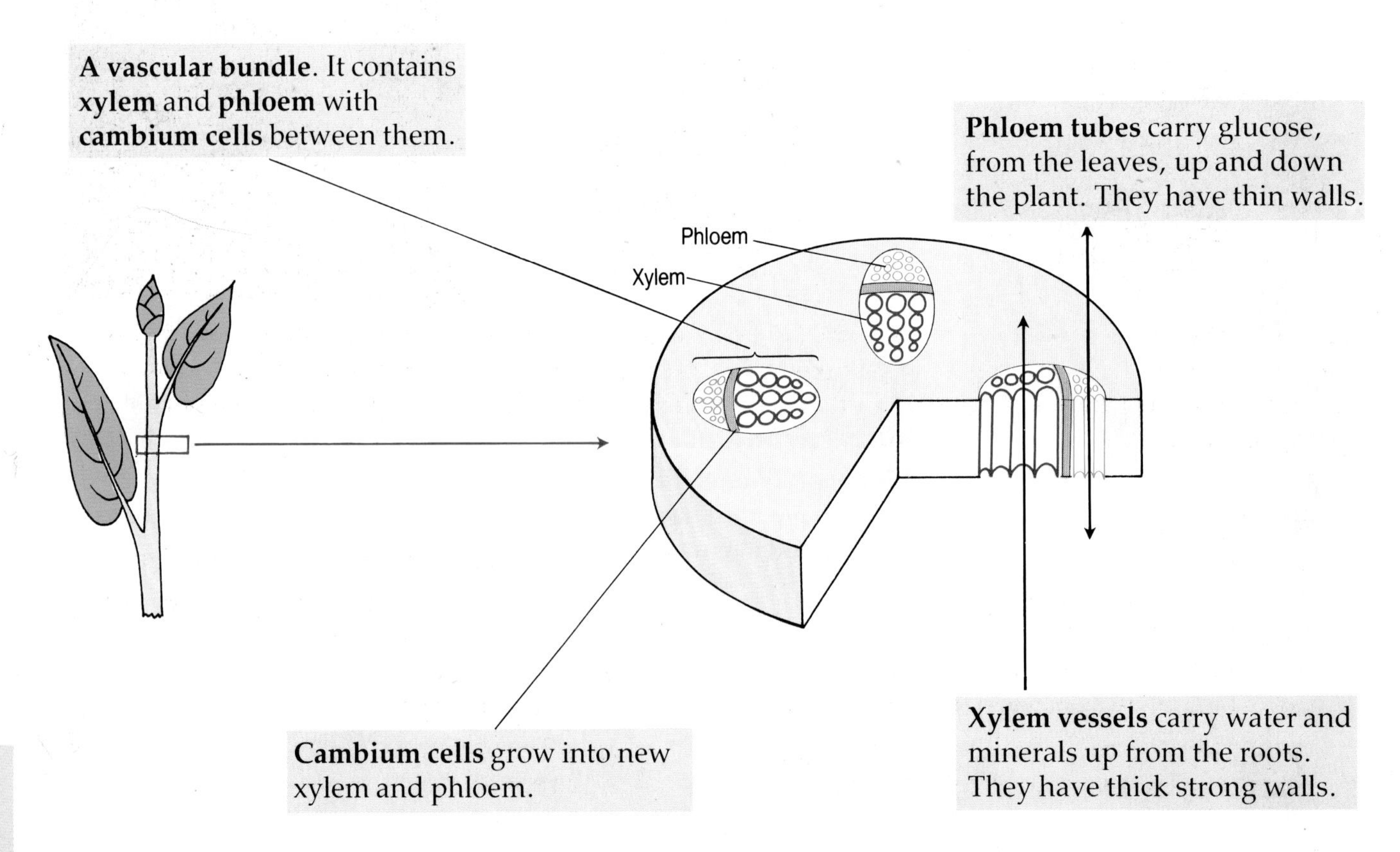

Inside a root

This is what the tip of a root would look like, sliced open and greatly magnified.

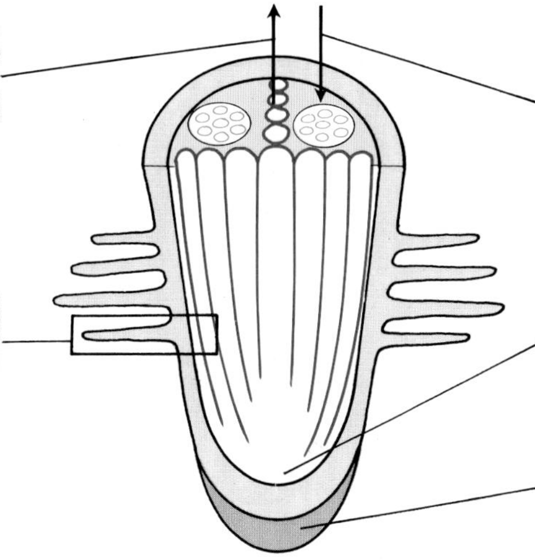

Support in plants

Xylem vessels have thick strong walls. So they help to hold up or support a plant. Bushes and trees have plenty of xylem to support them. In fact, tree trunks are mostly made of xylem.

The soft parts of plants, like leaves and flowers, are supported mainly by the water in their cells.

A plant with plenty of water has firm or **turgid** cells. Its stem is straight and its leaves and petals are firm.

A plant without water has soft or **flaccid** cells. Its stem, leaves and petals are soft. They droop, or **wilt.**

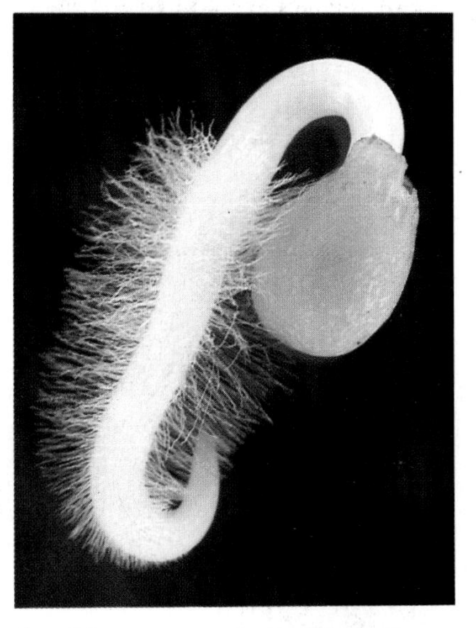

A white mustard seedling. It takes in water and minerals through its tiny root hairs.

Questions

1. **a)** Which tubes in a plant carry glucose?
   b) Where do they carry the glucose from?
   c) Where do they carry it to?
2. **a)** Which tubes carry water and minerals?
   b) Where do they carry them from?
   c) Where do they carry them to?
3. What does a vascular bundle contain?
4. Which cells grow into xylem and phloem?
5. Xylem vessels help support a plant. Why are they able to do this?
6. What is:
   a) a turgid cell?
   b) a flaccid cell?
   c) a root hair?

3.5 Transpiration

A plant loses water vapour through its leaves. This loss of water is called **transpiration**. The water first evaporates from cells inside a leaf. It then escapes through tiny holes in the leaf called stomata (one hole is called a stoma). In most plants stomata are on the undersides of the leaves.

The transpiration stream

Water is lost from the plant's leaves. At the same time more water flows up xylem vessels from the roots to replace it. This flow of water from roots to leaves is called the **transpiration stream**.

The transpiration stream flows fastest on a warm, dry, sunny days, because this is when water evaporates fastest from the leaves. It slows down on cold, dull, damp, days and when a plant is short of water.

Some water is used for photosynthesis, and some evaporates from the leaves.

Water flows up the stem.

Water is taken in by plant roots.

How plants survive in dry climates

If you moved a rhubarb plant to a hot, dry desert, it would quickly lose water through its large leaves with many stomata, and die.

Cactus plants are adapted to survive desert conditions by having features which reduce transpiration to a minimum.

They lost most of their stomata when, in the course of evolution, their leaves were reduced to spines. Photosynthesis is confined to green, succulent stems which have few stomata and store water soaked up during the desert's brief periods of rain.

As well as having leaves reduced to spines and stems which store water, cacti have a thick, waterproof cuticle to reduce evaporation. They also have a large root system to absorb water as quickly as possible if it rains.

Pines can live in dry, exposed, windy areas like mountains, because they have needle-shaped leaves with a small surface area for evaporation. Their stomata are sunk in deep pits, protected against the drying effect of wind.

Why transpiration is important

1 The transpiration stream carries water and minerals from the soil to the leaves. Water is needed for photosynthesis. Minerals are needed for making proteins.

2 Water is also needed to keep cells turgid (firm), so that they support the plant (Unit 3.3).

3 The evaporation of water from the leaves keeps them cool in hot weather.

How the transpiration stream flows

1 Root hairs take in water from the soil by osmosis (Unit 2.3).

2 Water passes into the root. It moves from cell to cell until it reaches the xylem vessels.

3 Water is sucked up the vessels (like lemonade sucked up a drinking straw). Water passes into the leaves through leaf veins. Water evaporates from leaf cells and escapes through stomata.

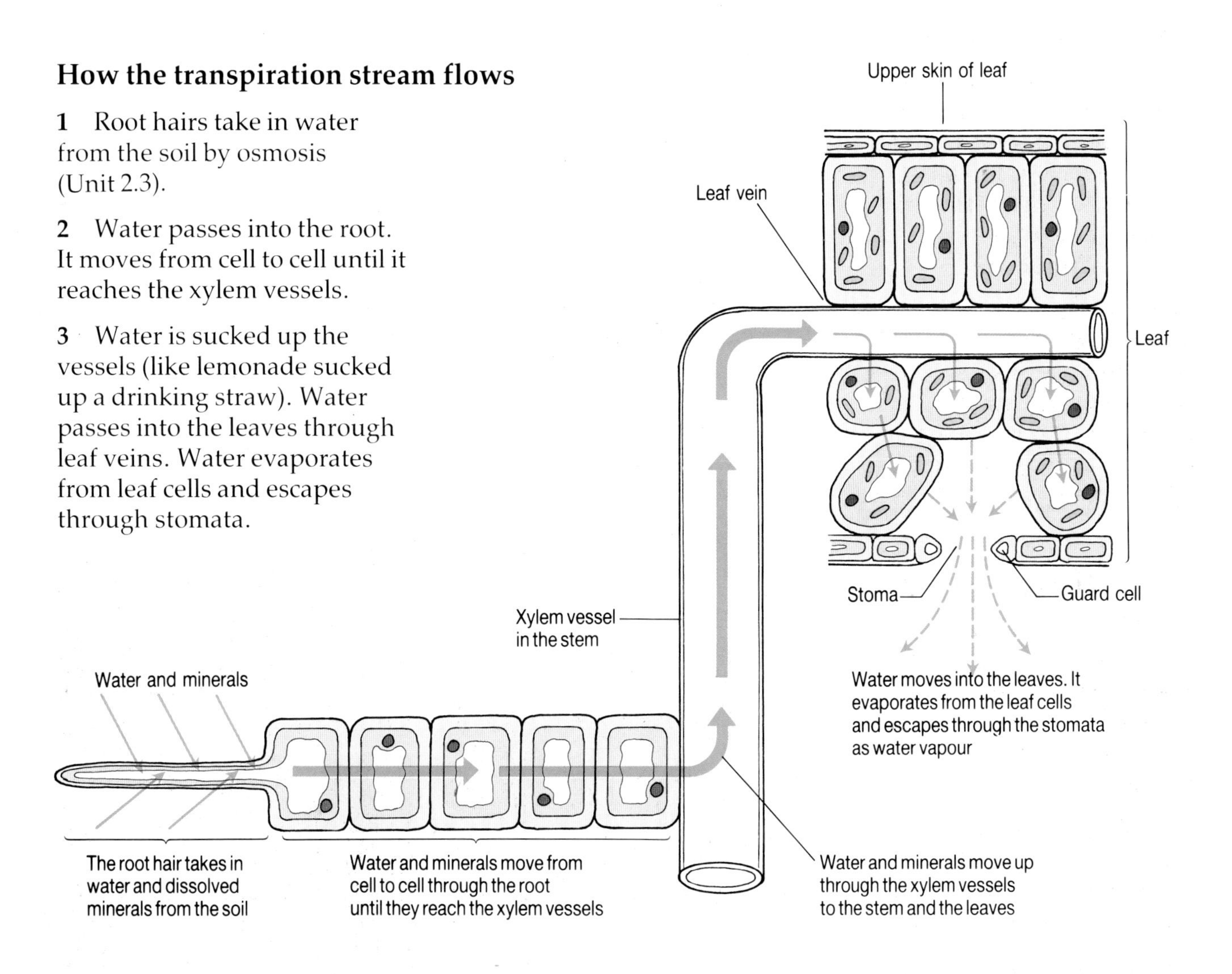

Questions

1 Does all the water which rises up a plant escape through the leaves?

2 **a)** What is the transpiration stream?
b) Give three reasons why it is important.

3 When is the transpiration stream fastest?

4 What happens if water is lost from the leaves faster than it is taken up through the roots?

5 Explain how guard cells work.

3.6 Flowers

Humans have male and female sex organs, and so do flowering plants. A plant's sex organs are in its flowers. If you slice a flower open you will see something like this:

This is a male sex organ. It is called a **stamen**.

This is a swelling called a **nectary**. It makes **nectar**, a sweet liquid that attracts insects.

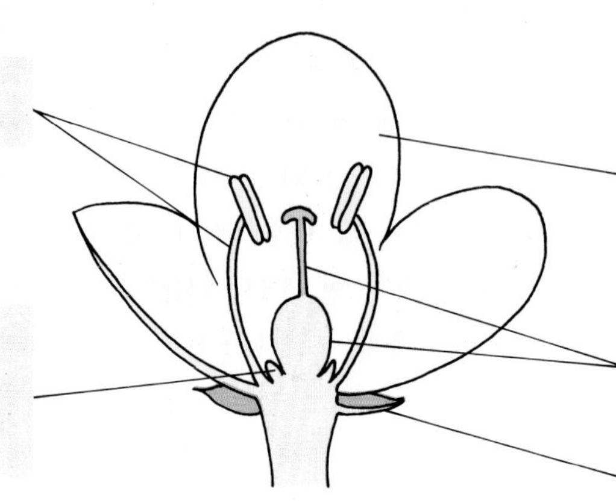

This is a **petal.** Some petals are coloured and scented to attract insects. Insects help plants to make seeds, as you will see later.

This is a female sex organ. It is called a **carpel**.

This is a **sepal**. Sepals protect the flower when it is in bud.

Parts of a stamen

The **anther**. It is made up of four **pollen sacs**. These are full of **pollen grains**. Pollen grains contain the male sex cells of the plant.

The stalk or **filament** of the stamen.

Anther cut open.

Pollen grains

The pollen sacs split open here

Pollen sacs

Parts of a carpel

The **stigma**. Pollen grains stick to this during pollination.

The **style**.

The **ovule** contains the female sex cell. It forms a seed after fertilization.

The **ovary** protects the ovule.

The **female sex cell**.

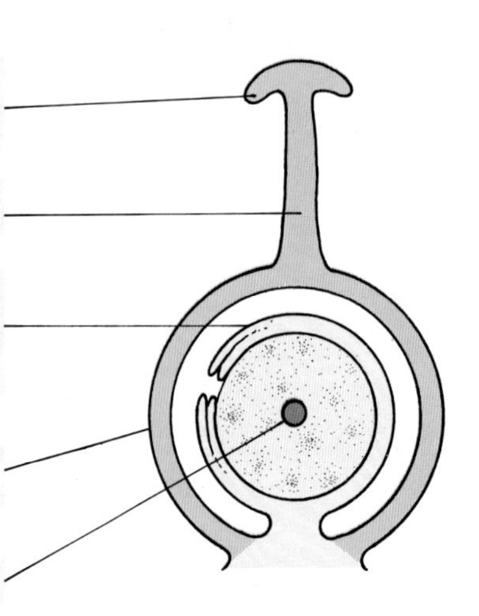

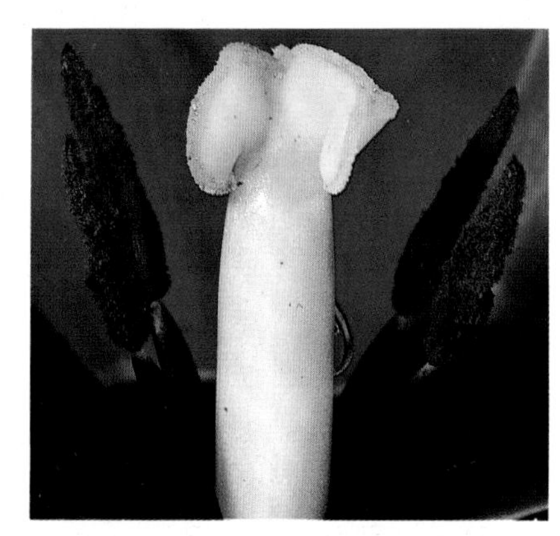

A tulip carpel surrounded by brown stamens. Note the large stigma at the top of the carpel. Why does it need to be sticky?

Some different flowers

There are many different kinds and shapes of flower. They have different numbers of petals, stamens, and carpels. Some flowers, such as grasses, do not have any petals or nectaries.

The petals on a plum flower are all the same shape. Note the stamens sticking out. The stamens have long filaments with the anthers at the end.

A grass flower doesn't have petals, scent, or nectar. You can see it has very large anthers.

Sunflowers are made up of hundreds of small flowers, called florets, gathered together.

The petals on a pea flower are of different shapes and sizes.

Questions

1 Why are some petals coloured and scented?

2 What is the name of:
  a) the male sex organ of a flowering plant?
  b) the female sex organ of a flowering plant?

3 What do sepals do?

4 What is nectar? Where is it made?

5 Copy and complete:
  The _____ is made up of four pollen _____ .
  These are full of _________ .

6 What is inside a pollen grain?

7 Draw a carpel and label the parts.

8 Draw a stamen and label the parts.

3.7 How fruits and seeds begin

A new seed begins when the male sex cell in a **pollen grain** joins up with a female sex cell in an **ovule**. The first step in making this happen is called **pollination**. Pollination takes place when pollen is carried from an anther to a stigma.

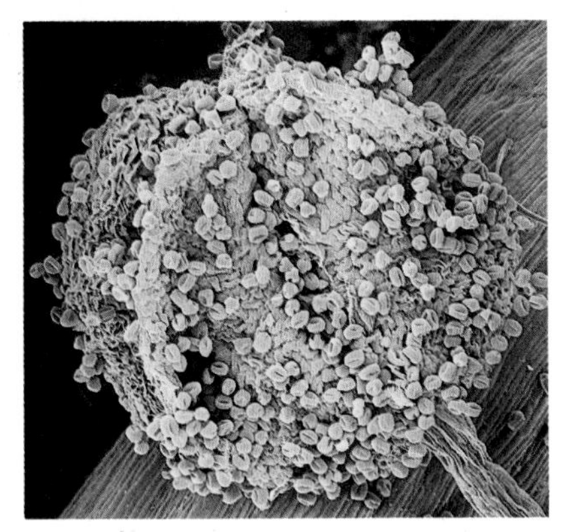

First the anther ripens. The pollen sacs split open releasing pollen grains.

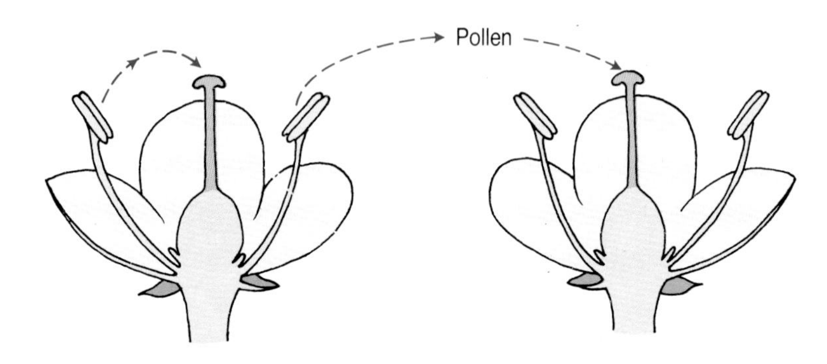

Pollen may be carried to a stigma in the *same* flower. This is called **self-pollination**.

Or pollen may be carried to a stigma in another flower. This is called **cross-pollination**.

Pollination by insects

Bees and other insects can carry pollen from flower to flower. That is what happens with buttercups, dandelions, and foxgloves.

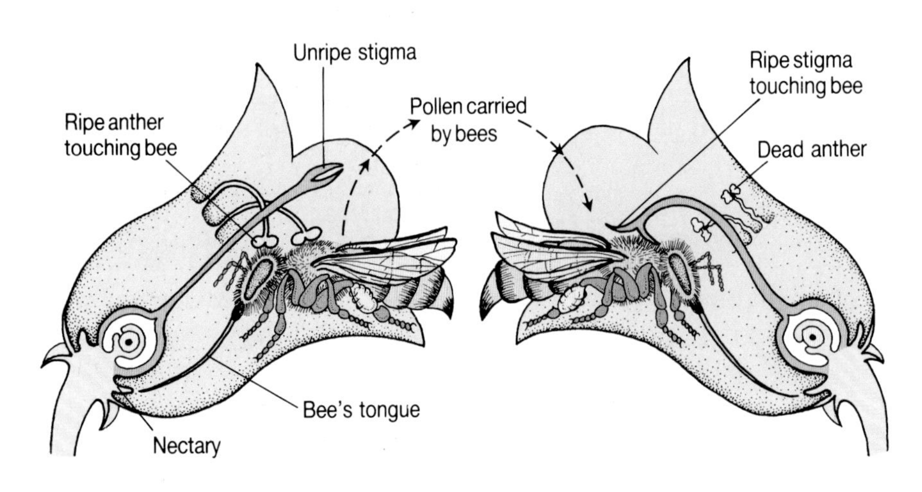

Insect-pollinated flowers have:

1 Large, coloured scented petals and nectar, to attract insects.
2 Large pollen grains that can stick to an insect's body.
3 Anthers and stigmas inside the flower, so that the insect can brush against them when it is drinking nectar.

A bee with foxglove pollen clinging to its body. Inside this foxglove, the pollen will rub off on the sticky stigma.

Pollination by wind

The wind can carry pollen from flower to flower. That is what happens with grass, stinging nettle, and catkins.

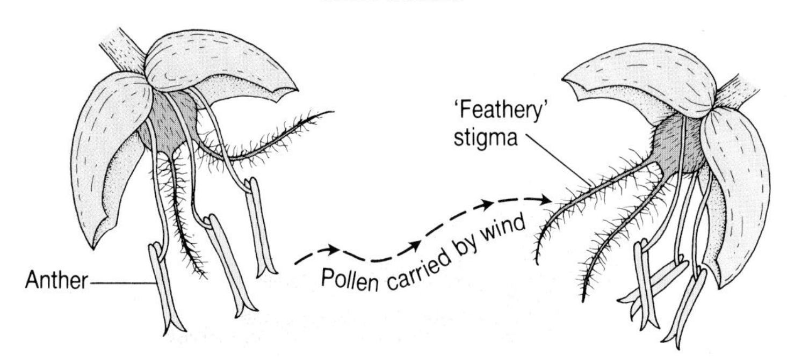

In catkins, the male and female sex organs are in separate flowers. Wind shakes the yellow pollen from the male catkin . . .

. . . and carries it to the female catkin's sticky stigmas.

Wind-pollinated flowers do not have large scented petals, or nectar, because they do not need to attract insects.

They do have:

1 Anthers that hang outside the flower, to catch the wind.
2 Large amounts of light, small pollen grains that blow in the wind.
3 Spreading, feathery stigmas to catch airborne pollen grains.

Fertilization

After a pollen grain lands on a stigma the next step is **fertilization**. Fertilization happens when a male sex cell joins up with a female sex cell.

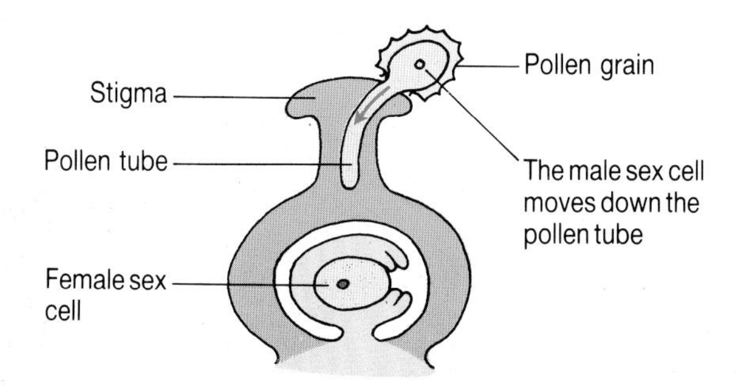

1 First a tube grows out of the pollen grain. It grows towards the female sex cell. A male sex cell moves down the tube.

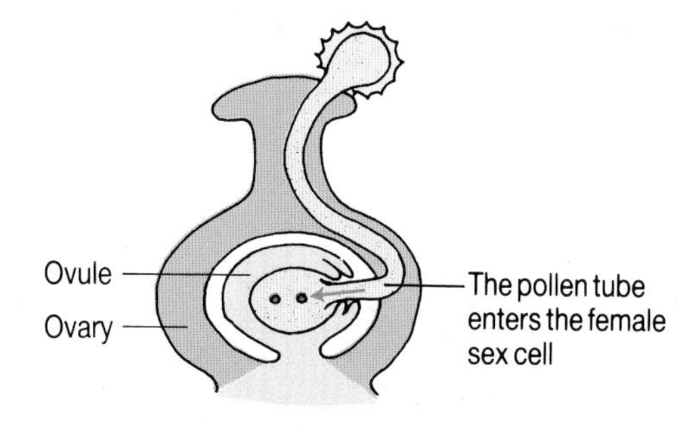

2 The tube enters the female sex cell. The tip of the tube bursts open. The male sex cell joins up with the female sex cell. The ovule then becomes a **seed**. The ovary becomes a **fruit** with the seed inside it. The petals die and drop off.

Questions

1 What are self-pollination and cross-pollination?

2 **a)** Describe the differences between insect-pollinated, and wind-pollinated flowers.
b) Give the reasons for these differences.

3 What happens after a pollen grain lands on a stigma?

3.8 More about fruits and seeds

In biology a **fruit** is any part of a plant that contains **seeds**.

Apples, oranges, lemons and tomatoes are all fruits. Think of the seeds inside them.

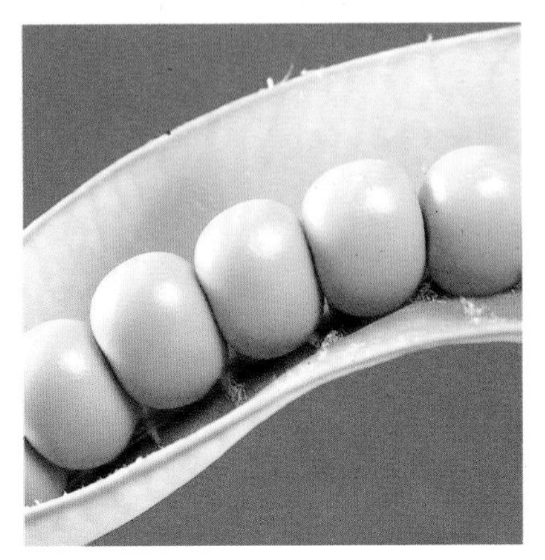

Pea pods are also fruits. The peas inside them are seeds.

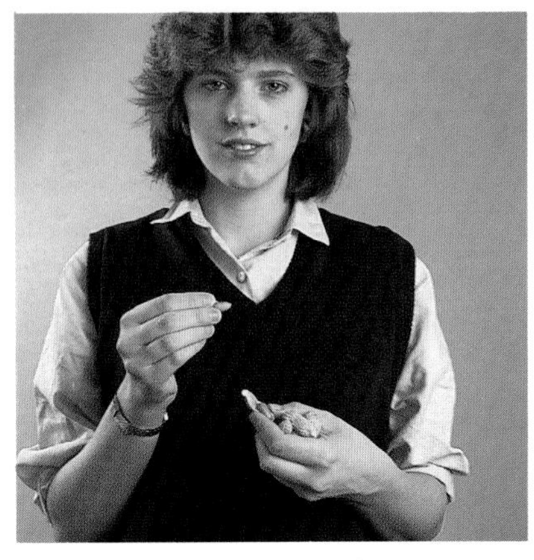

Nuts are fruits too. The part we eat is the seed.

How fruits and seeds get scattered

Most plants produce hundreds of seeds at a time. If they all fell close to the parent plant they would be too crowded. The new plants would have to compete with each other for water, minerals, and light.

Instead, fruits and seeds get scattered by the wind, or animals, or by the plant itself. This is called **dispersal** of fruits and seeds.

Dispersal by wind

These fruits and seeds are dispersed by the wind.

Dandelion and thistle seeds have a downy 'parachute' which floats in the wind.

Sycamore, ash, and lime seeds have wings which carry them long distances in the wind.

Poppy heads sway in the wind scattering seeds through holes in their sides.

Dispersal by animals

These fruits and seeds are dispersed by animals.

Blackberry, sloe, and hawthorn berries are eaten whole by animals. The seeds pass unharmed through their bodies.

Burdock has hooks which catch on animals' fur or on our clothes. The seeds can be carried long distances before falling off.

Acorns and beech nuts are carried away by birds and squirrels. Some are dropped before they are eaten.

Dispersal by plants

Some plants throw away their seeds.

Lupin seed pods suddenly burst open when dry. Their sides coil up, scattering the seeds.

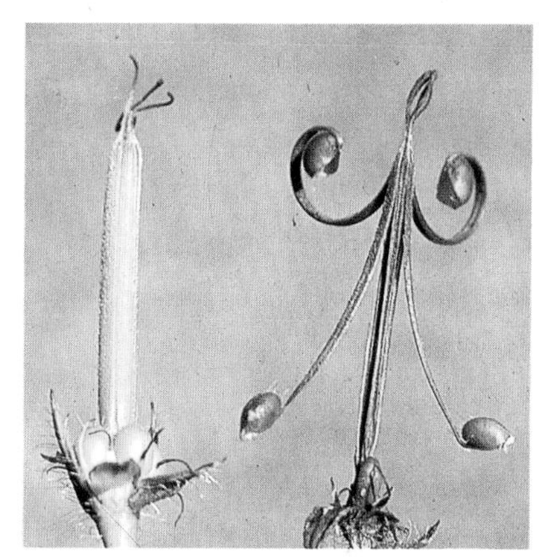

Geranium fruits also burst open when dry. Their sides spring upwards, scattering the seeds.

Some seeds are collected and stored and sold by man. They get dispersed by gardeners and farmers.

Questions

1 In biology, what is a fruit?

2 Why do seeds need to be dispersed?

3 **a)** Name two plants that use wind dispersal.
b) How are their fruits and seeds suited to this?

4 How do some plants disperse their own seeds?

5 Describe two ways in which animals can disperse fruits and seeds. Give an example of each.

6 Describe how you may have helped disperse seeds without knowing it.

3.9 How seeds grow into plants

A seed looks dead. But inside it there is a tiny plant called an **embryo**, and a store of food. If the seed is given water, air, and warmth it 'comes to life', and the embryo begins to grow. This is called **germination**.

The embryo in a seed has three parts. It has a root or **radicle,** a shoot or **plumule,** and one or more seed leaves called **cotyledons.** The embryo needs its stored food for growth, until it has enough leaves to make food by photosynthesis.

Not all seeds get a chance to grow into plants. Some are eaten first – like these baked beans.

Bean seeds

Bean seeds have two seed leaves, so they are called **dicotyledons** ('di' means 'two'). The seed leaves of beans are swollen, because they contain the stored food.

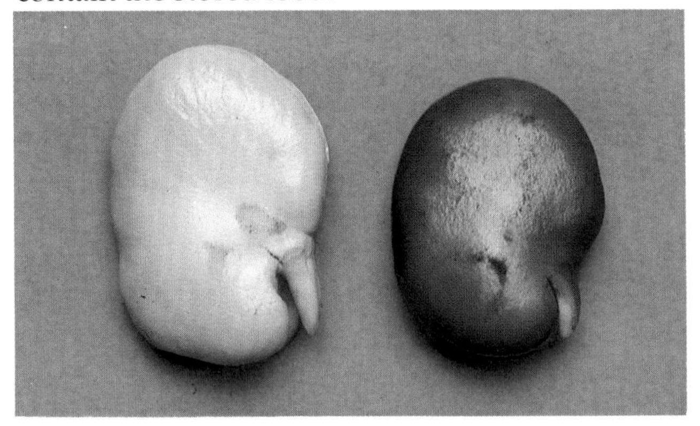

Broad beans. The one on the left has had its seed coat removed.

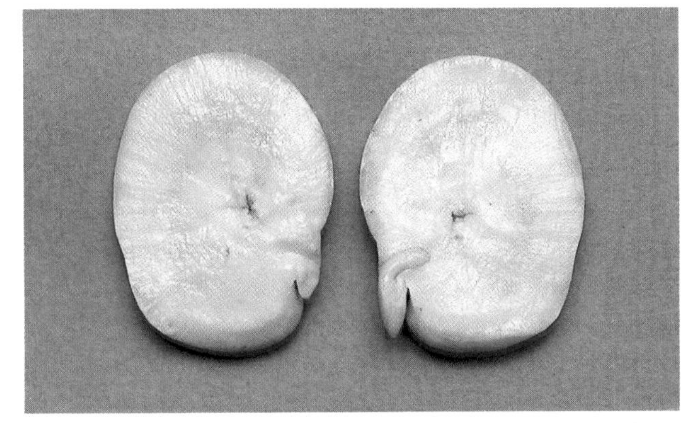

Broad bean cut in half to show cotyledons and embryo.

Germination of a French bean seed

When a French bean germinates, the cotyledons come above the ground. This is called **epigeal germination**. These pictures show a French bean germinating. The brown object in two of the pictures is the seed coat which has split off the seed.

The cotyledons push their way out of the soil.

They provide the food the plant needs, until its leaves grow.

The leaves are growing, so the cotyledons will soon drop off.

Maize and other cereal grains

Maize, wheat, rice, barley, and other cereal grains have only one cotyledon. So they are called **monocotyledons** ('mono' means 'one'). Their food is not stored in the cotyledon, but in a mass of cells called **endosperm** in a space above the cotyledon. When grains are milled, the endosperm spills out. We call it flour.

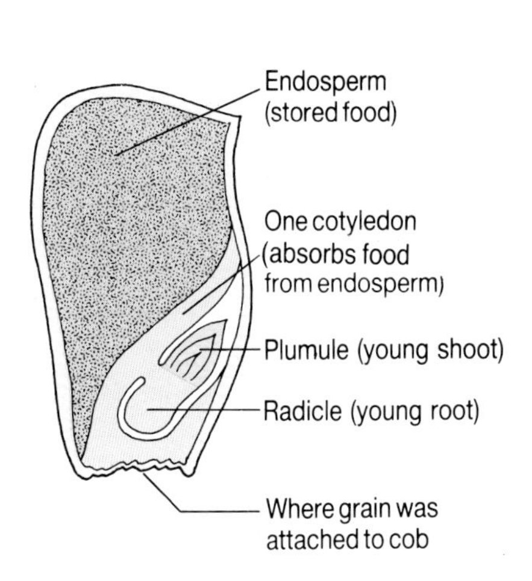

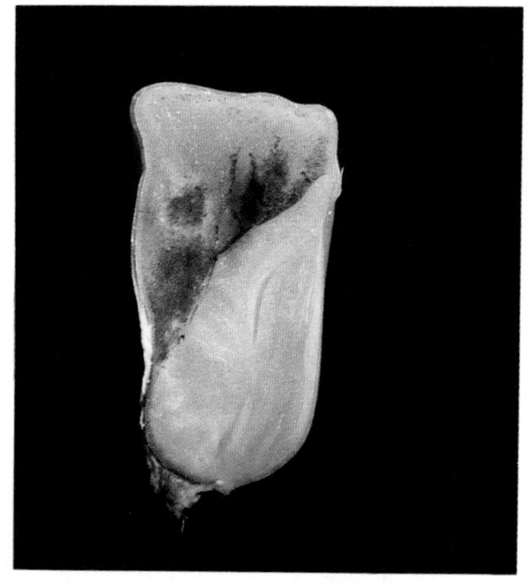

A maize grain cut in half. Compare it with the drawing.

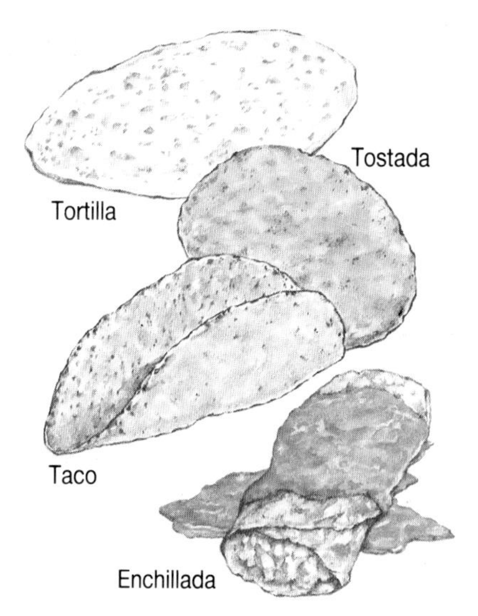

We make a lot of our food by grinding up the seeds of maize, wheat and other cereals, to get their food stores. Above are some Mexican foods made from maize flour.

Germination of maize grains

When a maize grain germinates, the cotyledon stays below the ground. This is called **hypogeal germination**. Other grains, and broad beans, also germinate in this way.

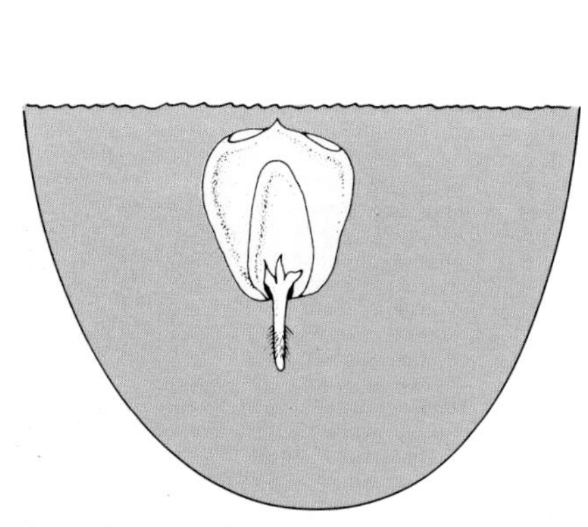

A maize grain germinating. Note the root growing downwards. Its tip is protected by a root cap.

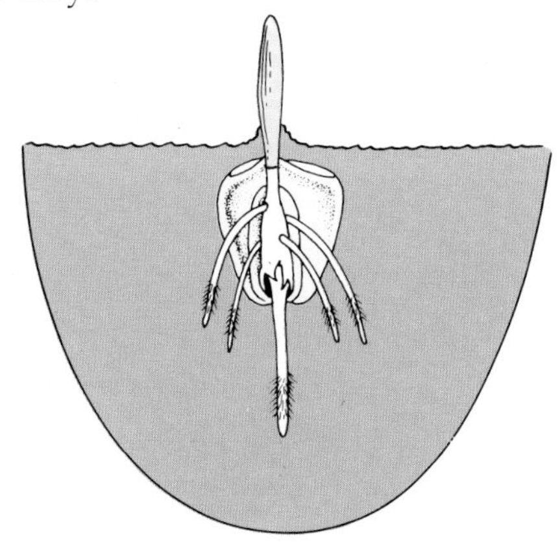

The young shoot pushes through the soil. The endosperm provides the food it needs.

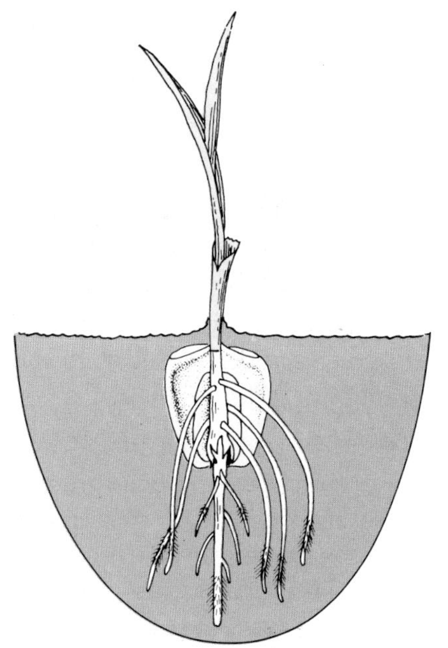

The new leaves can now make their own food. The cotyledon is still below ground.

Questions

1 What three things does a seed need to grow?

2 What is inside a seed?

3 What is:
   a) a radicle? **b)** a plumule? **c)** a cotyledon?

4 Why do seeds need a store of food?

5 Is maize grain a dicotyledon? Explain.

6 What is the main difference between epigeal germination and hypogeal germination?

3.10 Plant senses

Plants have no eyes or ears or noses. But they can still sense things. They can sense light, and the pull of gravity, and water. They respond to these things by growing slowly in certain directions. These responses are called **tropisms**.

Phototropism: response to light

Plants need light to make food. So they respond to light by growing towards it. They also turn their leaves to face the light. This makes sure leaves get as much light as possible.

These cress plants were grown with light coming from above.

These were grown with light coming from the right.

Geotropism: response to gravity

No matter how you plant a seed in the earth, it can sense which way is down, and which is up. It can sense this from the pull of gravity.

Roots grow down in response to gravity. This makes sure they find soil and water. Shoots always grow up. This makes sure they reach light.

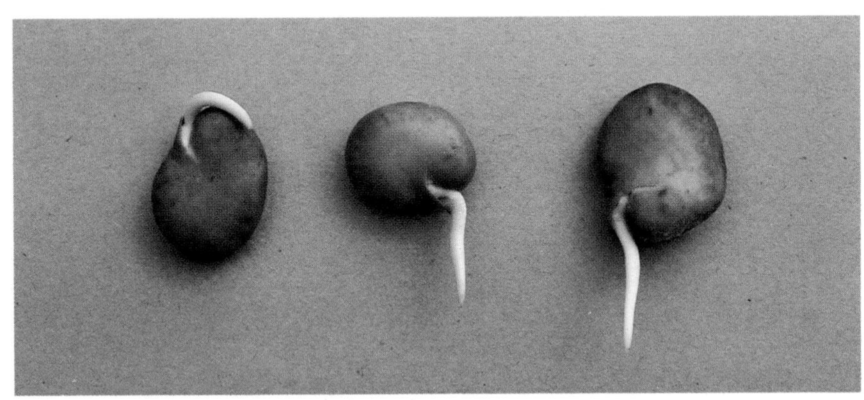

Roots responding to gravity. Which seed was planted upside-down?

Hydrotropism: response to water

Roots always grow towards water, even if this means ignoring the pull of gravity. If they have to they will grow sideways, or even upwards, to reach water.

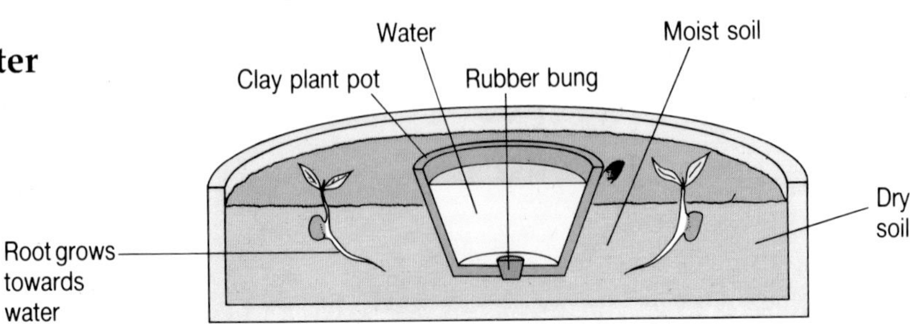

Auxin

A plant's responses to light and gravity are controlled by a chemical called **auxin**. Auxin is made by cells at the tips of stems and roots. It speeds up growth in stems. It slows down growth in roots.

Phototropism and auxin

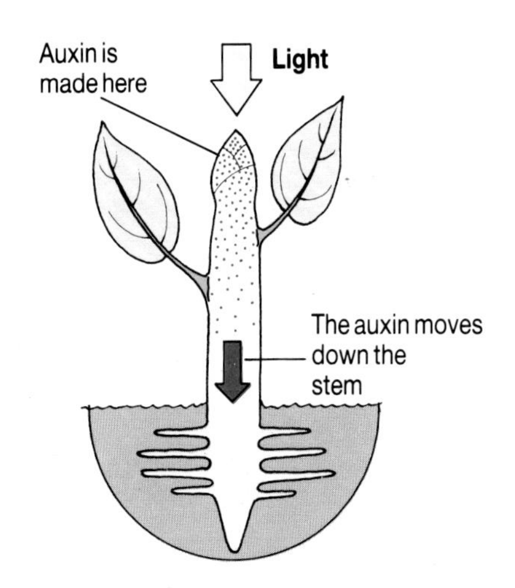

When light comes from above, auxin spreads evenly down the stem. The stem grows straight.

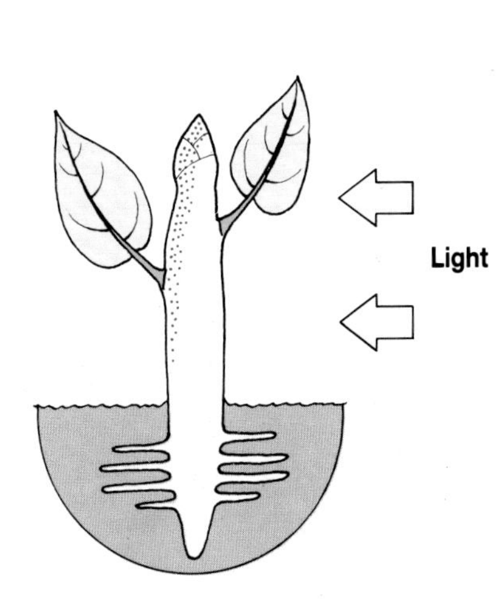

When light comes from one side, auxin spreads down the *shaded* side of the stem.

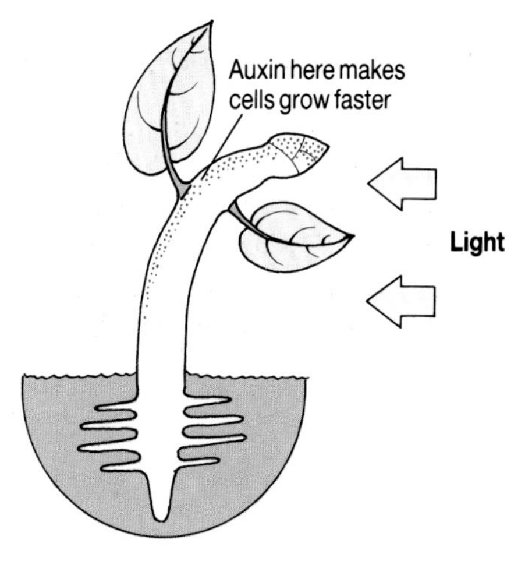

Auxin makes stem cells *grow faster*. This causes the stem to bend towards the light.

Geotropism and auxin

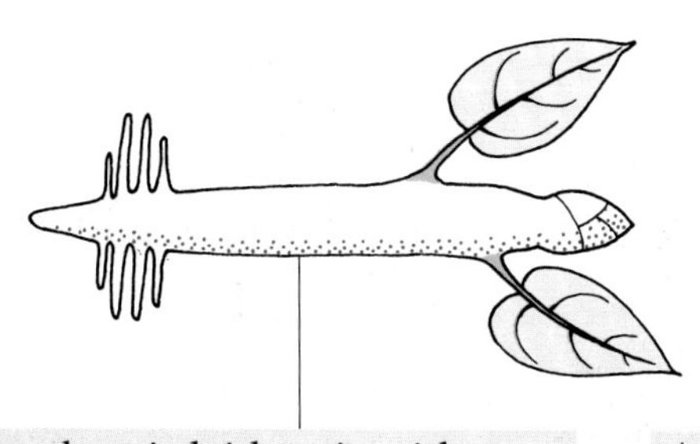

If a plant is laid on its side auxin gathers in the lower half of the stem and root.

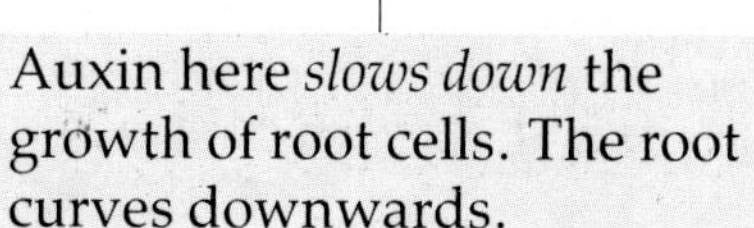

Auxin here *slows down* the growth of root cells. The root curves downwards.

Auxin here *speeds up* the growth of stem cells. The stem curves upwards.

Questions

1 Copy and complete:

a) A plant's response to light is called ________ .

b) A plant's response to gravity is called ________ .

c) A plant's response to water is called ________ .

2 What does a shoot do in response to light, and in response to gravity? How are these responses useful to the plant?

3 Name two responses controlled by auxin.

4 In which parts of a plant is auxin made?

5 How does auxin affect root and shoot cells?

Questions on Section 3

1 *Understanding/making predictions*
The diagram opposite shows an experiment to investigate how plants and animals interact. All the flasks were placed in bright light and kept at the same temperature.

a) Which flask would contain the most oxygen after one hour?
b) Explain your answer to (a).
c) Which flask would contain the most carbon dioxide after an hour?
d) Explain your answer to (c).
e) In which flask would the fish survive the longest?
f) Explain your answer to (e).
g) What are the controls in this experiment, and why are they necessary?
h) How does this experiment illustrate the oxygen cycle?

2 *Interpret data*
The diagram opposite shows apparatus designed to demonstrate a life process in plants. After the model was set up cotton wool in the tube gradually became red. The model was placed in different places and its weight loss recorded (see table opposite).

a) Name the parts of a plant represented by:
the plastic sheet;
pores in the plastic sheet;
the blotting paper;
the plastic tube.
b) What life processes are demonstrated when:
the cotton wool changes colour;
the apparatus loses weight.
c) Explain the experimental results.

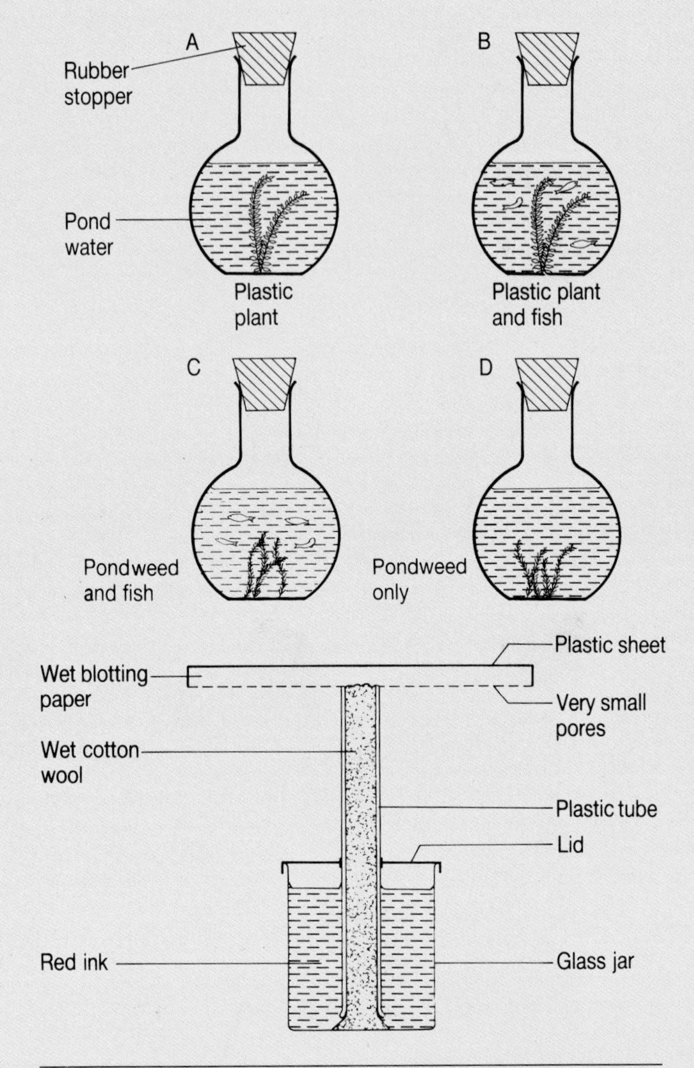

Experiment	Location	Weight loss
A	indoors	1.0 g
B	out of doors	1.8 g
C	inside plastic bag	0.1 g

3 *Understanding/making predictions*
The apparatus opposite is designed to investigate support in woody and non-woody (herbaceous) stems. Straight stems were placed in different liquids and a 5 g weight suspended from each.

a) After three hours what would be the appearance of stems in liquid A and liquid B?
b) Explain your predictions using the words turgid, flaccid and osmosis.
c) What could this experiment tell you about how woody and non-woody stems are supported?

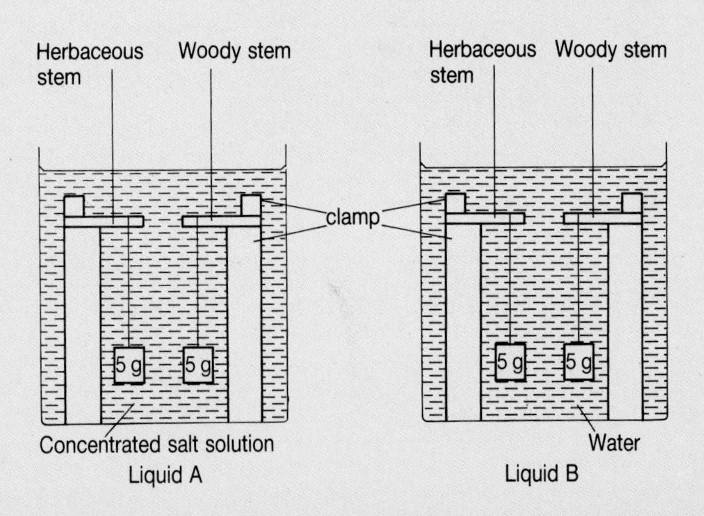

4 *Recall knowledge*
The diagram opposite shows the inside of a leaf.

a) Name the parts labelled A to G.

b) For each description below, write down the matching name and label:
cells which make food by photosynthesis;
layer of wax which makes a leaf waterproof;
hole which lets gases in and out of leaf;
contains tubes carrying sugar and water;
have chlorophyll inside them.

5 *Interpret data*
A pond contains many water plants. The graphs opposite show how the amounts of oxygen and carbon dioxide in the pond change.

a) When does the amount of oxygen start to increase?

b) When does the amount of oxygen stop increasing?

c) Why does the amount of oxygen increase and decrease at these times?

d) What happens to the amount of carbon dioxide in the pond, while the oxygen is increasing?

e) Why does the amount of carbon dioxide change in this way?

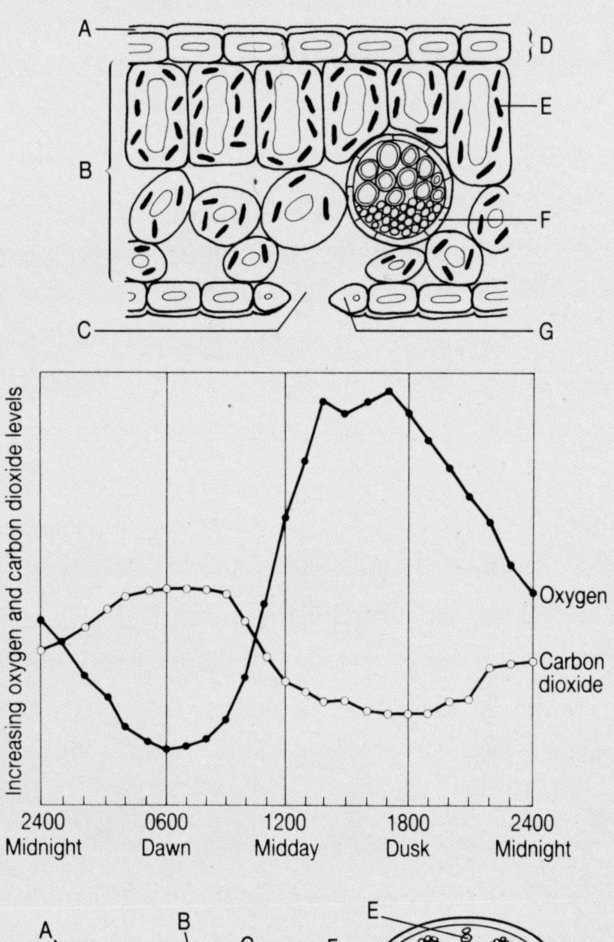

6 *Recall knowledge*
The diagrams opposite show a stem and root.

a) Name the parts labelled A to H.

b) For each description below, write down the matching name and label:
tubes which transport water from the roots;
tubes which transport sugar from the leaves;
cells which grow into xylem and phloem;
protects growing cells at the root tip;
absorbs water and minerals from the soil;
cells which produce auxin.

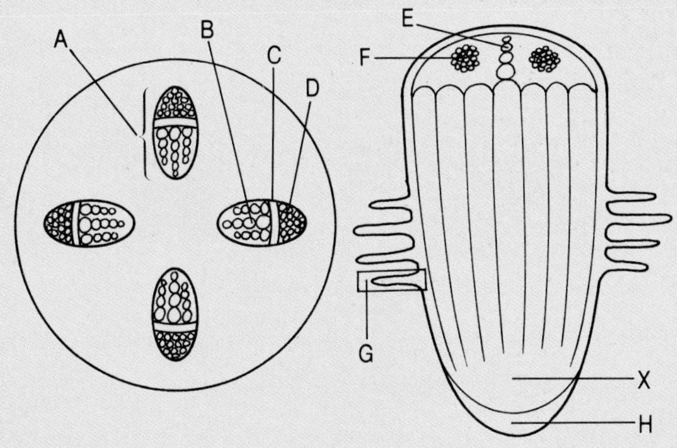

7 *Interpret data*
The experiment opposite used oat seedlings that were prepared in different ways. They are growing in a box with a hole cut at one end.

a) Which seedlings have grown towards the light?

b) Some seedlings have not grown towards the light. Explain why.

c) Explain how these results show that it is the tip of an oat seedling which allows it to bend towards the light.

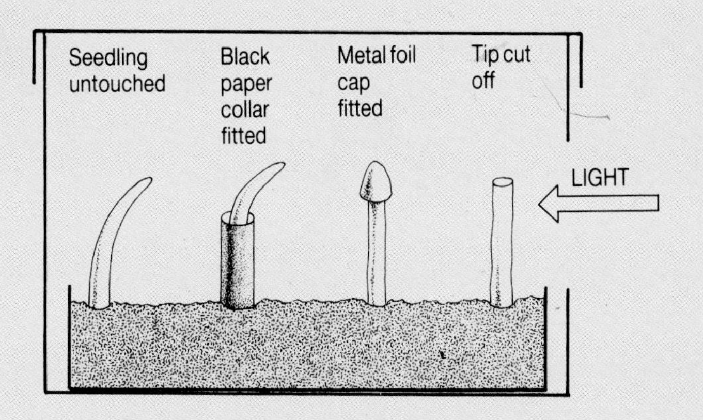

4.1 Skeletons

Jellyfish don't need skeletons since the water supports them. On land they just flop.

Some animals, like the jellyfish in this photograph, can live without a skeleton. They are supported by the water in which they float. But without your skeleton you would collapse in a heap. Like most creatures, especially land animals, you need a skeleton to support your body and give it shape.

Animals with liquid skeletons

Earthworms, caterpillars, and slugs are given shape and support by **liquid** inside them.

This liquid is mainly water. It fills the animal's cells and the spaces inside its body. Muscles squeeze against the liquid. This keeps it at a high pressure so the body stays firm.

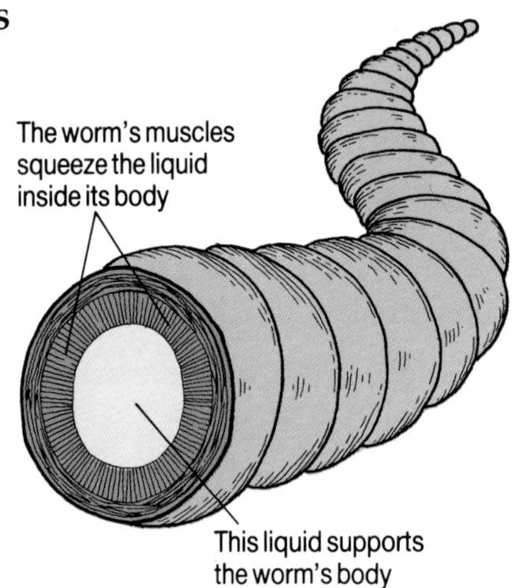

This will turn into a moth and get a hard external skeleton.

Animals with hard external skeletons

Insects, crabs, spiders, and other arthropods have a hard, tough skin over their bodies. This skin is like a suit of armour. It forms flat plates and hollow tubes which support and protect the body. This hard skin is called an **exoskeleton**.

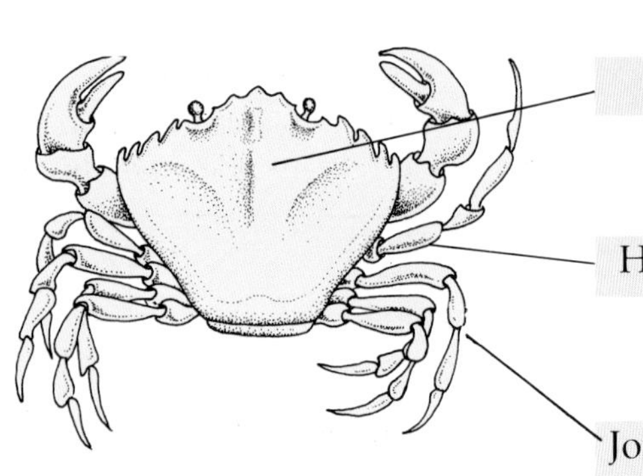

Flat plates cover the body.

Hollow tubes form the limbs.

Joints allow the limbs to move.

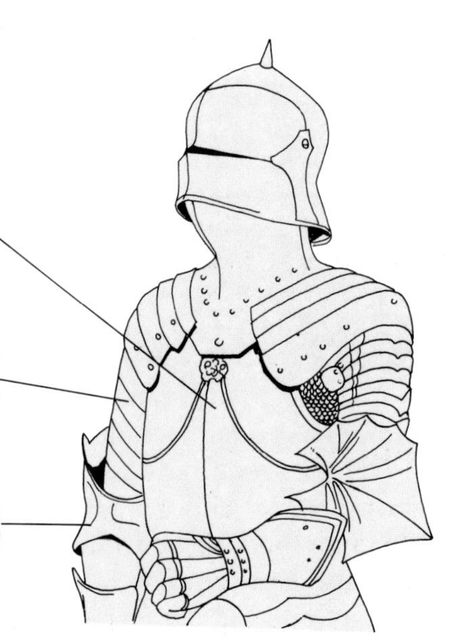

What an exoskeleton does

1 It **supports** the body and gives it shape.

2 It **protects** the soft insides from damage, dirt, and germs.

3 It is **waterproof** and stops the body drying up.

4 Its **colour** may help the animal to hide, or attract a mate.

5 Muscles are attached to it to move the joints.

A crab grows by shedding its exoskeleton. The skin beneath hardens to form a new one.

A male stag beetle has antlers as part of its exoskeleton. It uses them to fight for females.

Animals with hard internal skeletons

You are supported by a hard skeleton inside your body, called an **endoskeleton**. It is made of bone. Fish, amphibia, reptiles, birds, and mammals all have endoskeletons.

The **skull** protects the brain, eyes, and inner ears.

Ribs protect the heart, lungs, and main blood vessels.

Muscles are attached to bones. They can pull on the bones and make them move.

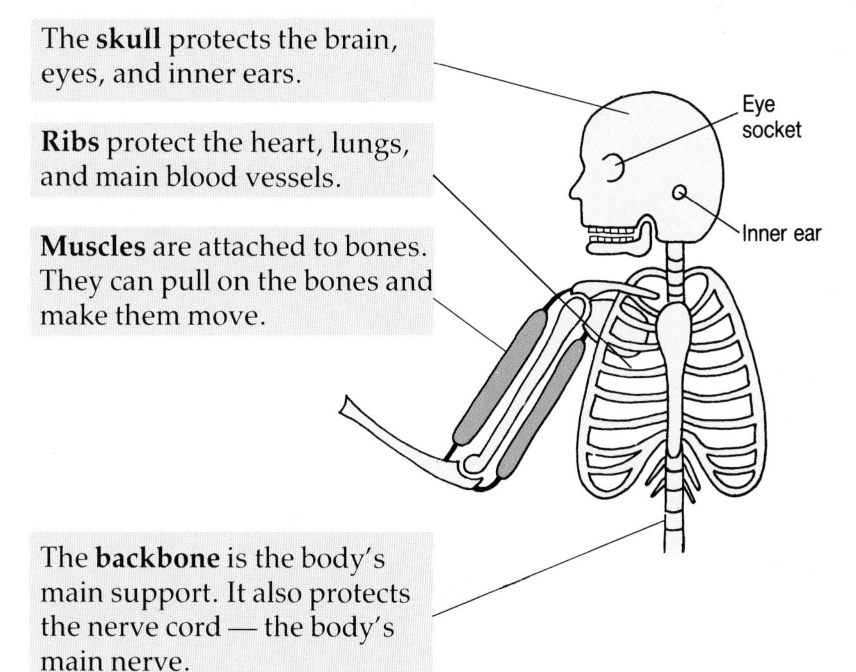

The **backbone** is the body's main support. It also protects the nerve cord — the body's main nerve.

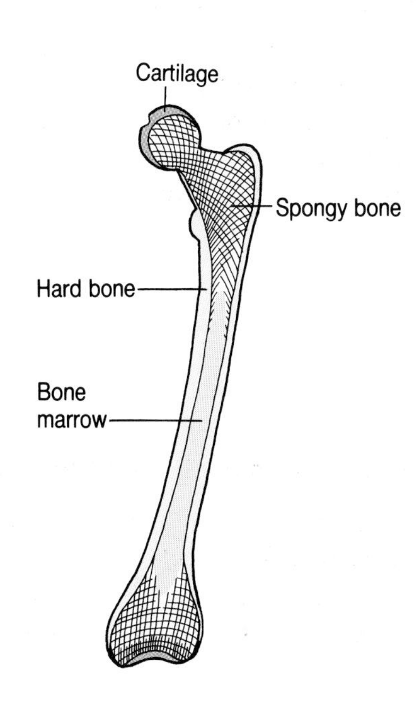

This is a leg bone cut in half. Bones are alive. They contain cells called **bone marrow** which make blood.

Questions

1 lobster mackerel housefly sparrow cat
Which of these have exoskeletons and which have endoskeletons?

2 What part of your skeleton protects:
a) your heart? **b)** your eyes?
c) your nerve cord?

3 Does this describe an endoskeleton or an exoskeleton?
a) It stops the body drying up.
b) It makes blood cells.
c) It keeps dirt and germs out of the body.
d) It is waterproof.

4.2 More about the human skeleton

There are about 200 bones in your skeleton. This is a simple diagram showing the main ones.

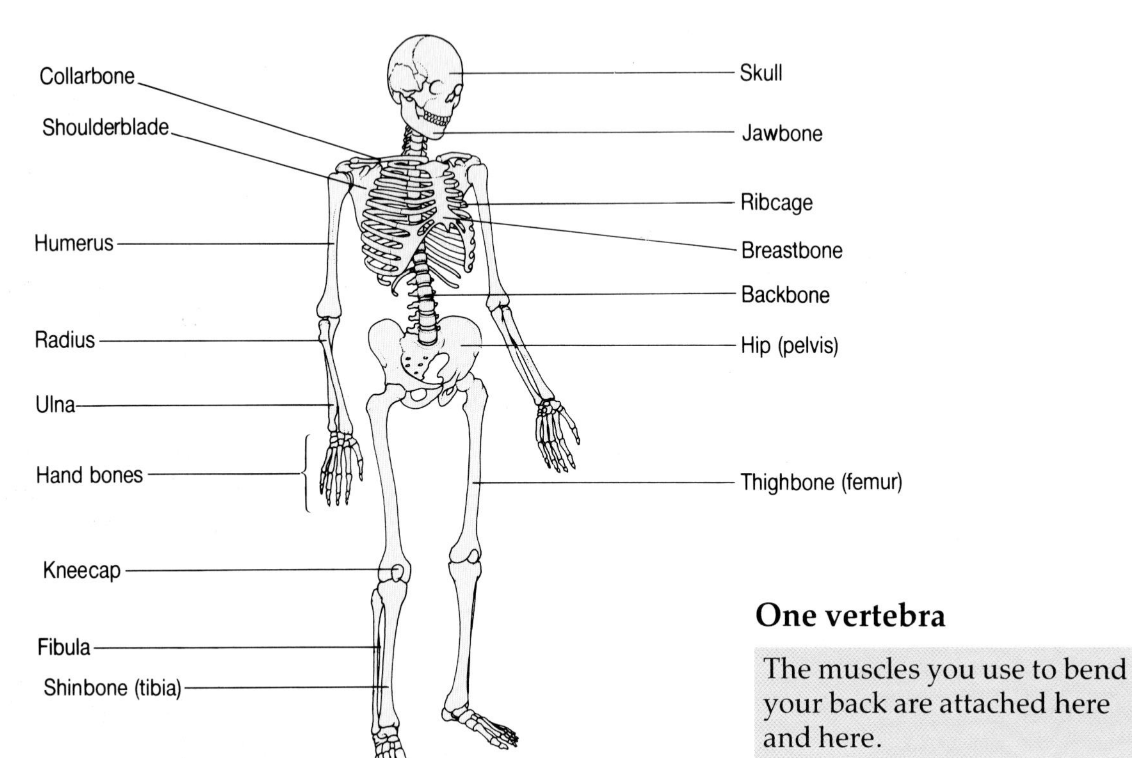

One vertebra

The muscles you use to bend your back are attached here and here.

This vertebra rubs against the next one here and here.

The vertebra has a hollow centre. The nerve cord goes through it.

The backbone

The backbone is called the **vertebral column**. It is made up of 33 small bones called **vertebrae**. These are joined in such a way that the backbone can bend and twist. This drawing shows part of a vertebral column seen from one side.

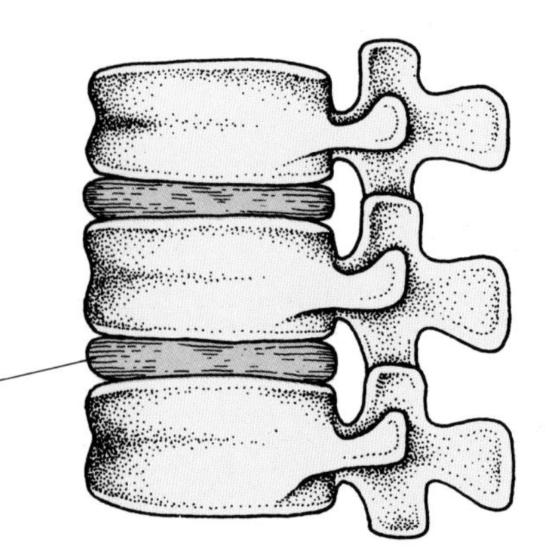

Intervertebral discs are pads of cartilage between the vertebrae. They stop vertebrae knocking against each other when you run or jump.

Joints

1 Fixed joints where the bones cannot move. There are fixed joints between the bones that make up the roof of your skull.

2 Slightly moveable joints where the bones can move a little. There are slightly moveable joints between your vertebrae.

3 Freely moveable joints where the bones can move easily. There are freely moveable joints at your knees and elbows.

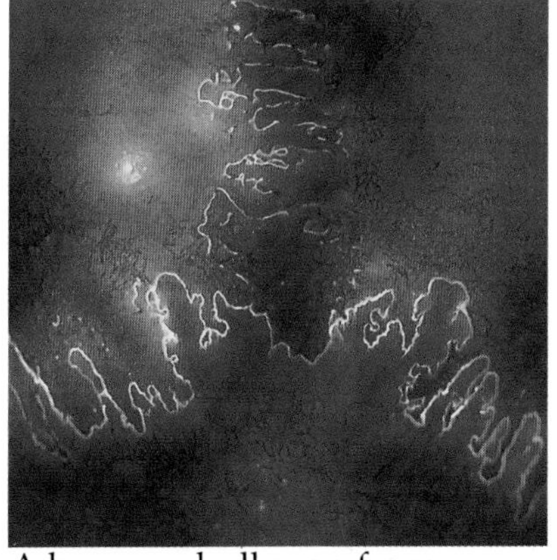

A human skull seen from above. The lines are the fixed joints.

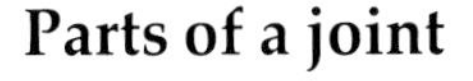

Parts of a joint

The bones at a joint are held together by strong fibres called **ligaments**.

Where the bones rub together they are covered by a slippery layer of **cartilage**.

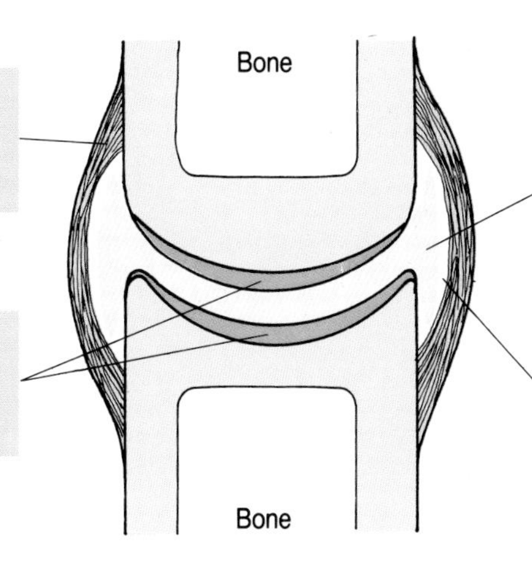

An oily liquid called **synovial fluid** helps the joint move smoothly.

The **synovial membrane** around the joint stops the synovial fluid from draining away.

Two types of freely moveable joint

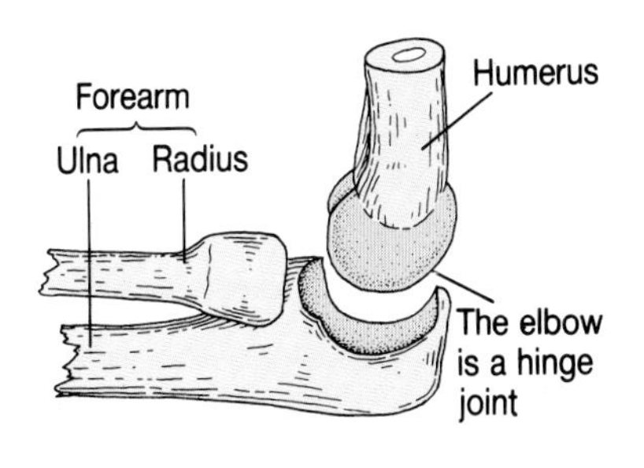

Hinge joints can bend in only one direction, like the hinge of a door. Elbows are hinge joints. joint.

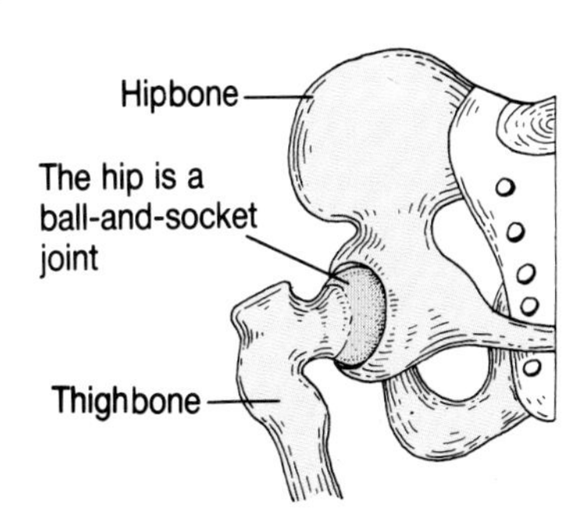

Ball-and-socket joints can bend in all directions. An example is the joint at your hip.

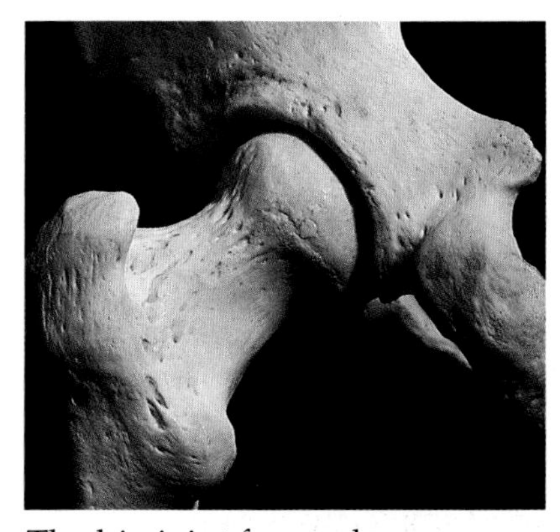

The hip joint from a human skeleton. Why is it called a ball-and-socket joint?

Questions

1 Write down the everyday names for these bones:
  a) pelvis **b)** vertebral column **c)** femur **d)** tibia

2 Where are your intervertebral discs, and what are they for?

3 What are ligaments for?

4 Name a hinge joint and a ball-and-socket joint on your body which are not mentioned here.

5 What is synovial fluid for?

6 What is cartilage for?

4.3 Muscles and movement

Muscles work by getting shorter, or **contracting**. When a muscle contracts it pulls what it is joined to, or squeezes something. There are three kinds of muscle in your body: **voluntary, involuntary** and **cardiac**.

Voluntary muscles

Voluntary muscles are muscles that you can control as you wish (voluntary means by free will).

The muscles attached to your bones are voluntary. When you decide to move they pull on your bones, which makes your joints bend. These muscles work quickly and powerfully, but they soon get tired.

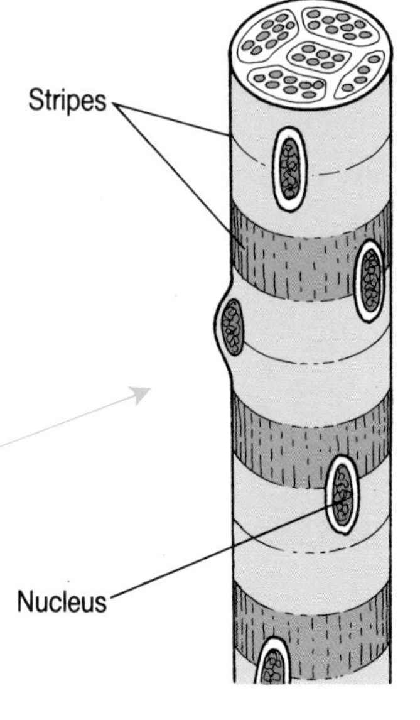

Voluntary muscles are made o fibres which are long and striped.

Involuntary muscles

Involuntary muscles are those you cannot control. They keep working on their own.

There are involuntary muscles in the walls of your gut. Their job is to push food along the gut. Involuntary muscles are also found in the walls of blood vessels, and the bladder. Involuntary muscles work slowly and do not get tired.

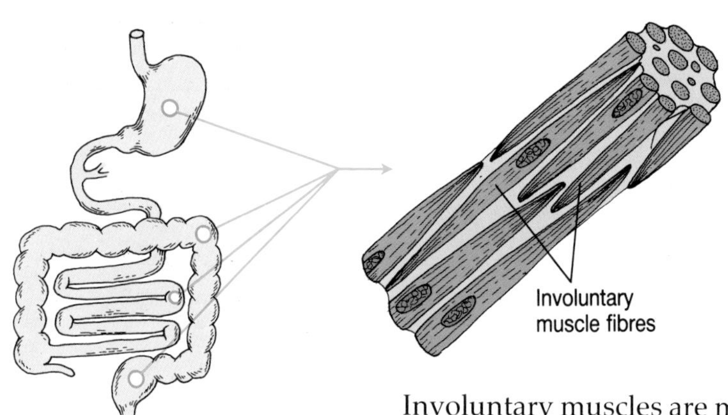

Involuntary muscles are made of short unstriped fibres.

Cardiac muscle

Your heart is made of cardiac muscle. This is a special kind of involuntary muscle.

Cardiac muscle works, without getting tired, all through your life, pumping blood around your body.

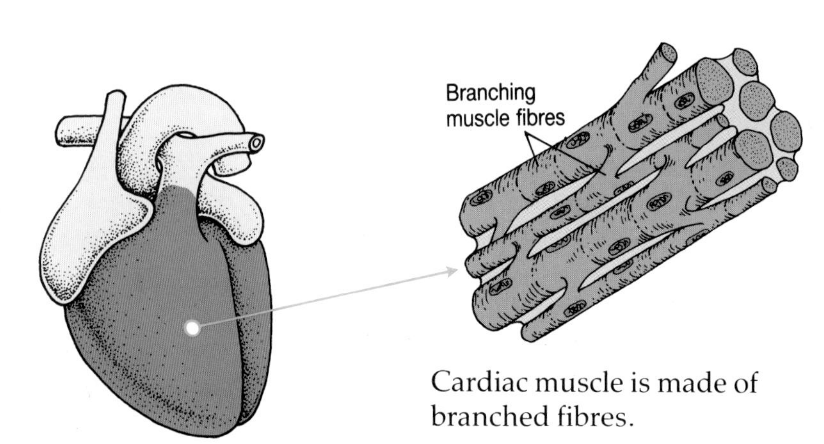

Cardiac muscle is made of branched fibres.

More about voluntary muscles

Muscles are attached to bones by strong fibres called **tendons**. A muscle has at least one tendon at each end.

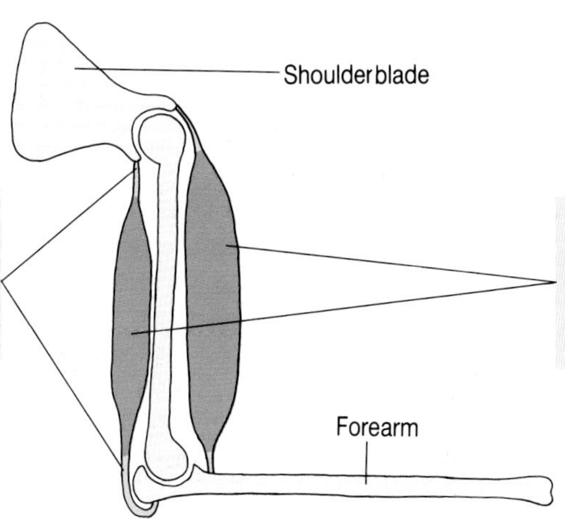

When one of these muscles contracts it pulls against the bones of the arm. This bends or straightens the elbow joint.

Voluntary muscles work in pairs

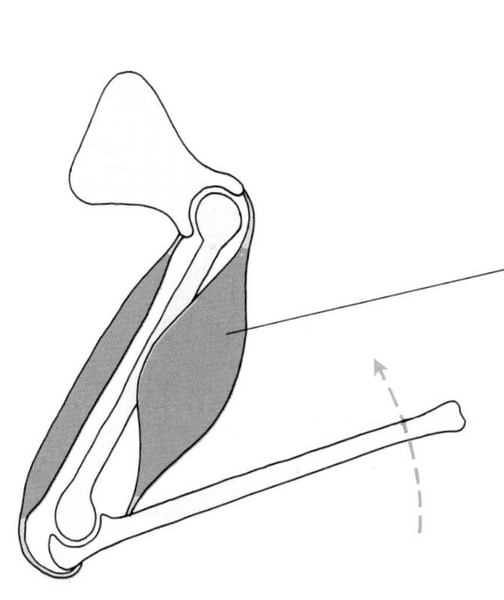

The muscles at a joint work in pairs. One muscle makes the joint bend, the other makes it straighten.

This muscle has contracted to bend or **flex** the elbow joint. Muscles which bend joints are called **flexor muscles**.

This muscle has contracted to straighten or **extend** the elbow joint. Muscles which straighten joints are called **extensor muscles**.

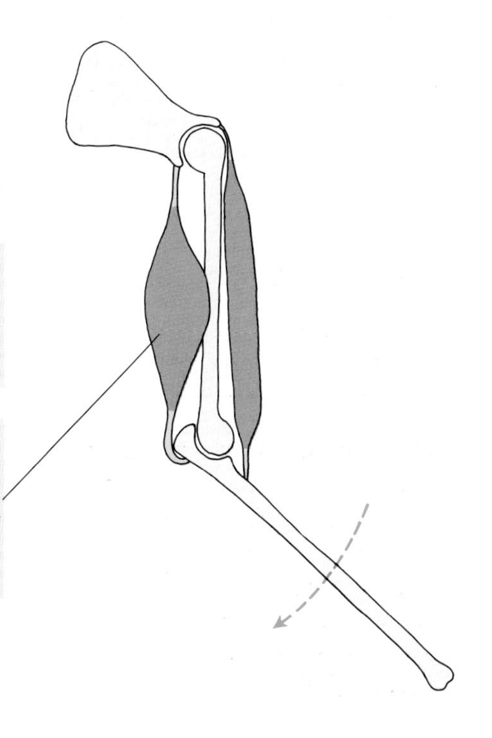

Questions

1 What kind of muscle:
   a) moves your arm when you turn this page?
   b) moves food from your mouth to your stomach?
   c) moves your legs when you run?
   d) moves blood around your body?

2 Why are at least two muscles needed at a joint?

3 What is the name for:
   a) muscles which straighten joints?
   b) muscles which bend joints?

4 Describe one way that heart muscle is:
   a) similar to stomach muscle.
   b) different from stomach muscle.

5 What is the difference between a ligament and a tendon?

4.4 Metabolism, enzymes, and respiration

Metabolism

Your body is a chemical factory. At any moment up to 1000 different chemical reactions are taking place inside every cell. Together these reactions are called **metabolism**. They each play some part in keeping you alive and well. So, metabolism is all the chemical reactions necessary for life.

Metabolism speeds up when you are active and slows down when you sleep. **Basal metabolism** is the slowest metabolism needed to keep you alive. There are two types of metabolism.

Catabolism is the breakdown of complex molecules into simpler ones, resulting in the release of energy. All living things obtain energy by a type of catabolism called **respiration.** This uses oxygen to break down glucose sugar into carbon dioxide, water and energy.

Anabolism is the opposite of catabolism. Anabolism *uses* energy from catabolism to build up complex molecules from simpler ones. For example, energy is needed to make starch out of glucose, and proteins from amino acids.

In the picture above, which person do you think has the lowest basal metabolism, and the highest rate of anabolism? Give reasons.

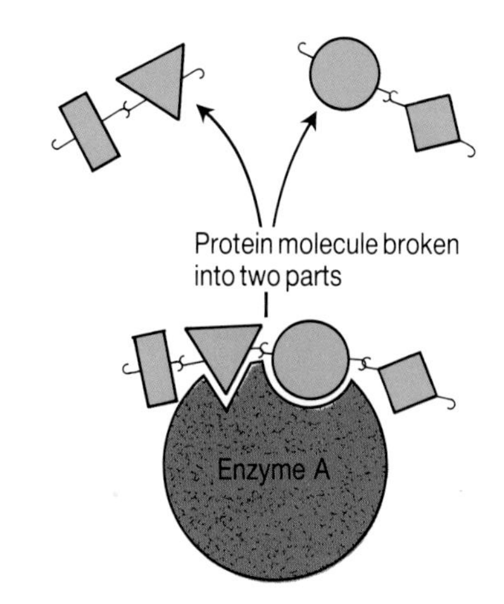

Enzymes

The chemical reactions of metabolism would go very slowly, or not at all, if it were not for **enzymes.** An enzyme speeds up a chemical reaction inside a cell, but the enzyme itself is not used up in the reaction. This allows the enzyme to do the same thing over and over again.

Digestive juices are enzymes which break down food into substances such as glucose and amino acids which can then be absorbed into the blood (p.125).

There is a theory that each enzyme has an 'active site' which combines briefly with a substance and changes it – either splitting it apart or linking two pieces together. The shape of the active site fits only one type of molecule so each enzyme can control only one type of chemical reaction.

Enzymes work best at a certain temperature. In humans that temperature is 37 °C. Some work best in acid conditions (e.g. the digestive enzyme pepsin), some in neutral conditions, and some in alkaline conditions.

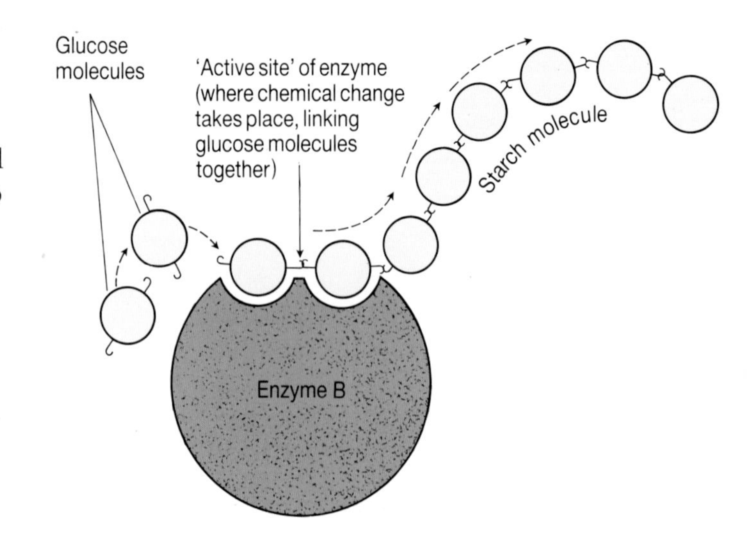

Respiration

You use energy to walk, and think, and digest your food. In fact, you use energy for everything that goes on in your body. The energy comes from your food during respiration, which can be described as the chemical breakdown of food to produce energy for life. This can also be called **internal respiration** because it takes place inside every cell of every living thing. Respiration usually needs oxygen and produces carbon dioxide as a waste product.

Even relaxing needs energy . . . and that means respiration. It goes on in all your cells even when you are quite still or asleep.

External respiration, or breathing

The word respiration used to mean breathing; that is, the movements which draw air into and push it out of the lungs. This is now called **external respiration** because it takes place *outside* cells and involves the exchange of oxygen and carbon dioxide between the body and the outside world.

The respiratory system in humans

You use your **respiratory system** to breathe in oxygen for respiration, and to breathe out carbon dioxide produced by respiration.

The voice box or **larynx**. It makes sounds used in speaking.

The windpipe or **trachea**. It is like the flexible hose of a vacuum cleaner or tumble drier. It is held open by rings of **cartilage**.

The **lungs** are soft and spongy.

The **ribs** protect the lungs.

The **intercostal muscles** help with breathing in and out.

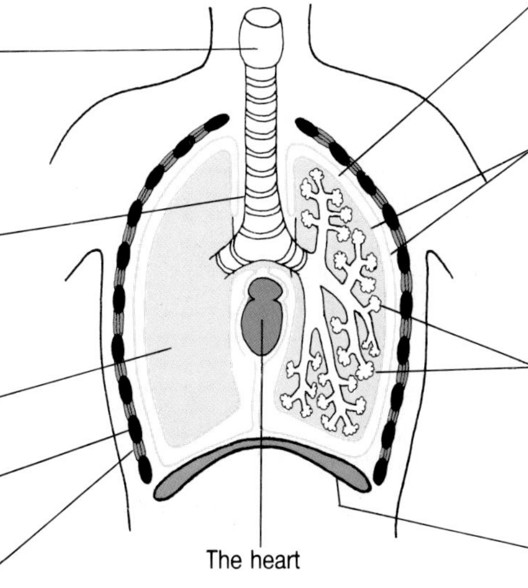

Lungs are in a space in the chest called the **thoracic cavity**.

This cavity is lined with a slippery skin called the **pleural membrane**. It protects the lungs as they rub against the ribs.

The windpipe has thousands of branches which end in tiny **air sacs**. This is where oxygen is taken in to the body, and carbon dioxide is got rid of.

The **diaphragm** is a sheet of muscle below the lungs. It helps with breathing in and out.

Questions

1 a) What are catabolism and anabolism?
  b) Which of the enzymes in the diagram on p. 88 is anabolic? Is the other catabolic? Give reasons for your answer.
  c) Name an example of catabolism.

2 Why is respiration essential for life?

3 What is the difference between respiration and breathing?

4.5 Aerobic and anaerobic respiration

Aerobic respiration

In most organisms internal respiration requires oxygen and involves the *complete* breakdown of food with the release of *all* the energy stored in it. This is called **aerobic respiration**. The complete breakdown (of one mole – 180 g) of glucose can be written as:

glucose+oxygen→carbon dioxide+water+**energy** (2898 kJ per mole)

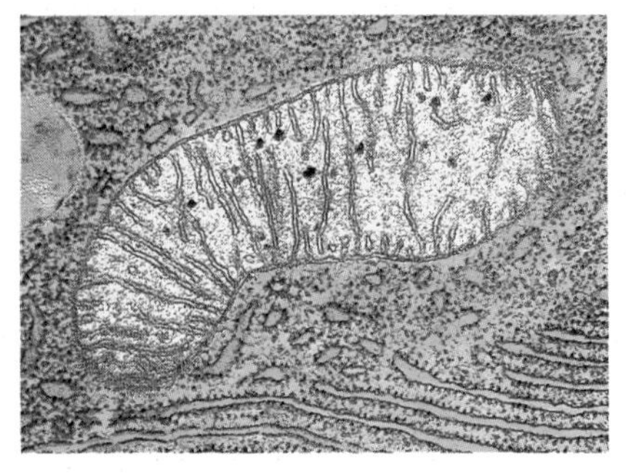

Respiration occurs in tiny sausage-shaped objects called **mitochondria**, found in all types of cell. They contain enzymes which release energy from food.

This is what happens:

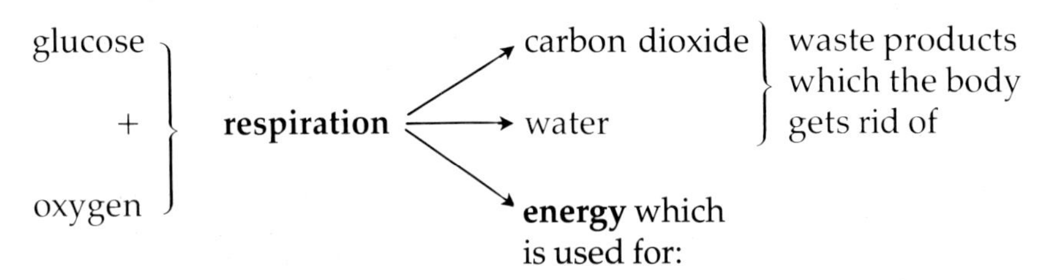

1 **Mechanical work** – contracting skeletal muscle, heart muscle, gut muscle.
2 **Chemical work** – in the liver, kidneys, nerve impulses.
3 **Growth and repair** – cell division for growth of new tissues, and repair of old and damaged tissues.
4 **Anabolism** – making proteins (enzymes and hormones), carbohydrates (e.g. glycogen), fats.
5 **Heat** – to maintain body temperature in warm-blooded animals.

Anaerobic respiration

Respiration can also take place without oxygen. It is then called **anaerobic respiration.** It produces less energy than aerobic respiration because food is not completely broken down into carbon dioxide and water with the release of all its stored energy.

Yeasts get energy by a type of anaerobic respiration called **fermentation.** It produces alcohol:

glucose → carbon dioxide + alcohol + **energy** (210 kJ per mole)

Alcoholic drinks are made using yeast to ferment the sugar in fruit juice. Cider is made from fermented apple juice. Beer is made by yeast which ferments malt obtained from barley seeds.

Fermentation is also used in baking. It takes place when yeast and sugar are added to dough. It produces carbon dioxide gas which fills the dough with bubbles and makes it rise. The yeast is killed in the hot oven when the dough is baked.

Which gas is produced when wine ferments?

Anaerobic respiration in muscle

Walking or jogging is called **aerobic exercise**, because the body can easily obtain enough oxygen for aerobic respiration to supply all the energy it needs. Food is broken down into water and carbon dioxide, which is breathed out so does not accumulate in the body. This is why walkers and fit joggers can continue for several hours.

Very fast running soon becomes **anaerobic exercise** because, no matter how fast you breathe, or how fast your heart beats, your body cannot obtain enough oxygen for aerobic respiration to supply all its energy needs. Under these circumstances, your body gets the extra energy from anaerobic respiration, because this does not need oxygen. But it produces lactic acid instead of carbon dioxide:

glucose → lactic acid + **energy** (150 kJ per mole)

After an exhausting race, these rowers gasp for breath to pay back the oxygen debt.

The oxygen debt

Lactic acid gathers in your muscles, making them ache. Within a minute, there is so much lactic acid in your muscles that they stop working altogether and you have to stop running. Fast, deep breathing as you recover soon supplies enough oxygen to combine with the lactic acid to make carbon dioxide and water, which is excreted.

The amount of oxygen needed to get rid of the lactic acid from muscles is called the **oxygen debt.**

How living things use and store energy

The energy released during respiration is not used directly. It is first used to build up a chemical called **ATP** (adenosine triphosphate) which acts as a temporary energy store. Think of ATP molecules as 'packets' of energy which are 'filled' during respiration. These packets can be 'emptied' to release instant energy whenever it is needed, without cells having to break down glucose molecules.

ATP delivers energy in precise amounts, and can transfer energy to other chemicals, turning them from inert substances into highly reactive ones.

ATP has three phosphate groups, the last of which is attached by a bond which needed much energy from respiration to create:

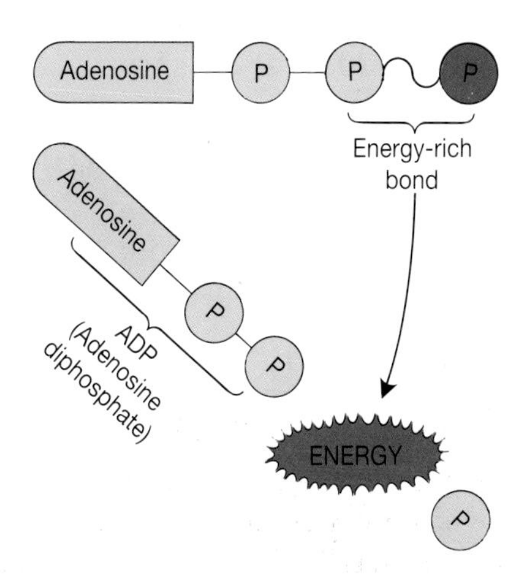

Energy is released when this bond is broken. More energy is needed to make ADP into ATP by adding another energy-rich phosphate bond.

Questions

1. Give at least two differences between aerobic and anaerobic respiration.
2. What kind of respiration is going on in your body now, and when you run very fast?
3. **a)** Why can marathon runners run for hours but sprinters have to stop after a few seconds?
   b) Which of the two builds up an oxygen debt?
4. What is ATP, and why is it important?

4.6 Lungs and gas exchange

Your body takes in oxygen from the air for respiration. At the same time it gives out the carbon dioxide produced by respiration. This exchange of gases takes place in your lungs.

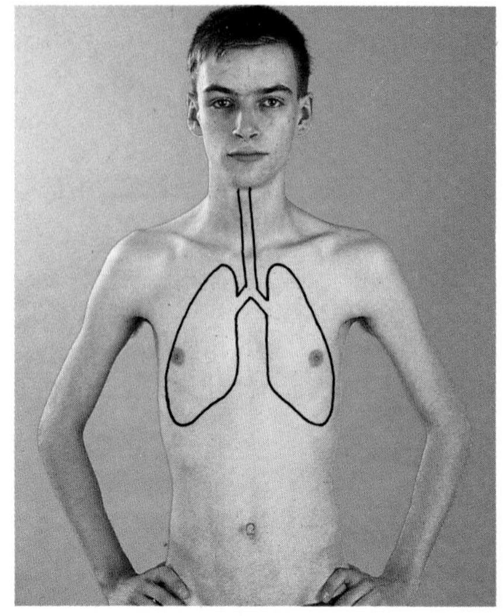

Where your lungs are . . .

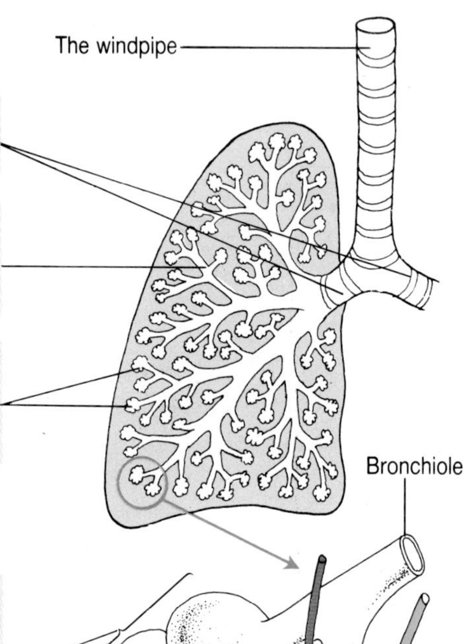

The windpipe divides into two tubes called **bronchi**.

The bronchi divide thousands of times into narrower tubes called **bronchioles**.

There are **air sacs** at the ends of the bronchioles. This is where gas exchange takes place.

A closer look at air sacs

This is one **air sac**. It looks like a bunch of grapes. The 'grapes' are called **alveoli**. They are smaller than grains of salt, and there are 300 million of them in your lungs.

Alveoli have thin moist walls, so that gases can pass through them easily.

Alveoli are covered with narrow blood vessels called **capillaries**. Oxygen passes into the capillaries from the alveoli.

Gas exchange in the alveoli

This diagram shows what happens in the alveoli.

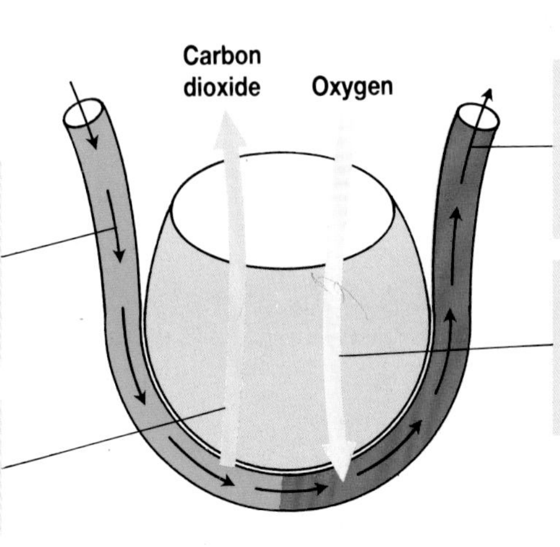

Blood flows to the lungs from around the body. It carries carbon dioxide produced by respiration in the cells of the body.

Carbon dioxide passes *from* the blood *into* the alveoli. Then it is breathed out of the body.

Blood carries oxygen away from the lungs to every cell in the body, where it is used for respiration.

Oxygen is breathed into the lungs. It dissolves in the water lining the alveoli. From there it passes *into* the blood.

When you breathe in

1 The **intercostal muscles** contract. These pull the ribcage upwards. So the chest increases in volume.

2 The **diaphragm** contracts. This makes it flatten out, so the chest gets even larger.

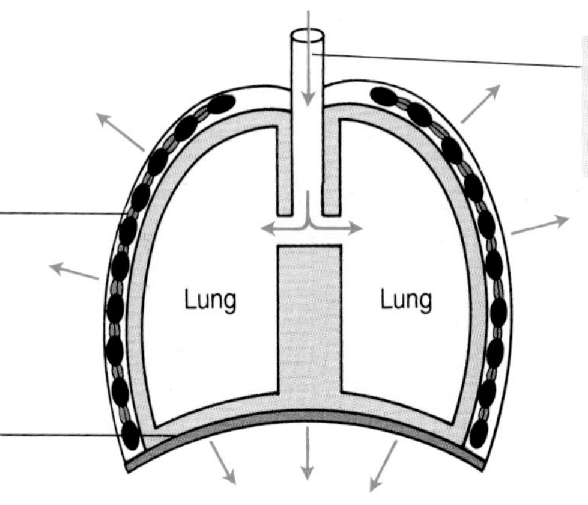

3 As the chest gets larger, air is sucked down the windpipe and into the lungs.

When you breathe out

1 The **intercostal muscles** relax, which lowers the ribcage. The chest decreases in volume.

2 The **diaphragm** relaxes, and bulges upwards. This decreases the volume of the chest even more.

3 Because the chest has got smaller, air is forced out of the lungs.

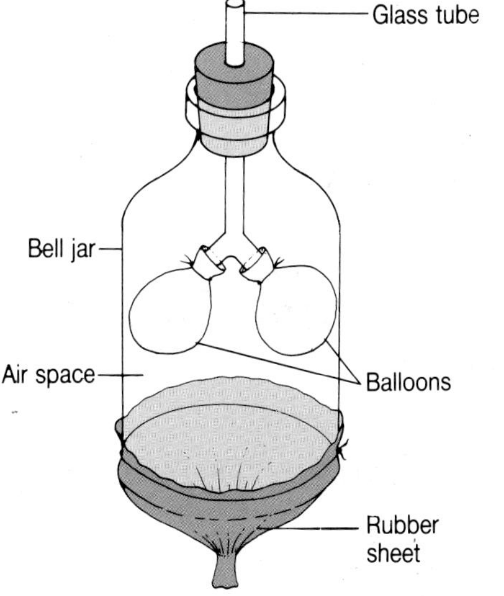

Watch what happens in a **bell jar** when you pull the rubber sheet downwards.

How air changes in your lungs

Gas	Amount of it in air you breathe in	Amount of it in air you breathe out
Oxygen	21%	16%
Carbon dioxide	0.04%	4%
Nitrogen	79%	79%
Water vapour	a little	a lot

Questions

1 Which gases are exchanged in your lungs?

2 Where in the lungs does gas exchange take place?

3 Explain what happens to intercostal muscles and the diaphragm as you breathe in and out.

4 Explain why you breathe out:
 a) more carbon dioxide than you breathe in
 b) as much nitrogen as you breathe in.

5 Do you use *all* the oxygen you breathe in?

6 What are alveoli?

Questions on Section 4

1 *Recall knowledge/understanding*
The diagram opposite shows bones and muscles in the right leg of a runner waiting for the starting gun.

a) Give the name and letter of a ball-and-socket and a hinge joint.
b) What holds bones X and Z together?
c) What do flexor and extensor muscles do?
d) Which letters point to flexor muscles and which point to extensor muscles?
e) Which muscles will contract when the starting gun fires?

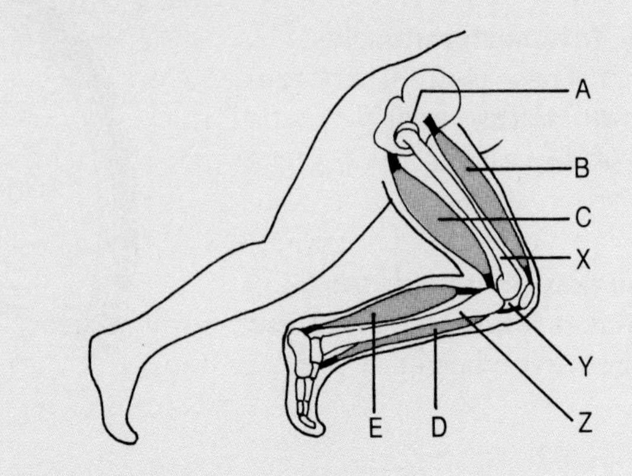

2 *Recall knowledge*
This diagram shows a freely moveable joint. Match each label on the diagram with one or more of the descriptions below:

a) an oily liquid;
b) slippery gristle;
c) fibres which hold the bones together;
d) synovial membrane;
e) lubricates the joint;
f) ligament.

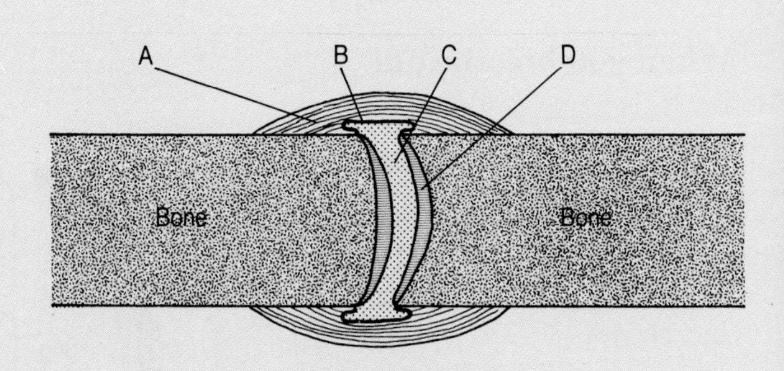

3 *Recall knowledge*
This diagram shows the parts of the human respiratory system.

Match each label on the diagram with one or more of the descriptions below:

a) holds open the windpipe;
b) the trachea;
c) bulges upwards when it relaxes;
d) ribs;
e) bronchus;
f) contain millions of alveoli;
g) tube connecting the mouth with the lungs;
h) lungs;
i) diaphragm;
j) lift the ribcage when they contract.

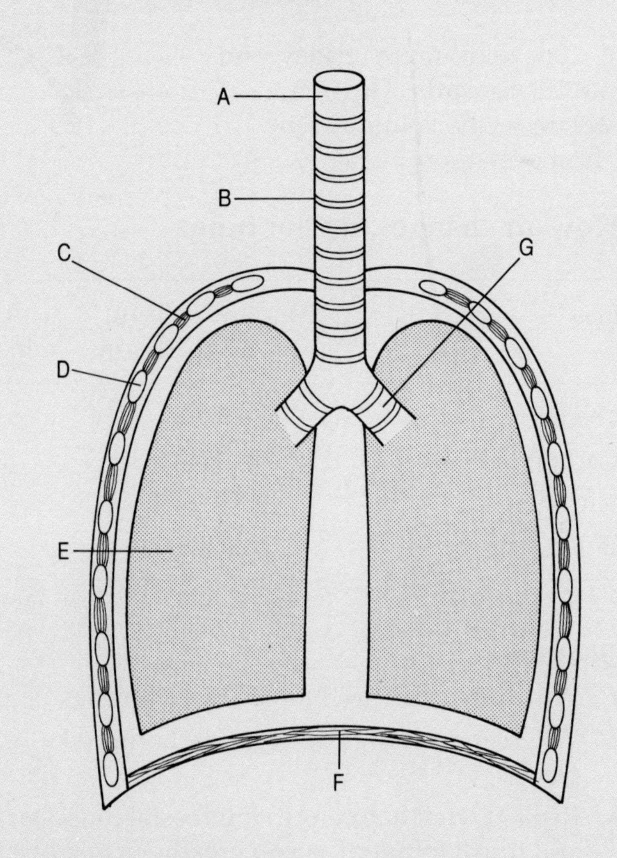

4 *Recall knowledge*

a) Name the gas your muscles produce when they are working.
b) How does your body get rid of this gas?
c) Name the gas your muscles need more of, during exercise.
d) How does this gas get to your muscles?

5 *Interpret data (extract information)*
The graphs opposite show the breathing rates of a runner before and just after a race. The distance between a peak and a trough on the graph is the volume of one breath.

a) Calculate the number of breaths per minute at rest and after running.
b) Calculate the volume of air breathed out each minute at rest and after running.
c) Explain the difference between these two volumes.
d) If unbreathed air is 20 per cent oxygen, and breathed air is 16 per cent oxygen what is the volume of oxygen taken in per minute while at rest and after running?

6 *Understanding/interpret data*
Study the apparatus in the diagram opposite. All the joints are airtight.

a) What is the purpose of this apparatus?
b) Explain the change in level of the liquid in the glass tube.
c) Why are the containers surrounded with cotton wool?
d) What is the functions of the clips?
e) What is the function of the part labelled A?
f) What control is necessary?

7 *Interpret data/understanding*
A student was given various mixtures of gas to breathe and his rate of breathing was measured. The graphs opposite show the results.

a) What is the main cause of an increase in breathing rate?
b) In view of this, why is mouth-to-mouth resuscitation normally a better method of artificial respiration than pressing down on the chest?

8 *Understanding/interpret data*
Study the apparatus in the diagram opposite. Soda lime absorbs carbon dioxide from air.

a) How far does the oil drop travel in apparatus A?
b) In view of the function of soda lime, give one reason why it moved in this way.
c) What is the apparatus measuring?
d) What is the function of apparatus B?

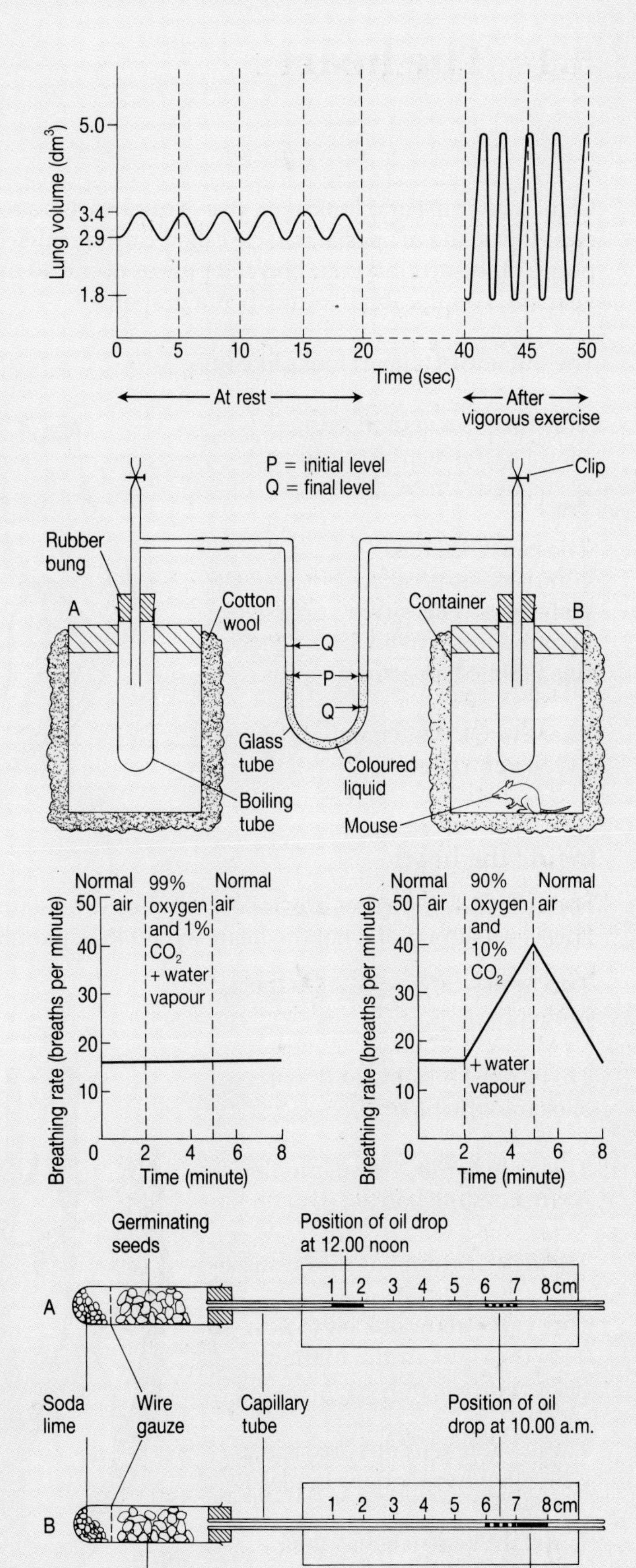

5.1 The heart

Clench your fist and look at its size. Your heart is about the same size. It is made of special muscle called **cardiac muscle**. Its job is to pump blood around your body. It pumps about 40 million times a year and weighs about as much as a grapefruit.

The outside of a heart looks like this.

This tube is a **vein**. It brings blood to the heart from all parts of the body *except* the lungs.

The heart has four compartments called **chambers**. These two upper chambers are called **atria**. (Each one is called an **atrium**.)

These two lower chambers are called **ventricles**.

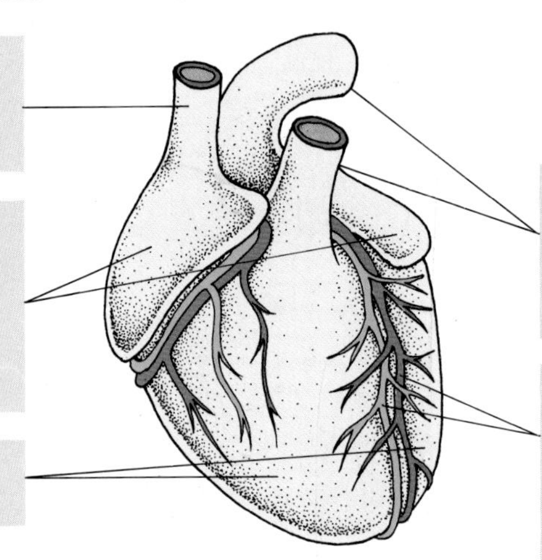

These two tubes are called **arteries**. They carry blood away from the heart to all parts of the body.

The heart has its own blood supply, carried by the **coronary** artery and vein. These tubes carry food and oxygen to the heart, and carry wastes away.

Inside the heart

The atria and ventricles are hollow, so they can fill up with blood. This is a diagram of what the heart would look like sliced open.

This **artery** carries blood to the lungs.

This **valve** stops blood flowing back into the heart.

This **vein** carries blood into the heart from the body.

The **right atrium** has thin walls.

This **valve** allows blood to flow from the right atrium to the right ventricle only.

These **valve tendons** are strings which hold valve flaps in place.

The **right ventricle** has thick walls. It pumps blood to the lungs.

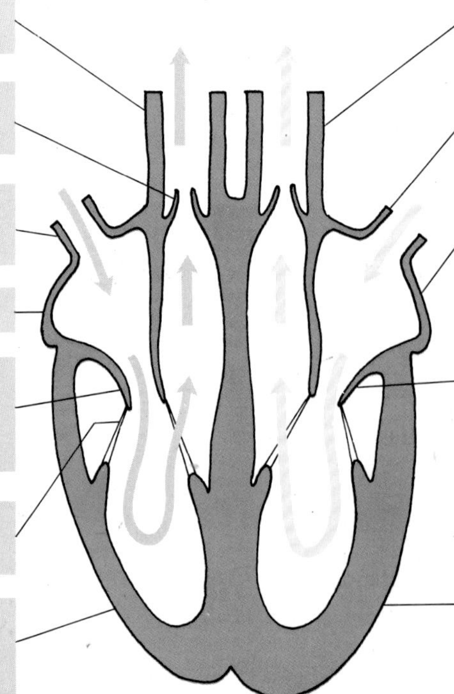

This **artery** carries blood away from the heart to the body.

This **vein** carries blood into the heart from the lungs.

The **left atrium** has thin walls.

This **valve** allows blood to flow from the left atrium to the left ventricle only.

The **left ventricle** has very thick walls. It pumps blood to all parts of the body, except the lungs.

How the heart pumps blood

The heart pumps blood by tightening, or **contracting.** That makes it smaller, so blood gets squeezed out into the arteries. Then it **relaxes** again, and fills up with blood from the veins.

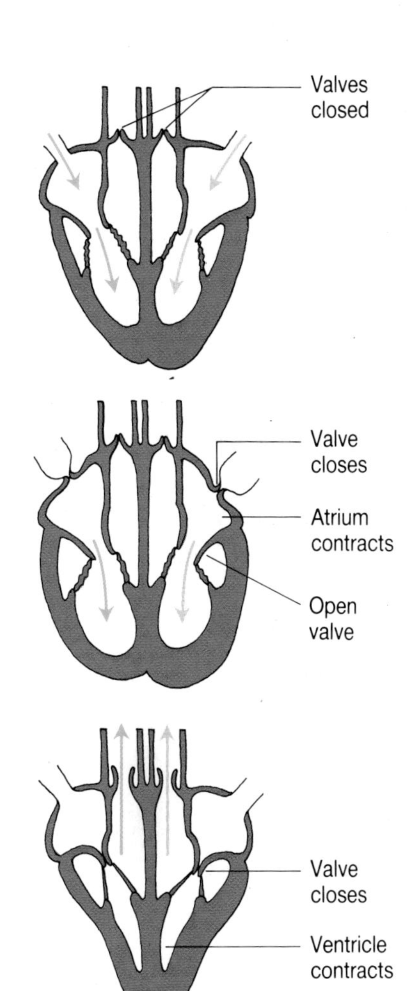

1 When the heart is relaxed, both sides fill up with blood from the veins.

2 Then the **atria** contract. The veins also contract where they join the atria. So blood is forced into the ventricles through the valves.

3 A fraction of a second later, the **ventricles** contract. The valves between the atria and ventricles close. So the blood is squeezed into the arteries.

4 The heart relaxes again, and fills up with blood.

One complete contraction and relaxation is called a **heartbeat**. It takes less than a second. Each beat pumps out a cupful of blood. The heart usually beats around 70 times a minute when you're resting.

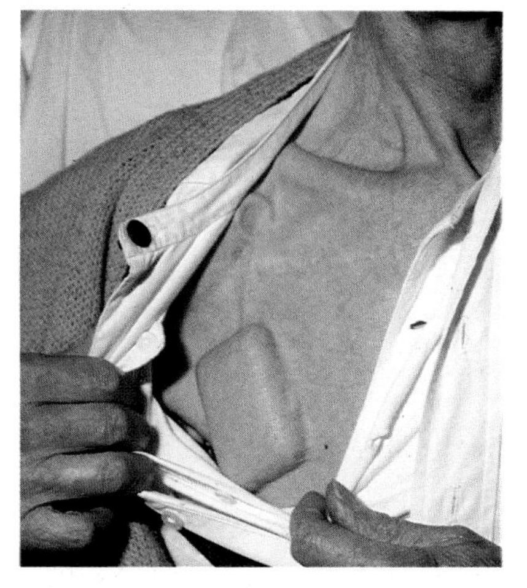

This is an electronic heart **pacemaker.** It can be used instead of the heart's own built-in pacemaker if this stops working after a heart attack. Both pacemakers keep the heart beating by sending tiny, regular electric shocks through the heart muscle.

The heart's own pacemaker is a bundle of special muscle found in the wall of the right atrium near the opening of the main vein.

The electronic pacemaker is battery-operated and small enough to be inserted under the skin. It sends shocks to the heart muscle through wires.

Questions

1. What is the heart made of?
2. What are the upper parts of the heart called?
3. Which ventricle has the thickest walls?
4. What stops blood flowing backwards through the heart?
5. What are valve tendons for?
6. What are the coronary arteries and veins for?
7. The heart can **contract.** What does this mean?
8. What happens when the atria contract?
9. What happens when the ventricles contract?
10. What is a heartbeat?
11. How fast does the heart usually beat?

5.2 What makes blood flow

When the powerful muscles of your heart contract they force blood out into tubes called **arteries**. The arteries branch into tiny little tubes called **capillaries**. The capillaries join together to form **veins**. The veins carry blood back to the heart.

This is an **artery**. There are elastic fibres in its walls. The blood pumps into the artery very fast, at high pressure, so the elastic fibres stretch. Then they contract, and that squeezes the blood towards the capillaries.

This is a **capillary**. Its walls are so thin that liquid from the blood can pass through them. This liquid takes food and oxygen to the cells of the body. It also takes away carbon dioxide and other wastes.

This is a **vein**. Veins are wider than arteries and have thinner walls. The blood flows through them more slowly. They have valves to stop it flowing backwards.

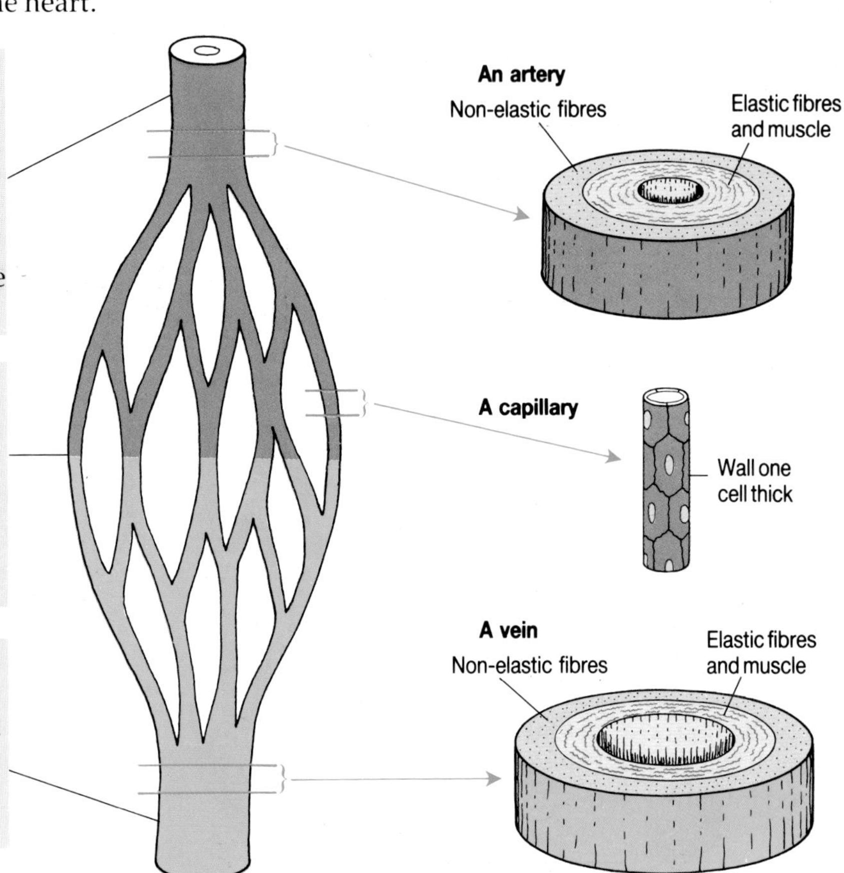

Vein valves

There are many large veins inside muscles of the legs and arms. When these muscles contract they squeeze the veins. This squirts blood toward the heart. The vein valves stop it flowing in the opposite direction.

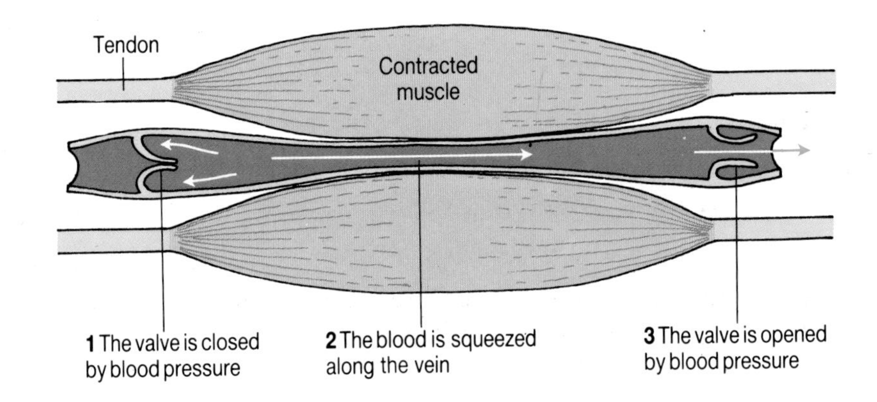

The circulatory system

The tubes that carry blood are called **blood vessels**. The heart and blood vessels together make up the **circulatory system**. This system has two parts:

1 The right side of the heart pumps blood to the lungs and back again. In the lungs it loses carbon dioxide and picks up oxygen.

2 The left side of the heart pumps blood to the rest of the body and back. On its way around the body the blood loses oxygen to the body cells and picks up carbon dioxide.

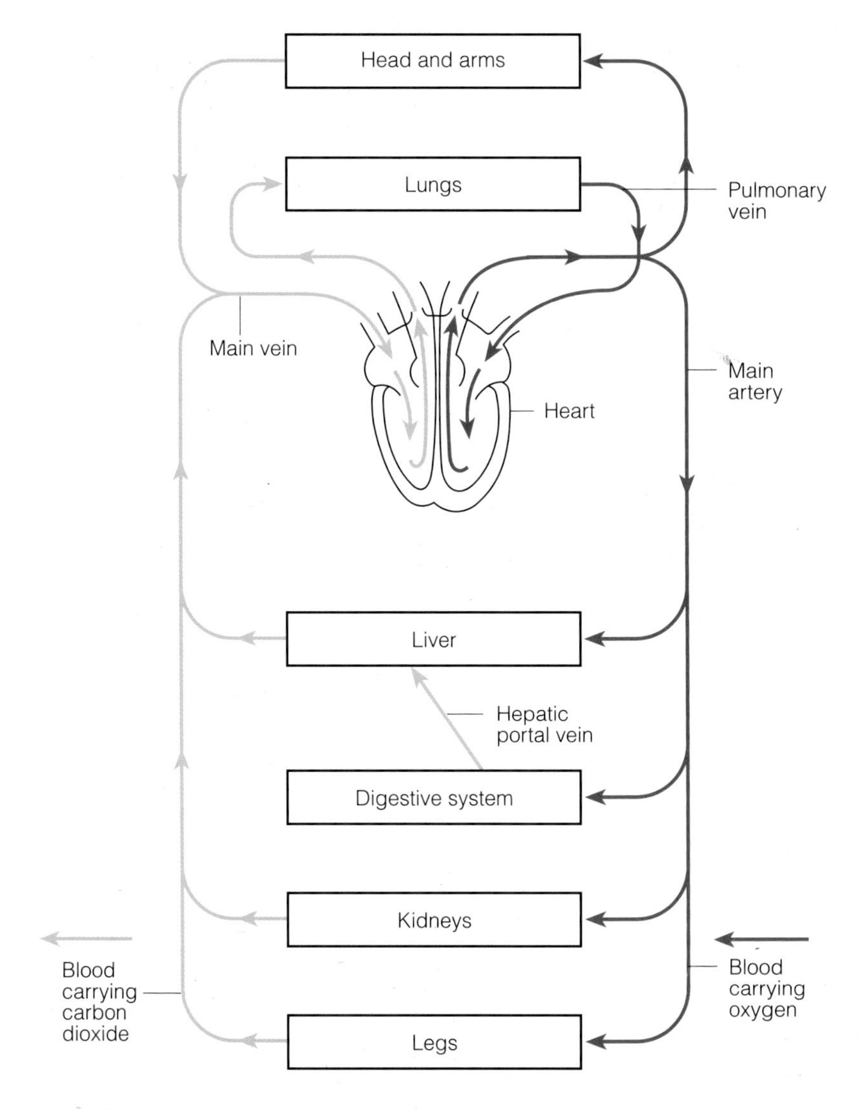

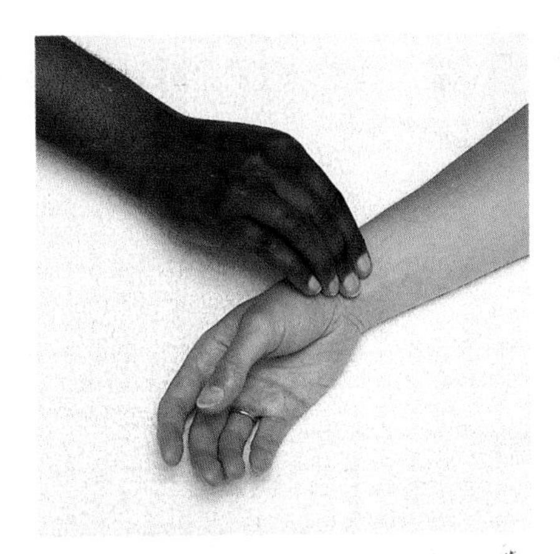

When the heart pumps blood into the arteries at high pressure, it causes a pulse. Use your fingers, as shown here, to find the pulse at your wrist.

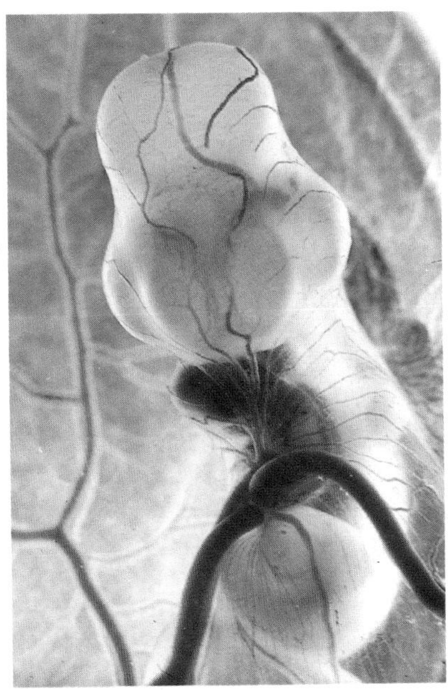

This photograph shows the network of capillaries and arteries in a chick embryo. There are enough capillaries in your body to stretch one-and-a-half times round the world!

Questions

1 What happens in the arteries?

2 What happens in the capillaries?

3 How are arteries different from veins?

4 a) Blood is at its highest pressure just where it leaves the heart. Why do you think this is?

b) The blood in an artery flows faster and at higher pressure than the blood in a vein. Can you explain why?

5 In the diagram above find:

a) a vein with blood full of oxygen;

b) an artery containing blood full of oxygen;

c) a vein with blood full of digested food.

5.3 Heart disease

To keep pumping, your heart needs food and oxygen. It gets them from its own blood supply, carried in the **coronary arteries**. If these arteries get blocked the result is **heart disease**.

How arteries get blocked

A fatty substance called **cholesterol** can stick to the walls of an artery. The arteries become narrower. So blood gets slowed down.

Cholesterol can make artery walls rough. This causes blood to clot, as it flows past. A blood clot can block an artery completely. The blockage is called a **thrombosis**. The blood flow is stopped.

Bits of cholesterol can break off into the blood stream, and block narrow blood vessels.

A thrombosis in blood vessels in the brain is called a **stroke**. Brain cells die. A person suffering from a stroke may get paralysed or even die.

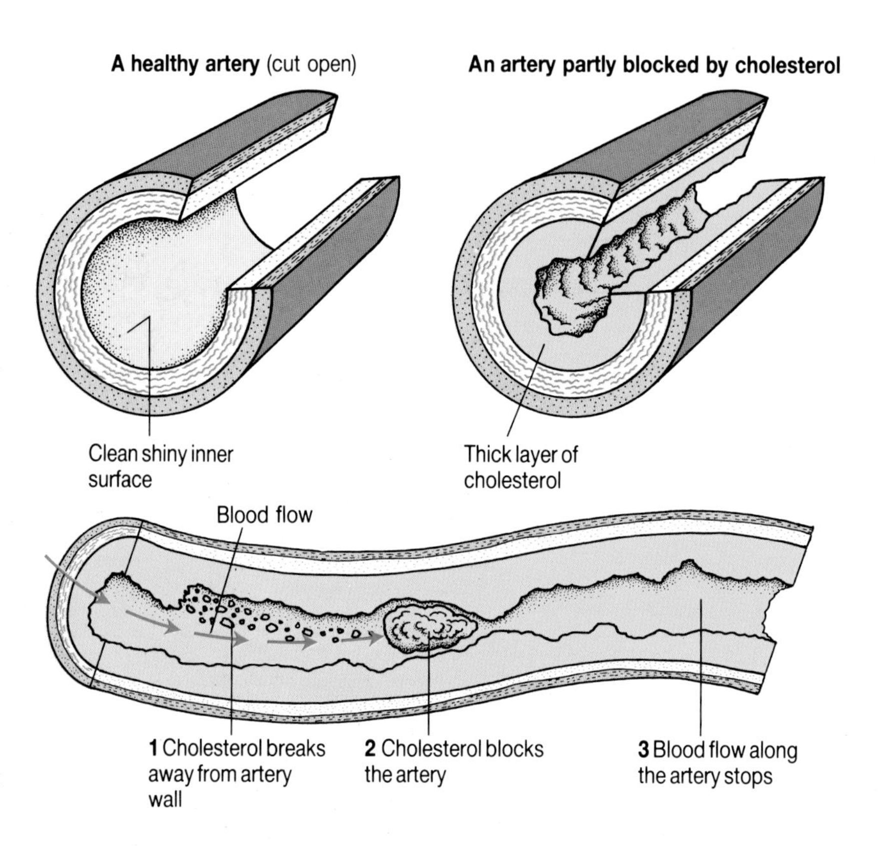

Blocked arteries, and heart transplants

If a coronary artery is partly blocked, the heart muscle gets too little food and oxygen, causing severe chest pains called **angina.** The effect of a thrombosis (blood clot) in the coronary artery is called a **heart attack.** The heart stops beating. But the right treatment can start it beating again.

If a heart becomes very badly diseased, it can be replaced with a healthy donor heart by **transplant surgery.** The patient is put on a heart-lung machine which charges blood with oxygen and pumps it back into the body. Most of the diseased heart is removed, leaving only the back wall of the atria. The donor heart is then sewn in place.

Transplant surgery is made possible by drugs which stop the patient's immune system treating transplants as if they were germs and destroying them.

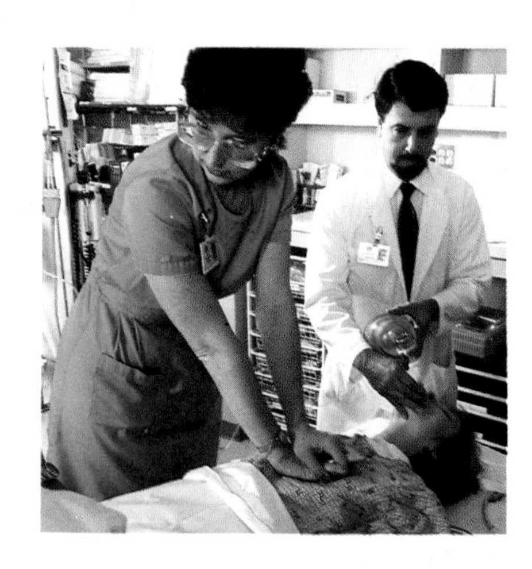

Heart massage can start a heart beating again. The green cylinder contains oxygen.

Things that can lead to heart disease

Heart disease is a big problem. In Britain it kills about 400 people a day. Most doctors agree that your chances of suffering from it are greatest:

1 If you eat lots of cream, butter, eggs, fat, and fried foods. These can lead to a high level of cholesterol in the blood.
2 If you smoke cigarettes.
3 If you are overweight.
4 If you take too little or no regular exercise.
5 If your way of life involves stress (worry, anger, fear, and so on).

One way to avoid heart disease is to take regular exercise.

How you can avoid heart disease

1 Cut down on fried food. You can grill, boil, or steam, rather than fry. If you do fry food use corn, soya, or sunflower oils.

2 Eat less red meat. When you do eat it cut off any fat you can see.

3 Eat less dairy foods (eggs, butter, milk, and cream).

4 Eat more poultry and fish, because these are less fatty.

5 Eat more fresh fruit and vegetables.

6 Do not smoke.

7 Take exercise regularly.

8 Take time to relax before you go to bed.

If you follow these guidelines you will be less likely to suffer from heart disease. Your general health will be better too.

Questions

1 Name the fatty substance that can block arteries.
2 What is a thrombosis? What causes it?
3 What is a stroke?
4 What causes the disease called angina?
5 Explain what causes a heart attack.
6 **a)** Why is the man in the drawing above likely to develop heart disease?
  b) Name two things, not shown in this drawing, which can also cause heart disease.
7 How can you avoid heart disease?

5.4 What blood is

When blood is left standing for a time, the solid particles start settling to the bottom of the plasma.

You have nearly half a bucket of blood (5.5 litres) in your body! Blood is a liquid called plasma, with red cells, white cells, and platelets floating in it. Let's look at each of these things in turn.

Plasma

Plasma is a yellow liquid. It is mainly water, with digested food, hormones, and waste substances dissolved in it.

Red cells

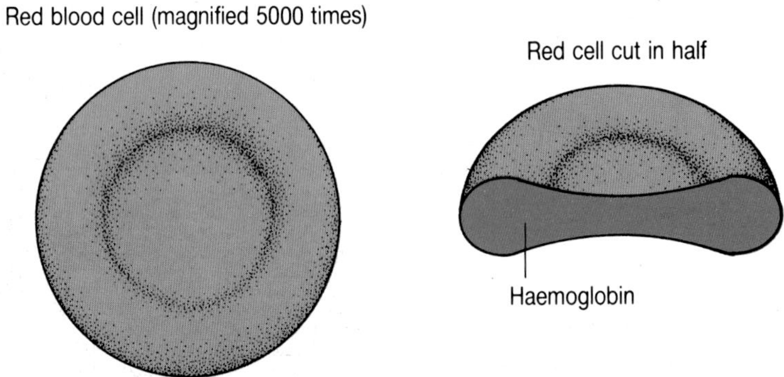

Red cells are wheel-shaped. They have no nucleus. They are red in colour because of the **haemoglobin** inside them.

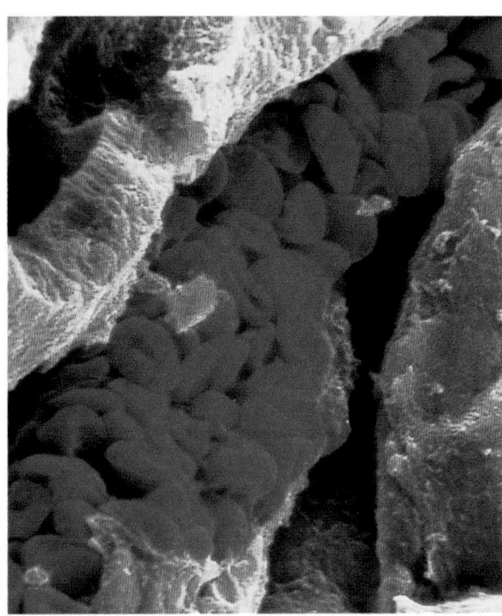

There are about five million red cells in one drop of blood. This photograph shows them moving through a capillary.

How the red cells carry oxygen

Red cells are the body's oxygen carriers. They carry oxygen from the lungs to all the cells of the body.

1 The red cells pick up oxygen as blood is pumped through the lungs.

2 The oxygen and haemoglobin join to form **oxyhaemoglobin**. This is bright red.

3 As the blood passes around the body, the oxyhaemoglobin breaks down and releases oxygen to the body cells.

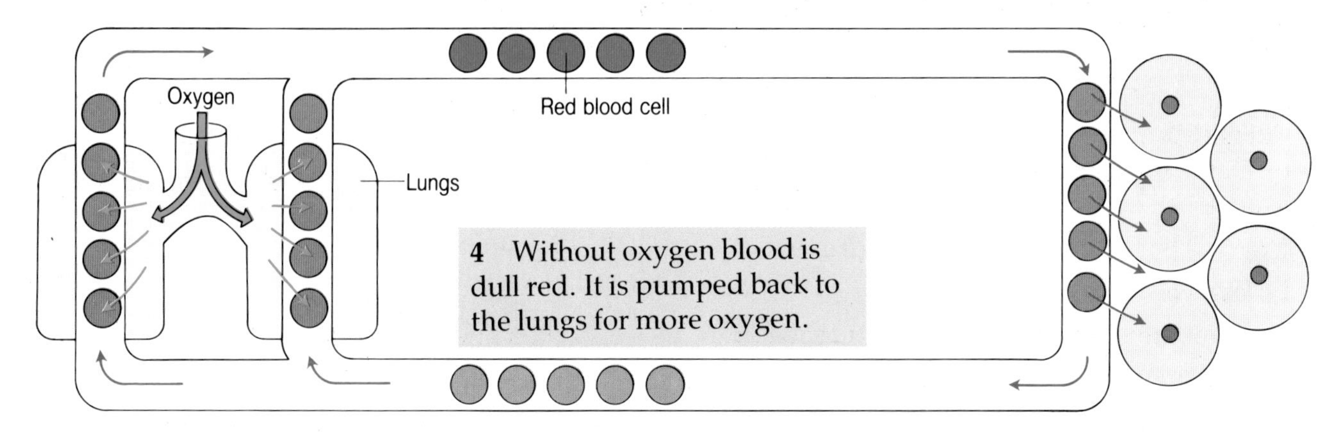

4 Without oxygen blood is dull red. It is pumped back to the lungs for more oxygen.

White cells

White cells are larger than red cells. They all have a **nucleus.** They can change shape. White cells protect us from disease. White cells called **phagocytes** can eat up the germs that cause disease.

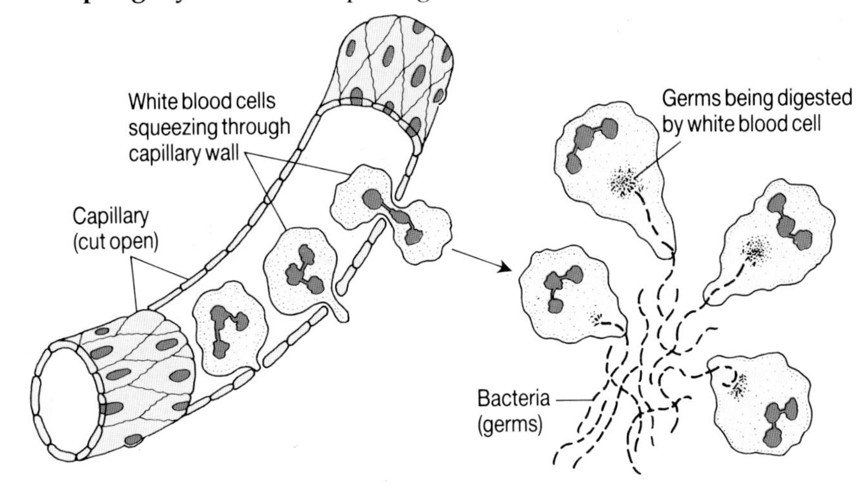

1 Phagocytes can squeeze through capillary walls.

2 They move towards germs and surround them. They then digest them.

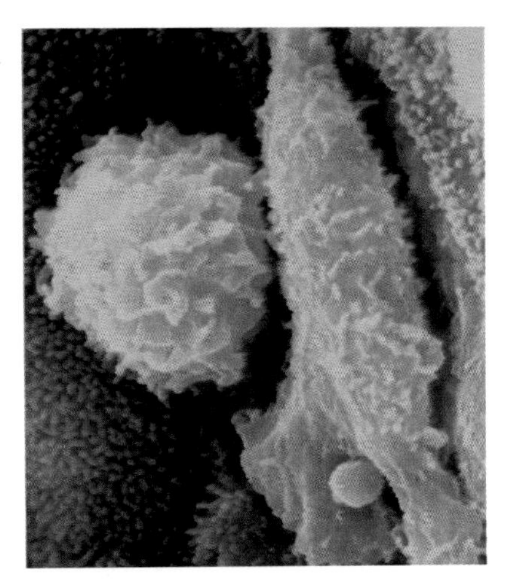

Two phagocytes in a human lung. One has grown long and thin, and is about to destroy a small particle. Phagocytes keep lungs clear of dust and pollen.

Other white cells make chemicals called **antibodies.** These chemicals destroy germs that get into the body by making them stick together, or by dissolving them. They also destroy **toxins** (poisons) that germs make. There is a different antibody for each kind of germ.

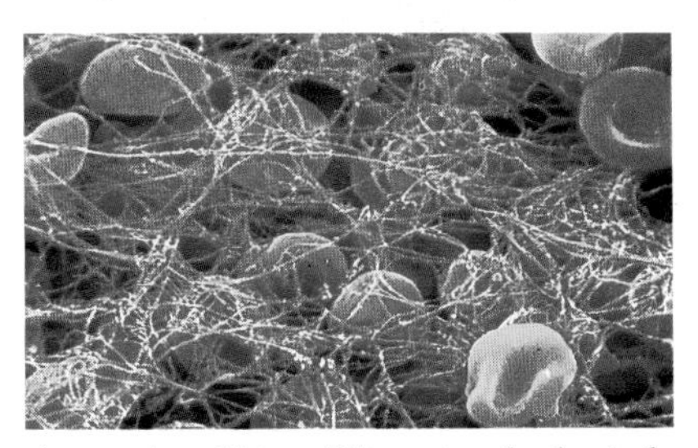

A cut healing. The tangled pink fibres were produced by the platelets. The orange bits are trapped red cells.

Platelets

Platelets are fragments of cells formed in bone marrow. Their job is to help stop bleeding from cuts.

1 Bleeding washes out dirt and germs from the cut. Then the platelets produce tiny fibres. Red cells get trapped in these fibres and the blood changes into a thick red jelly called a blood clot.

2 The clot hardens to a scab. This keeps the wound clean while new skin grows. Then the scab breaks off.

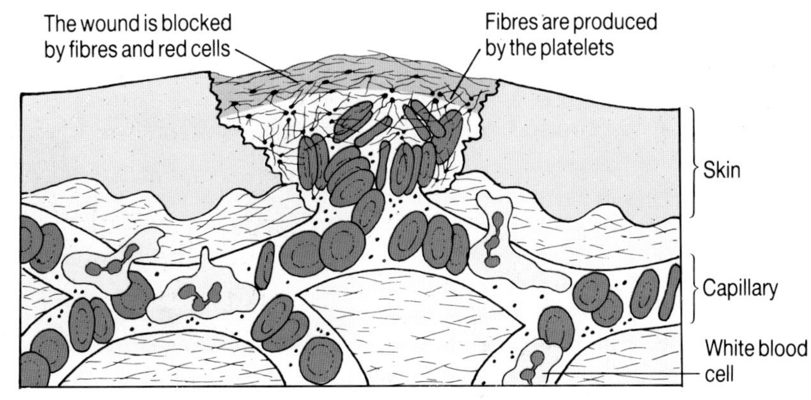

Questions

1. Name the four things that make up blood.
2. Name the liquid part of blood.
3. What do red cells do?
4. What are phagocytes?
5. What are antibodies?
6. How do platelets help blood clot?

5.5 What blood does

Blood does two jobs. It carries things round the body, and it protects us against disease.

Blood carries things

What it carries	How carried
1 Oxygen from the lungs to the rest of the body	In the red cells
2 Carbon dioxide from the body to the lungs	Mainly in plasma
3 Digested food from the gut to the rest of the body	In the plasma
4 Wastes from the liver to the kidneys	In the plasma
5 Hormones from hormone glands to where they are needed	In the plasma
6 Heat from the liver and muscles to the rest of the body so it is all at the same temperature	In all parts of the blood

Blood protects us

1 White cells called **phagocytes** eat germs.

2 Other white cells make **antibodies** to fight disease. Antibodies destroy the germs that cause disease. Antibodies also destroy toxins that germs make. There is a different antibody for each disease.

Once your white cells have made a particular kind of antibody, they can make it faster next time. Also it may stay in your blood for a while. This makes you **immune** to the disease. You will not catch it again, or else you will have only a mild attack.

You can get **vaccinated** against some diseases. Specially treated germs are injected into you, to give you a mild attack of the disease. Your body makes antibodies, so you become immune for the future.

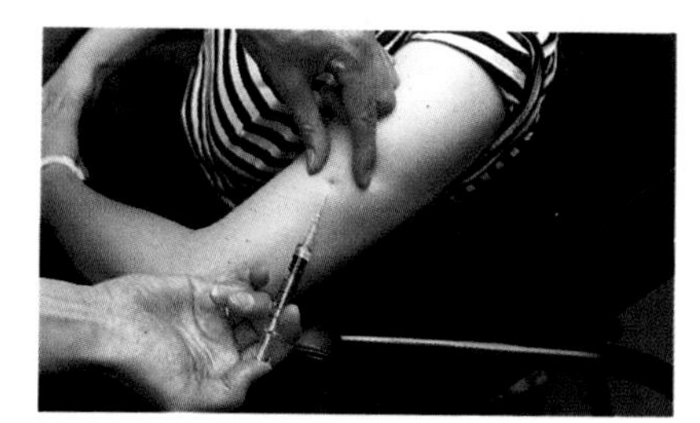

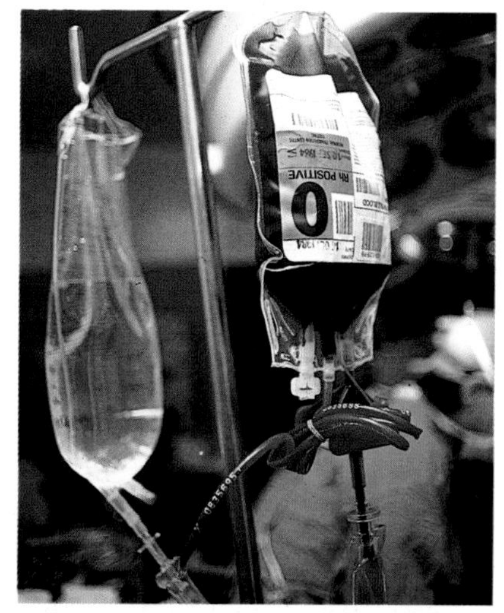

Often, ill people need blood transfusions. The blood comes from donors. It is sent out to hospitals in plastic packs.

Questions

1 What part of the blood:
   a) carries digested food?
   b) helps protect you against disease?

2 Describe two ways in which your body can become immune to a disease.

3 What do antibodies do?

5.6 How oxygen and food reach cells

Blood carries oxygen and food around the body. But it never comes into contact with body cells. So how does food and oxygen get from the blood to the cells where they are needed?

Tissue fluid

The answer is that capillary walls are so thin that they leak. A liquid called **tissue fluid** leaks through them from the blood, into tiny spaces between the body cells.

Tissue fluid carries oxygen and food from the blood to the cells, and washes away their wastes.

Most tissue fluid then seeps back into the blood vessels. But some drains into **lymph capillaries**, carrying germs and bits of dead cells with it. It becomes **lymph**.

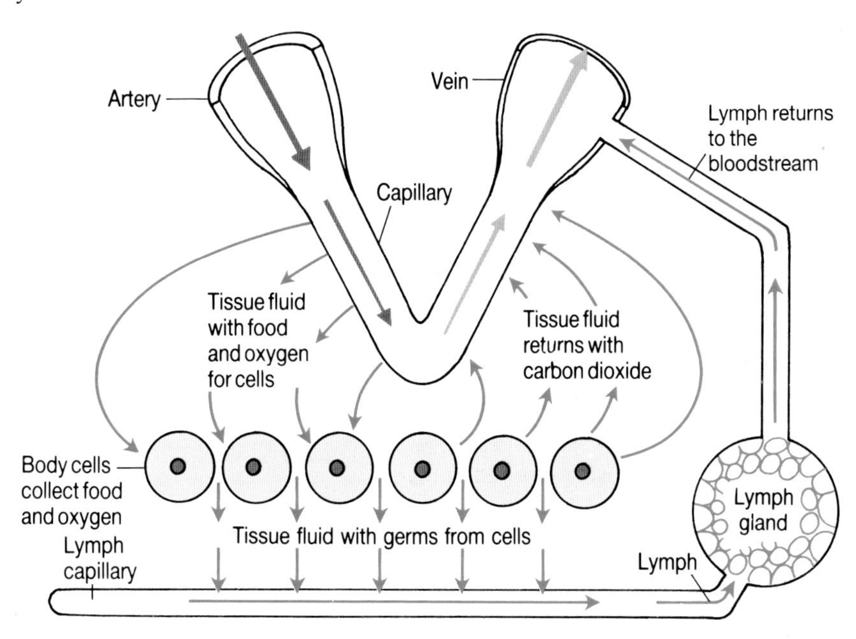

Where does the lymph go?

The lymph capillaries join to make larger tubes. These tubes drain into **lymph glands**.

Here the lymph is cleaned up by white cells called **lymphocytes**. These eat the germs and dead cells, and also make antibodies.

Clean lymph is returned to the bloodstream through tubes which join a vein in the neck.

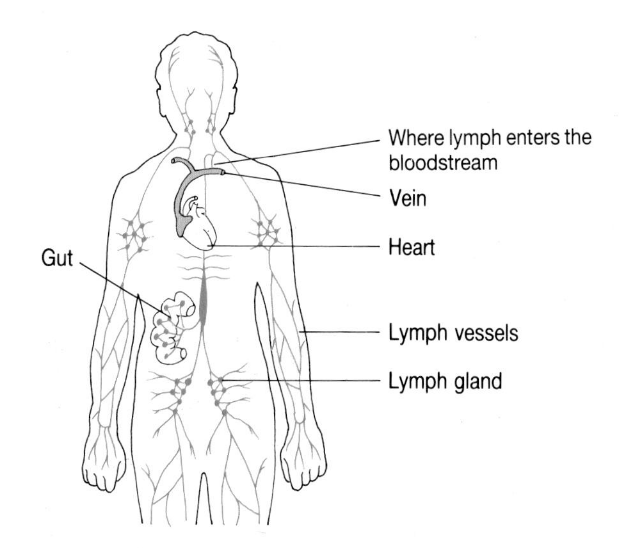

Questions

1 How does food and oxygen reach your body cells?

2 What happens to tissue fluid after it has been to the cells?

3 What happens inside lymph glands?

4 How does lymph get back into the blood-stream?

5 Where in your body are the lymph glands?

5.7 Homeostasis

Tissue fluid – your internal environment

Imagine you are an *Amoeba* in a pond. Water temperature will vary between very warm and freezing cold. Some days the water will be pure, and others it could be polluted with pesticide from a nearby farm. Its oxygen content could be more than enough one day, or so little the next, that you nearly suffocate.

In the same way, the cells of your body are also bathed in a kind of pond water. But unlike *Amoeba*, their pond consists of a liquid called **tissue fluid** (Unit 5.6). Tissue fluid forms the **internal environment** in which your body cells live.

But the internal environment of your body does not change all the time, like pond water. Many organs work day and night, making constant adjustments to keep it as near perfect an environment as possible for health, growth, and efficient functioning of cells. This maintenance of a constant internal environment is called **homeostasis.**

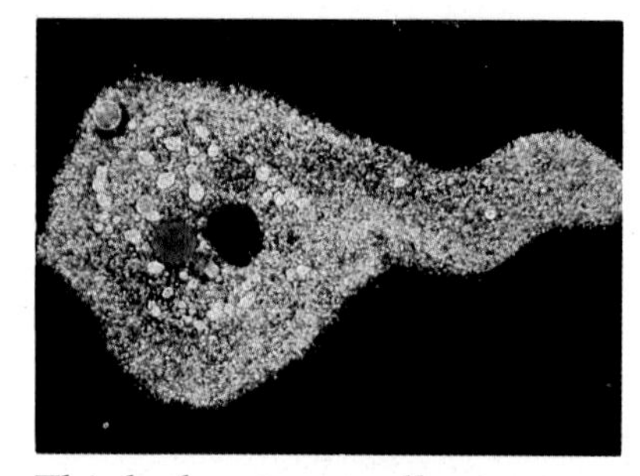

Think about one cell somewhere in your body. Will the temperature, oxygen content and cleanliness of the surrounding tissue fluid vary like pond water around this *Amoeba*? If not, why not?

Homeostasis and feedback mechanisms

Homeostasis is controlled by **feedback** mechanisms, which consist of three parts. A **sense organ** detects a change somewhere in the body's internal environment, and then sends information about the change to a **control centre** (usually the brain). The control centre sends messages to a **responding organ**, which does whatever is needed to bring the internal environment back to normal.

The sense organ detects that the internal environment is returning to normal and so **feeds back** this information to the control centre, which then tells the responding organ to stop what it is doing.

These non-stop adjustments happen all your life, without you being aware of them.

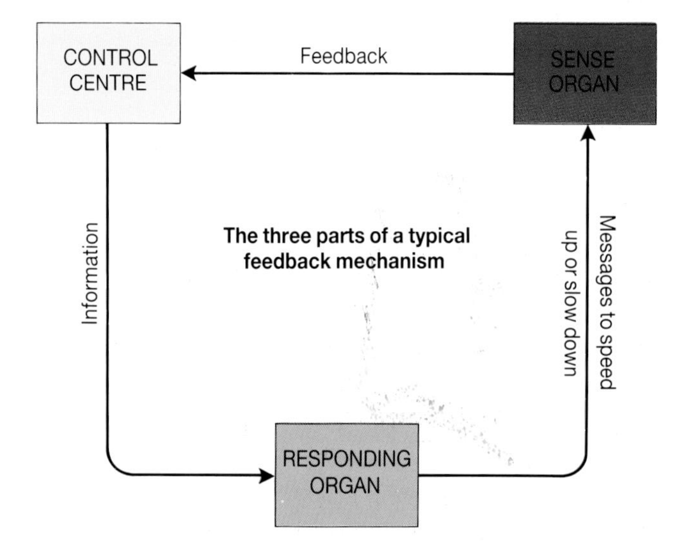

The three parts of a typical feedback mechanism

The boy on the pogo stick in the photo on the right illustrates a simple example of feedback in action. He would fall sideways if he did not make constant adjustments to his balance. Messages from his eyes and balance organs travel to his brain and tell it about the position of his body. In response, his brain sends messages to the muscles which control his movements. Then his senses feed back more information to his brain so that it can check that this response was correct, and so on.

The organs of homeostasis

Your skin helps keep your body within 1 °C either side of 37 °C (Unit 5.8). Sweat glands let your body lose excess heat and cool down. A layer of fat under your skin helps keep heat inside your body, and the shivering action of muscles generates extra heat to warm you up in cold weather. The fur of hairy animals traps a layer of warm air next to the skin like the clothes you wear. Without clothes, your body can only keep at 37 °C if the air temperature stays above 27 °C.

Your lungs control the amount of carbon dioxide and oxygen in tissue fluid (Unit 4.6).

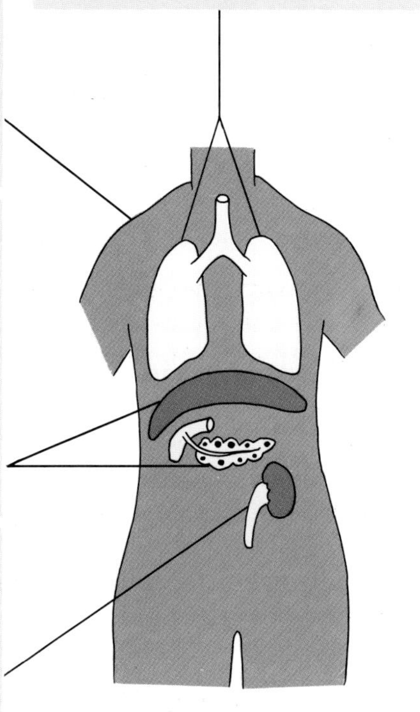

Your liver and pancreas together control the amount of glucose in your blood and tissue fluid. If the glucose level falls below normal, glands in the pancreas produce the hormone **glucagon** which makes the liver release glucose into the blood. If there is too much glucose in the blood, these glands release the hormone **insulin** which makes the liver remove glucose from the blood. The liver also controls the amounts of amino acids, and removes poisons from blood and tissue fluid. And it makes heat to help keep the body warm.

Your kidneys keep blood and tissue fluid 'clean' by removing urea, excess water and other wastes, and excreting them (Unit 5.9). This controls the amounts of dissolved substances in the blood and tissue fluid, a process called **osmoregulation**. This is important for two reasons. If tissue fluid becomes too concentrated, cells will lose water by osmosis and become dehydrated. And if it becomes too dilute, cells will take in excess water by osmosis (Unit 2.3).

Osmoregulation

If you drink a lot of liquid the water content of your blood rises. This is detected by special sense organs and your kidneys respond by increasing the amount of water in urine. As a result, a greater amount of more dilute urine is excreted until the water level in tissue fluid returns to normal.

The opposite happens if you are thirsty: your kidneys excrete only a small amount of concentrated urine to keep water loss to a minimum, until you can find something to drink.

Sugary meal eaten
Pancreas
Which hormone?
Glycogen (stored in the liver)
Which direction should arrow point?
High/low level of ? in blood

12 hours since last meal
Pancreas
Which hormone?
Glycogen (stored in the liver)
Which direction should arrow point?
High/low level of ? in blood

Questions

1. What is homeostasis, and what are the main organs of homeostasis?
2. Explain why tissue fluid is called the internal environment of the body.
3. Study the diagram above then copy it after answering the questions it contains.
4. What is osmosis, and osmoregulation?

5.8 Skin and temperature

What skin does

1 It **protects** your body against damage, dirt, and germs.

2 It contains millions of tiny **sense organs**, which are sensitive to touch, temperature, and pain.

3 It **excretes** water and salts from your body, as **sweat**.

4 It helps keep your body at a **steady temperature**.

Skin highly magnified. You can see the dead cells flaking off. What do you think the spike is?

The structure of skin

This is a drawing of a tiny piece of skin, greatly magnified.

The **outer layer** of your skin is a tough protective layer made of dead, flat cells. Dead cells continually flake off. But they are replaced by cells from below, which grow and flatten out as they move to the surface.

Sebaceous glands produce an oily substance called **sebum**. This makes skin and hair supple and waterproof. It also slows the growth of germs.

The **hair root** is the only living part of a hair. It is deep inside the skin in a tube – the **hair follicle**. Hair protects your head from direct sunlight and, in hairy animals, keeps the body warm.

Hair

Pore

Capillaries near surface of the skin

Blood vessel

This layer of cells is full of **fat and oil**. It helps keep the body warm.

Sweat glands produce liquid called sweat, to cool the body when it gets too hot.

Skin colour

Some skin cells make melanin, the brown pigment responsible for skin colour. Skin makes more melanin in sunshine, because melanin protects it from the Sun's harmful ultraviolet rays.

Dark skin contains more melanin than pale skin does.

Skin and body temperature

The outside of your body may feel hot or cold. But inside, it always stays at about the same temperature, 37°C. Your skin helps to keep it at this temperature.

When you get too hot:

1 You sweat a lot.-The sweat evaporates and cools you down.

2 The blood vessels below your skin **expand**. So a lot of blood flows near your body surface and loses heat – like a radiator.

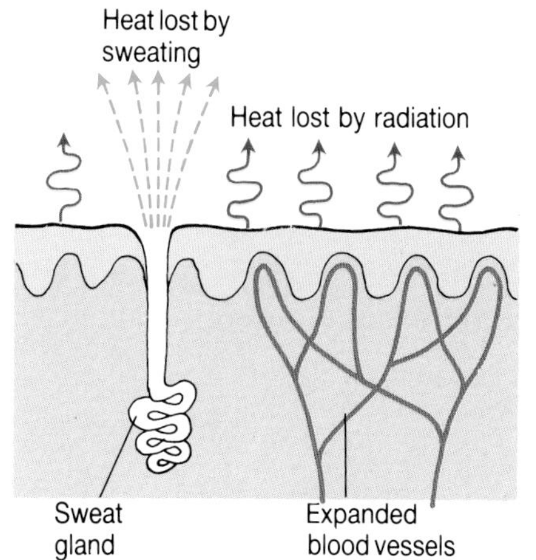

But your skin alone won't keep you cool in hot weather...

When you get too cold:

1 You stop sweating.

2 Blood vessels below the skin **contract**. So only a little blood flows near the surface and loses heat.

3 Muscles start rapid, jerky movements we call **shivering**. This produces extra heat, which warms up the body.

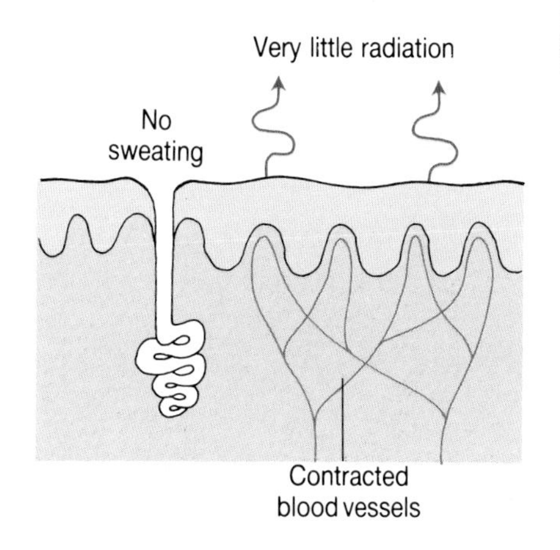

...or warm in cold weather. You help it out by wearing the right kind of clothes.

Body hair

Many animals have a thick coat of hair, or fur. This traps a layer of warm air around the body.

In cold weather, tiny muscles make the hairs stand up. So more air is trapped. In warm weather the hair muscles relax. So less air is trapped.

A seal pup, all dressed up for arctic life.

Questions

1. Your skin is being worn away all the time. How is it replaced?
2. How is skin protected by a suntan?
3. What changes happen in your skin when you get too hot?
4. What changes happen in your skin when you go from a warm room out into a cold winter night?
5. How does fur help arctic animals to keep warm?

5.9 Excretion

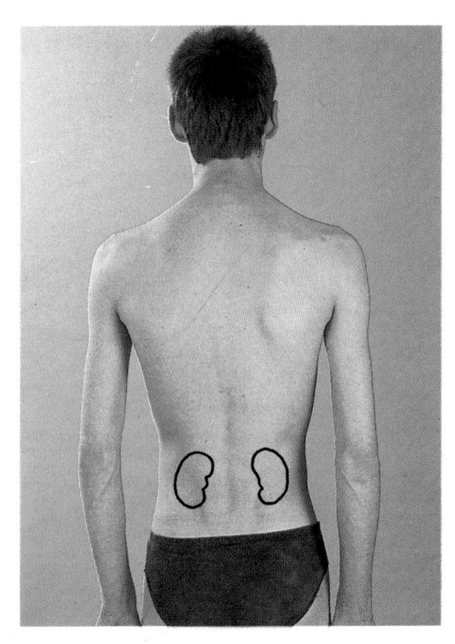

Where your kidneys are . . .

All day long, even while you sleep, your body produces waste substances. These wastes include carbon dioxide and urea.

Your body must get rid of these things as they are poisonous. Waste removal is called **excretion**. Your lungs excrete carbon dioxide. Your kidneys excrete urea. Your kidneys are your main **excretory organs** – they are towards the back of your body just above your waist.

What kidneys do

Your kidneys remove urea, water, and other unwanted substances from your blood. Urea is a waste produced by your liver.

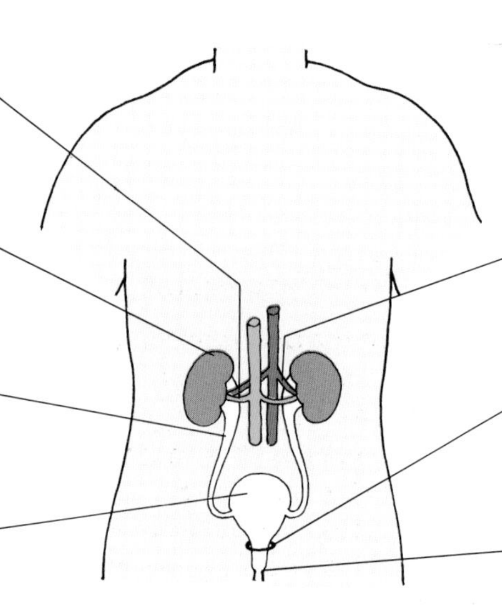

The **renal vein** carries 'clean' blood away from the kidneys.

The **kidneys** remove urea and other wastes from blood, and excrete it in a liquid called **urine**.

Ureters are tubes which carry urine to the bladder.

The **bladder** is a bag which stores urine until you go to the toilet.

The **renal artery** carries 'dirty' blood to the kidneys.

The **sphincter** is a ring of muscle which keeps the bladder closed until you go to the toilet.

The **urethra** is a tube which carries urine out of your body.

How kidneys clean blood

Kidneys clean blood by filtering it. They filter all your blood 300 times a day. The filtering is done by over a million tiny tubes packed into each kidney. These tubes are called **nephrons**.

Some people's kidneys are not very good at filtering blood. Kidney machines help by filtering their blood for them. The little girl in the photograph is attached to a kidney machine.

Surgeons can sometimes replace faulty kidneys with healthy ones. This operation is called a **kidney transplant**.

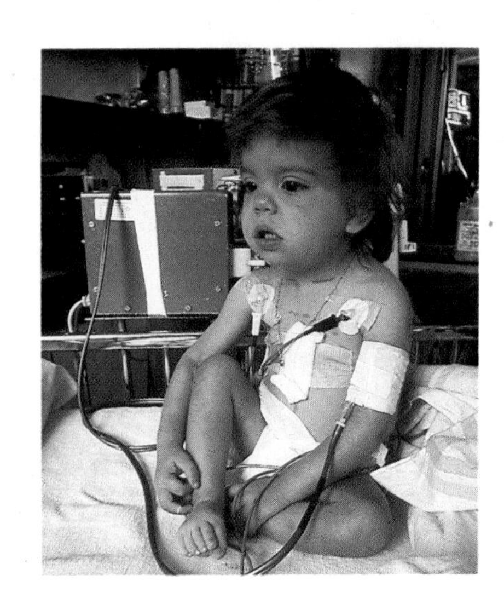

The parts of a kidney

This diagram shows a kidney cut in half, with one nephron magnified.

A **nephron** begins with a cup-shaped bag containing a bunch of capillaries called a **glomerulus**. 'Dirty' blood enters a glomerulus and is filtered.

Filtered liquid enters this tube-shaped part of a nephron. Here it is turned into **urine**.

The urine from many nephrons drains into a **collecting duct**.

The urine drains into the **ureter**.

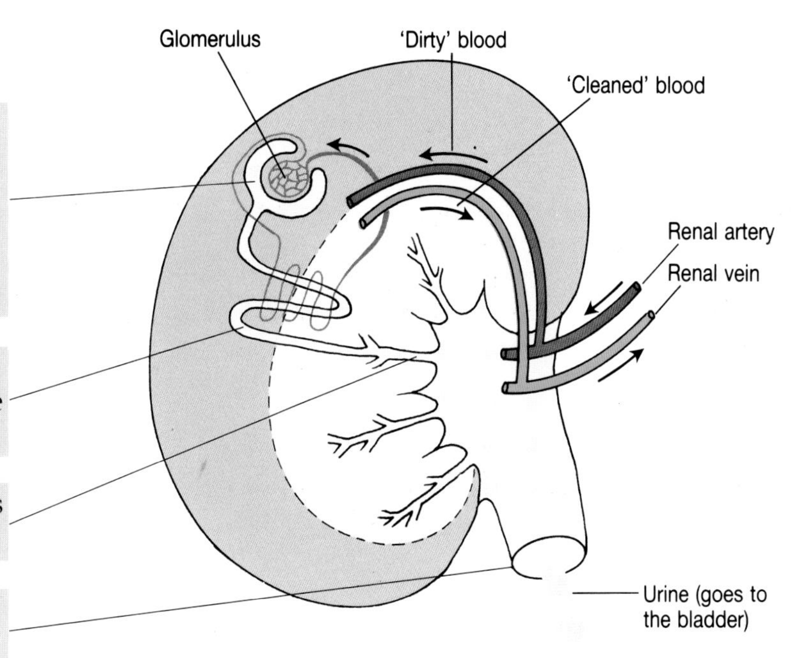

How a nephron works

The diagram below shows one nephron straightened out.

Blood is **filtered** by the glomerulus. Nearly all the blood except red cells filters through into the nephron.

Liquid in the nephron contains useful substances such as glucose, as well as urea. Useful substances pass back into the blood (called **selective reabsorption**, because only certain substances are selected in this way).

The liquid left in the nephron is **urine**. It contains urea and water plus other unwanted substances. It goes through the collecting ducts and ureters to the bladder.

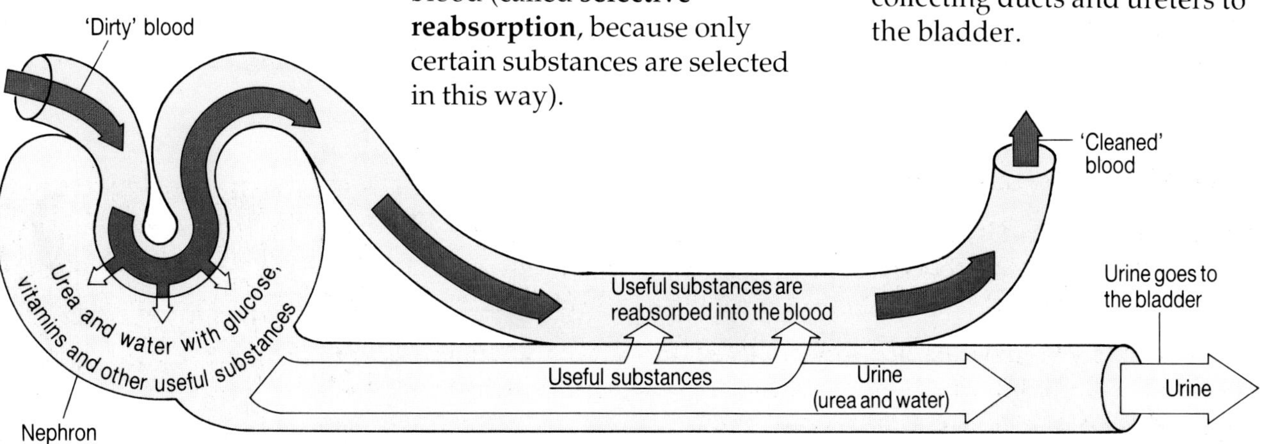

Questions

1 Why would you die if excretion stopped?
2 What do lungs and kidneys excrete?
3 What do the bladder and sphincter do?
4 Where is 'dirty' blood filtered?
5 What happens to stop useful substances being lost from the body as urine is produced?

Questions on Section 5

1 *Recall knowledge*
This diagram shows the inside of the heart. Match the labels on the diagram with the descriptions below:

a) right atrium;
b) blood on its way around the body;
c) valve tendon;
d) right ventricle;
e) a vein carrying blood from around the body;
f) left ventricle;
g) valve which stops blood flowing back into the left atrium;
h) cardiac muscle:
i) blood on its way to the lungs;
j) blood returning from the lungs;
k) valve which stops blood flowing into the left ventricle.

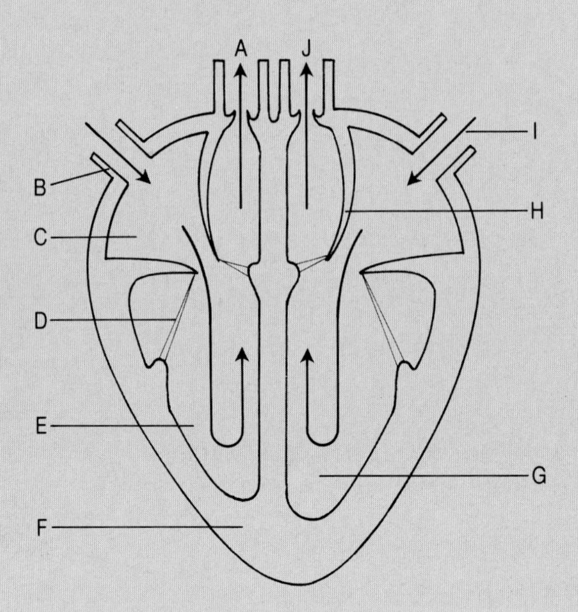

2 *Recall knowledge/understanding*
Opposite is a simpler version of the diagram from page 91, showing the heart and main blood vessels.

Which of the blood vessels labelled 1 to 4:

a) contains blood at the highest pressure in the whole body?
b) contains blood at very low pressure?
c) contains blood full of carbon dioxide?
d) contains blood full of oxygen?

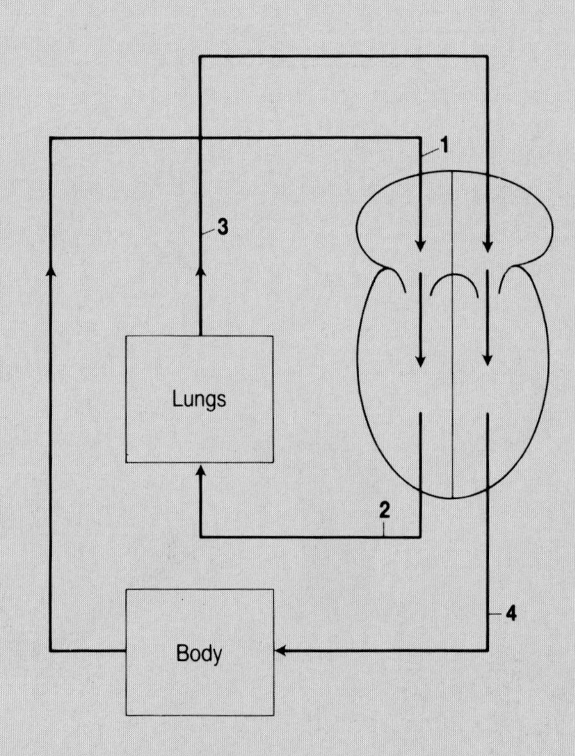

3 *Recall knowledge*

a) Explain what these medical words mean: angina, thrombosis, heart attack.
b) Why is it important not to eat too much fatty food?
c) Name ten foods which contain lots of fat.
d) Apart from fatty food, what else can cause heart disease?
e) List all the things you should do to avoid heart disease.

4 *Recall knowledge*
This is a drawing of human blood cells.

a) Name each cell.
b) Which of the cells:
contains haemoglobin?
is a phagocyte?
can pass through capillary walls?
transports oxygen?
eats germs?

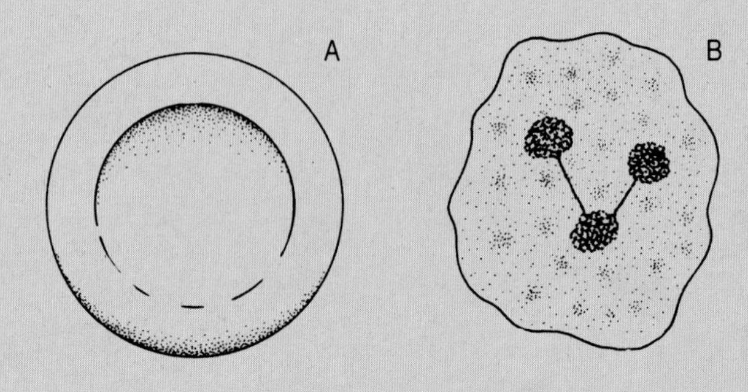

5 *Recall knowledge*
This diagram shows part of the human skin:

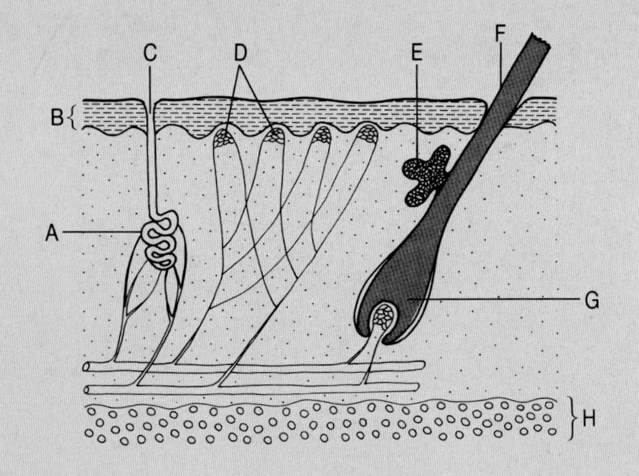

a) Name the parts labelled A to H.
b) Name the parts which help to control body temperature.
c) What do these parts do when we get too hot?
d) What do these parts do when we get too cold?
e) Explain how part B is replaced as fast as it wears away.
f) Why does layer H grow thicker in wild animals in autumn?

6 *Understanding/interpret data*
A student was put in the same environmental conditions and given the same food and drink for two days. She rested completely on one day and performed regular exercises on the other. The water she lost in sweat and urine was measured (see table opposite).

	Loss of water	
	Sweat	**Urine**
Rest day	100 g	1900 g
Exercise day	5000 g	500 g

a) Which environmental conditions should be kept the same for the two days?
b) Why should food and drink be kept the same for the two days?
c) Explain the differences in sweat production for the two days.
d) On the exercise day what is the advantage of producing more sweat and less urine?

7 *Understanding/interpret data*
The graph opposite shows how body temperature in humans and lizards changes with external temperature.

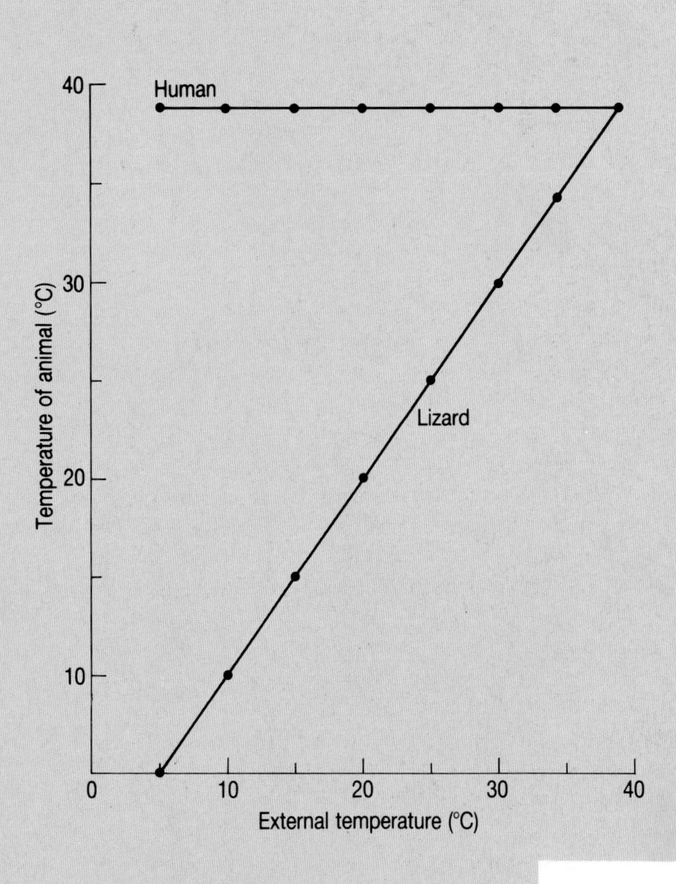

a) What is the body temperature of humans and lizards when the external temperature is 5°C and 30°C?
b) Which is likely to be the most active in cold weather? Explain your answer.
c) How does the human body maintain the temperatures shown on the graph?
d) What are the advantages of this ability?
e) Think of some reasons why lizards do not have this ability?
f) What could a lizard do to prevent its body overheating?

6.1 Food

Why we need food

For growth and repair. Your body grows by forming new cells. You also need new cells to replace dead ones. Cells are built from substances in food.

For energy. You need energy to work your muscles, and all other organs. The energy in food is measured in **calories** or in kilojoules.
(4.2 kilojoules = 1 Calorie)

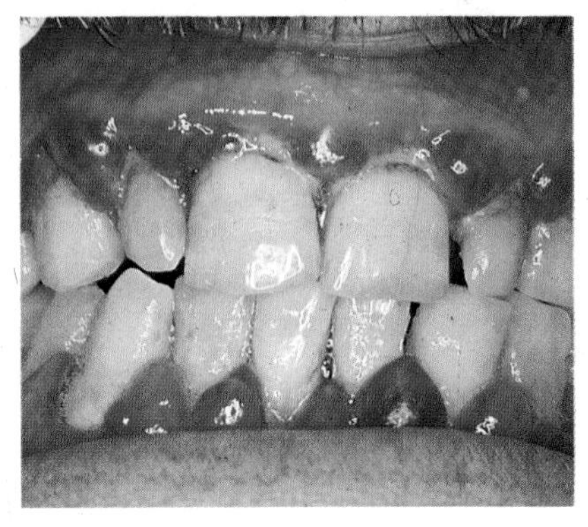

To stay healthy. You need vitamins and minerals in food. This shows what happens to gums if you don't get enough vitamin C.

What's in food

Food contains a mixture of substances. The main ones are proteins, carbohydrates, fats, oils, vitamins, minerals, fibre, and water.

These eggs contain:

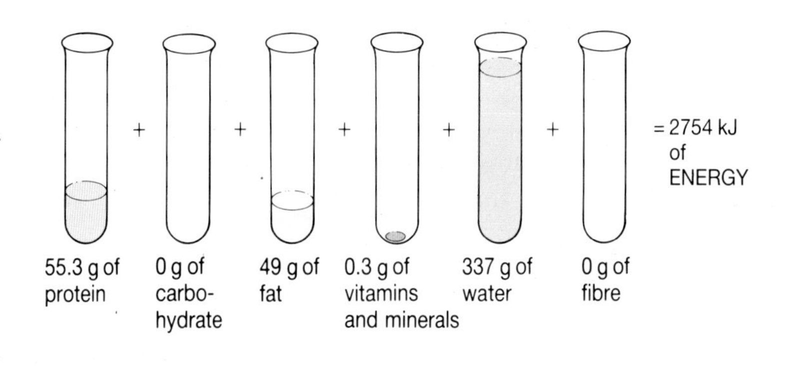

These beans contain:

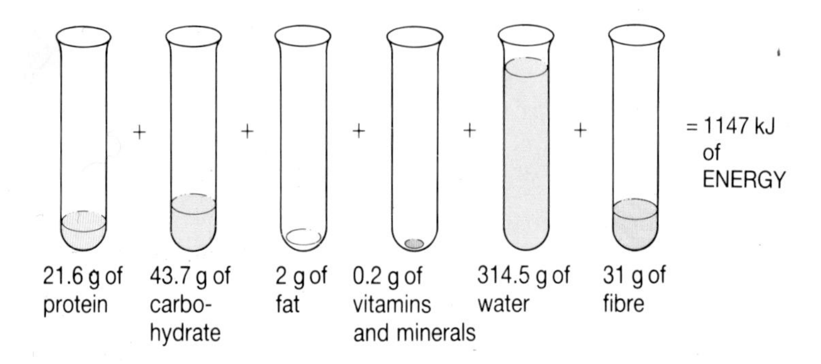

What foods do for you

Carbohydrates give energy. Sugar and starch are two types of carbohydrate.

You cannot digest **dietary fibre**. But it prevents constipation and cleans your bowels.

Proteins are body-building foods. You need proteins for growth and repair.

Fats and oils give twice as much energy as other foods. They are stored for warmth and future use.

The main vitamins and minerals you need

Substance	Where you find it	Why you need it	Shortage can cause
Vitamin C	Oranges, lemons, grapefruit, green vegetables, potatoes	For healthy skin and gums and to heal wounds quickly	**Scurvy.** Gums and nose bleed. The body bleeds inside
Vitamin D	Milk, butter, eggs, fish, liver (also made by skin in sunshine)	For strong bones and teeth	**Rickets.** The bones become soft and bend
Calcium	Milk, eggs	For strong bones and teeth	**Rickets.**
Iron	Liver, spinach	For making red blood cells	**Anaemia.** The person is pale and has no energy

Questions

1 Name three foods rich in:
- **a)** dietary fibre;
- **b)** sugar;
- **c)** protein;
- **d)** starch;
- **e)** fats and oils.

2 Name foods needed:
- **a)** for strong bones and teeth;
- **b)** to avoid scurvy;
- **c)** to avoid constipation;
- **d)** for warmth;
- **e)** for growth.

6.2 More about food

Food and energy

All foods except vitamins and minerals contain stored energy. This energy is released and made available to living things during respiration (Unit 4.4).

When food is burned, it releases the same amount of energy, in the form of heat, as it does when respired. So, to discover the energy value of a food, it is burned and the heat given off used to heat a known quantity of water. Since it takes 4.2 joules of heat to raise the temperature of 1 g of water by 1 °C, the energy value of a food can be calculated by the formula:

energy = temperature rise in °C × 4.2 × mass of water in g

A food's energy value is usually quoted in **kilojoules**: 1 kilojoule (1 kJ) = 1000 joules.

These measurements give only the *potential* and not the *actual* energy value of a food. If you eat bread and butter with an energy value of 1000 kJ, you do not get 1000 kJ of energy. First, the bran in bread and some of the fat in butter passes undigested through the gut so its energy is lost. Second, up to 85 per cent of the energy released during respiration is lost as heat. Study the photograph and caption opposite before reading on.

Which of these foods is a source of instant energy? Explain your answer. One of these will yield about 15 per cent of its potential energy during respiration, and the others will yield even less. Which is which? Explain your answer.

Types of food and their composition

The six types of food are: carbohydrates, proteins, fats and oils, minerals and vitamins.

Carbohydrates These are sugary and starchy foods such as sweet fruits, honey, jam, bread, cakes, potatoes, rice, and spaghetti. Carbohydrates are the main source of energy for living things. One gram yields about 17 kJ of energy. Most carbohydrates are converted by the body into glucose before they are respired.

Plants store large quantities of starch in their seeds and in storage organs like potato tubers. In animals, the main carbohydrate food reserve is **glycogen**, which is similar to starch. Animals can store only limited amounts of glycogen. When this limit is exceeded, any excess carbohydrate is changed into fat or oil, and is then stored in the body.

After reading all of this Unit, sort these foods into those which are: a source of energy, form the main food reserve in animals and plants, and supply the raw materials for growth and repair.

Fats and oils These are also called **lipids**, and include butter, lard, suet, dripping, and olive and cod-liver oil. These are important sources of energy because 1 g of lipid can yield up to 38 kJ of energy when respired. But they are less easily digested than other foods, which reduces their energy value a little.

Fats are very important food reserves in animals and plants, firstly because each gram contains twice the energy of other foods, and secondly because layers of fatty tissue, especially under the skin of animals, insulate the body against loss of heat.

Proteins The most important sources of protein are meat, liver, kidney, eggs, fish, and beans. These foods supply the raw materials living things need for growth, and for repair of damaged and worn-out tissues. Proteins are not usually respired for energy, but when this happens they can yield up to 17 kJ of energy per gram.

Proteins are digested into chemicals called **amino acids**. There are about twenty-six of these but only ten, called the **essential amino acids**, are needed by humans. Our bodies can make the rest. Animal proteins contain all ten essential amino acids and so are called **first-class proteins**.

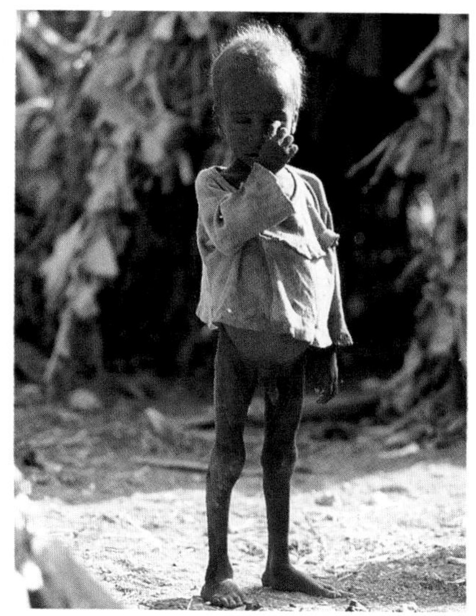

This child is suffering from **marasmus** – a condition caused by lack of proteins and energy-giving carbohydrates in the diet.

A note on vegetarian diets No plant protein contains all the essential amino acids, so they are called **second-class proteins.** But all ten can be obtained by eating a wide variety of plant foods. A vegetarian diet can be extremely healthy because it does not usually contain much fat, oil, sugar, or salt, and is high in dietary fibre.

Vitamins Vitamins have no energy value but are essential for growth and health (see Unit 6.1). Some are needed in very small amounts – less than one millionth of a gram a day. Vitamins take part in vital chemical reactions in the body, usually in conjunction with proteins.

Minerals You require about fifteen different minerals in your diet. Like vitamins, they have no energy value, but are essential for health. Most essential minerals are supplied by a diet of meat, eggs, milk, green vegetables and fruit (Unit 6.1).

Question

Copy the chart opposite and fill it in using information in this Unit and in Unit 6.1.

	Carbohydrates	Fats and oils	Proteins	Vitamins	Minerals
Fried cod					
Orange juice					
Prevents rickets					
Toffee					
Fish and chips					
Body-building food					
Liver					
Heal wounds					
For strong bones					
Milk chocolate					
Contains amino acids					

6.3 Diet and health I

If you ate only sweets, or cream buns, or crisps you would stay alive for a time. But you would not stay healthy, because you are not eating a balanced diet.

A **balanced diet** is one which contains the right amounts of protein, carbohydrate, fat, vitamins, and minerals to satisfy all your body's needs.

How much should you eat?

The amount of food you need depends on the energy you use in a day. And that depends on:

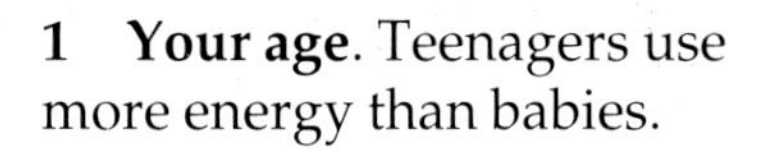

1 **Your age**. Teenagers use more energy than babies.

2 **Your work**. A footballer uses more energy than a snooker player.

3 **Your sex**. Males use more energy than females of the same age, even for the same work.

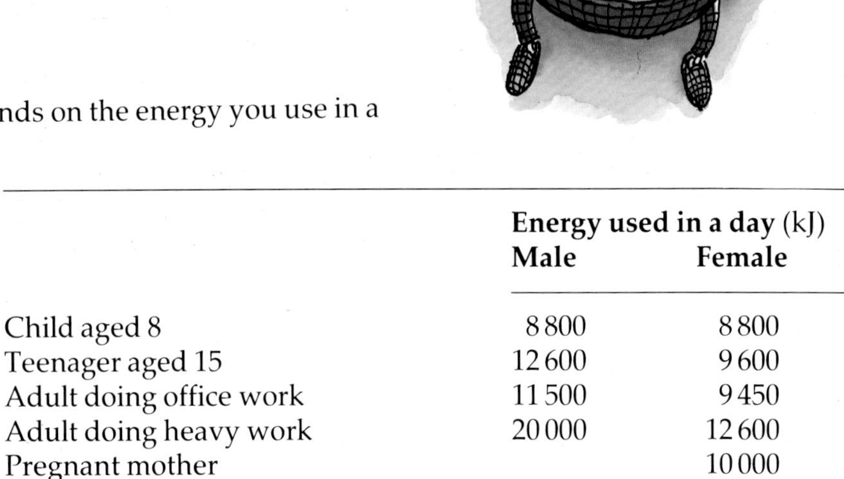

	Energy used in a day (kJ) Male	Female
Child aged 8	8800	8800
Teenager aged 15	12600	9600
Adult doing office work	11500	9450
Adult doing heavy work	20000	12600
Pregnant mother		10000
Breast-feeding mother		11300

This table shows the amount of energy used in a day.
Energy used by the body is measured in kilojoules (kJ for short).

Getting the balance right

The food you eat each day should supply just enough energy to get you through that day.

If you eat too much, your body stores the extra as fat. So you get overweight, or **obese**. Obese people are more likely to have heart attacks than slim people.

If you eat too little, you lose weight. You feel weak and have no energy. Some people eat so little that they suffer from **anorexia**. This happens if a person starts a slimming diet and doesn't know when to stop. But we have to eat to live – so get the balance right.

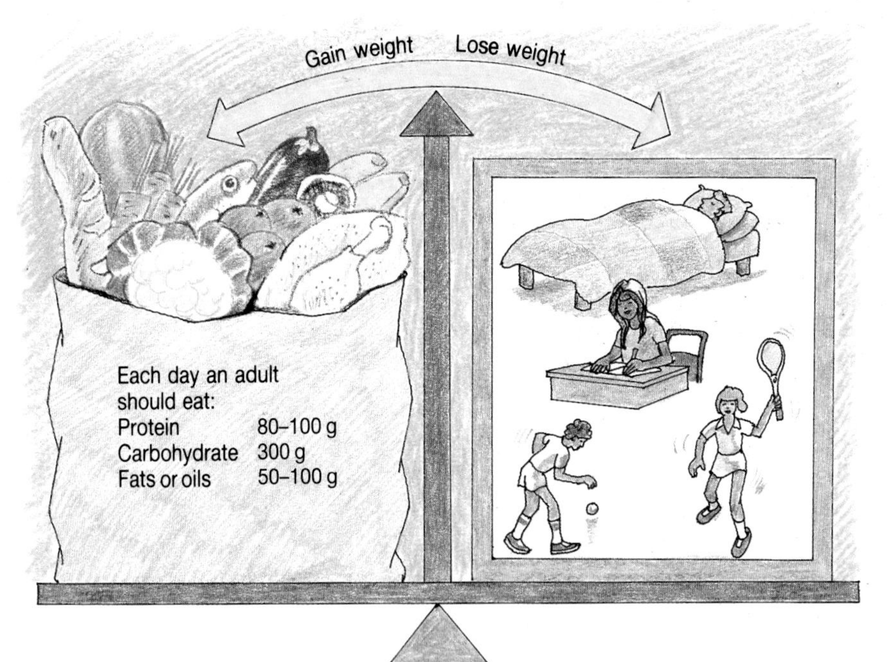

Food eaten must **balance** energy used

Healthy eating

Some foods are bad for you if you eat too much of them. So it makes sense to eat healthy foods, and avoid those which can damage your health.

A little fat gives you a *lot* of energy. So it is easy to eat more than you need. Fats make you fat, and can cause heart disease.

Sugar gives you energy but nothing else. It rots your teeth and also makes you fat. Cakes, chocolate, and ice-cream also contain hidden fat.

During processing in factories, food loses dietary fibre and other goodness. Salt, sugar, colouring, and chemicals are added. These may do you harm.

These foods contain protein but not much fat. They build you up without making you fat.

These give you carbohydrate, protein and fibre. They fill you up without making you fat.

These foods contain dietary fibre, vitamins, and minerals to keep you healthy.

Questions

1. Name the five types of food which make up a balanced diet.
2. Why does a teenager need more food than a baby?
3. What happens if you eat more than you need?
4. What happens if you eat less than you need?
5. Explain why you should:
   a) eat less cakes and ice-cream;
   b) cut most of the fat from meat;
   c) eat fresh food instead of processed food;
   d) eat more fruit and vegetables.
6. Name five foods that contain fibre.

6.4 Diet and health II

The Seventh Day Adventists are a religious sect based mainly in the USA. They never smoke, nor drink alcohol, and eat a mainly vegetarian diet including just a little meat and some dairy produce.

Before reading this Unit, use these facts and the graph opposite to formulate hypotheses to account for the Seventh Day Adventists' remarkable health record, compared with the general population.

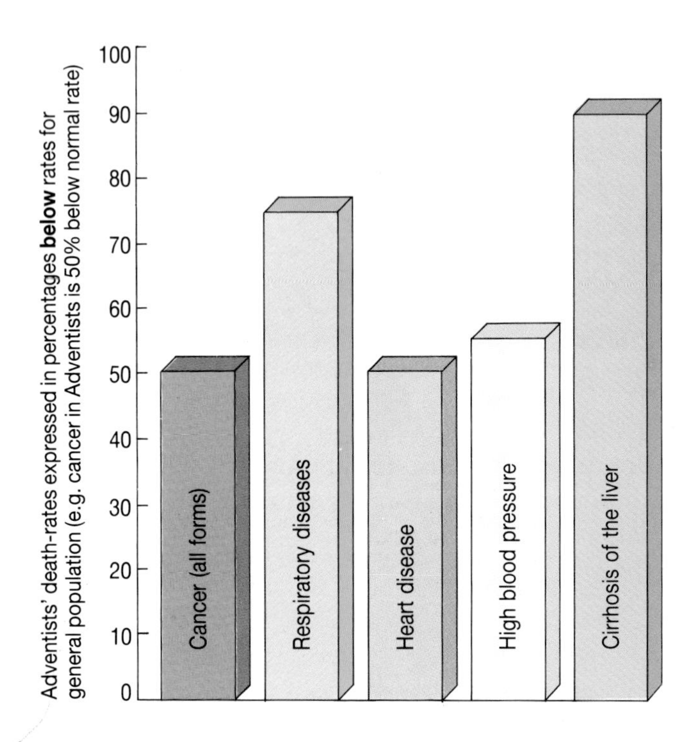

Malnutrition

Malnutrition means bad nutrition, but the word is usually used to refer to the effects on the body of eating too much, too little, or the wrong kinds of food.

Eating too much

You are eating too much if the energy value of the food you eat each day is more than the amount of energy you use in that time (Unit 6.2). Eating too much can lead to weight increase and obesity (fatness).

There is clear evidence that obesity contributes to many diseases: heart disease, high blood pressure, diabetes, gall bladder disease, cancer of the bowel, and also breast and womb cancer in women. Eating too much of the wrong kinds of food, such as fatty, sugary and salty foods, is especially dangerous.

Eating too much fat Excess animal fat in the diet is dangerous because it leads to the formation of cholesterol (Unit 5.3). It has also been estimated that diets high in animal fats cause one-third of all cancers.

Remember that, in addition to butter, cream and fatty meat, there is much hidden fat in foods such as ice-cream, burgers, sausages, crisps, peanuts, and chocolate.

What does the graph opposite tell you about the dangers of obesity?

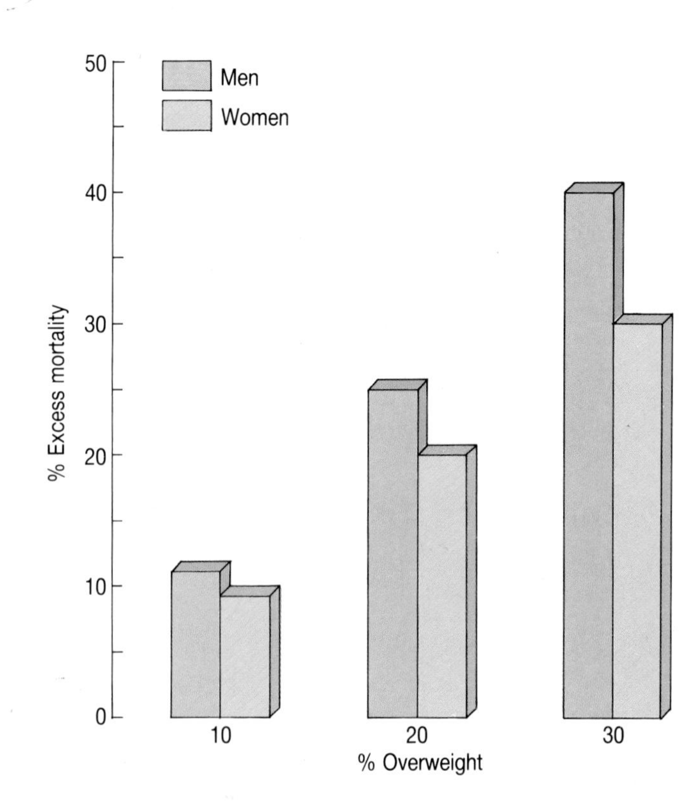

Eating too much sugar Sugar causes tooth decay (Unit 6.6), is linked with diabetes, and can cause obesity. Ice-cream and chocolate are among the most fattening foods, as they contain both sugar and fat.

Eating too much salt This can cause high blood pressure (a major factor in heart disease and strokes) and stomach cancer. On average, the British eat 2.5 teaspoons of salt daily (12 g). About 88 per cent of this is already present in food, especially factory-made (processed) foods (see graph opposite). So add less salt to food, eat fewer packet soups, foods preserved in brine, bacon, and salted potato crisps.

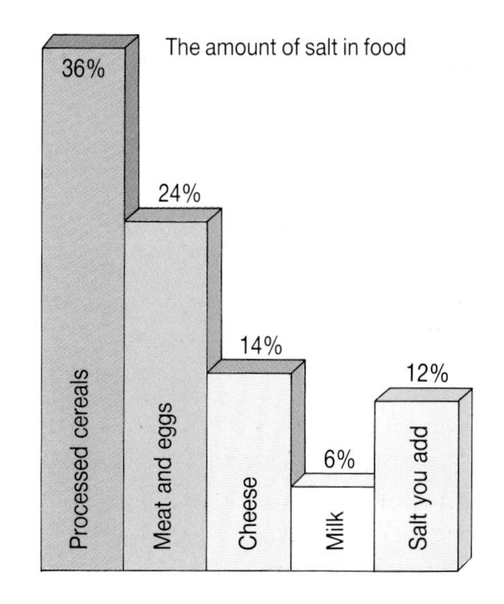

Dietary fibre (roughage)

Fibre is the bran of cereals and the cellulose walls of plant cells. It passes unchanged through the gut because it is not digested.

High-fibre foods Wholemeal products contain up to 8.5 per cent more bran than white flour products. Bran breakfast cereals are 25 per cent bran. Other high-fibre foods are potatoes in their skins, berries, nuts, bananas, apples, oranges, and vegetables such as sweetcorn, spinach and broccoli.

Why fibre is important Low-fibre foods (e.g. meat and eggs) can cause constipation, heart disease, diabetes, and cancer of the bowel. This is because these foods pass slowly through the gut, forming sticky lumps which cause constipation. High-fibre foods have more bulk, which allows gut muscles to grip and move them along quickly by peristalsis. Fibre absorbs poisonous wastes from digesting food and adds indigestible matter. This satisfies your appetite but can't make you fat.

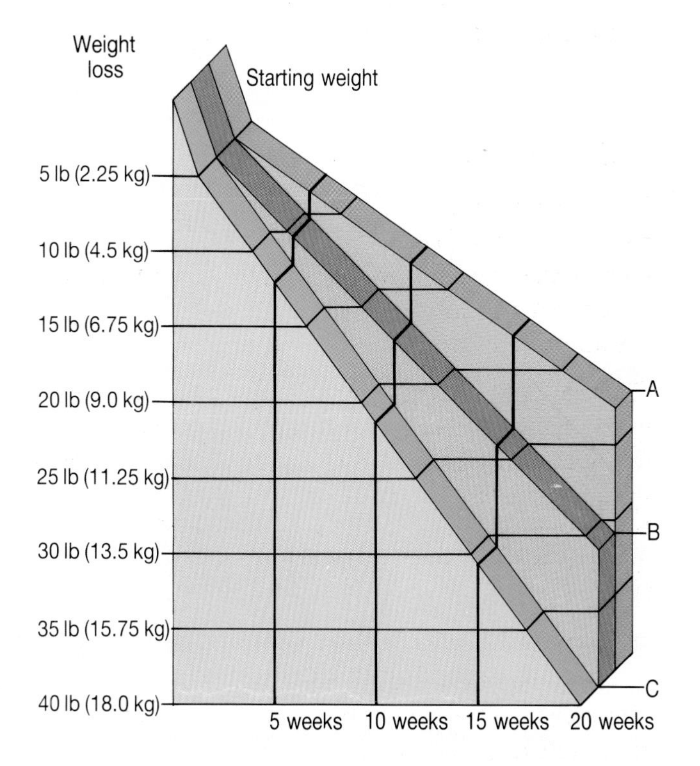

Questions

1 Five chocolate biscuits (about 100 g) supply enough energy for a man to cycle for 90 minutes at 15 km/h. The **same weight** of wholemeal bread supplies only enough energy for 30 minutes' cycling at this speed. So why might scientists say that 'Some foods supply your daily energy needs long before they satisfy your appetite', and 'It is easy to eat too much without realizing it.'

2 An **energy gap** is caused by a daily food ration which give **less energy** than you use each day. In the graph above:
A = an energy gap of 2100 kJ (500 Calories)
B = an energy gap of 2900 kJ (700 Calories)
C = an energy gap of 4200 kJ (1000 Calories)
Explain the weight losses shown in the graph. What effect would extra exercise (e.g. jogging) have on the energy gap, and weight loss?

3 **a)** The word 'malnutrition' is mostly used to describe the effects on the body of starvation. Give a better definition.
b) Name foods which can cause obesity, tooth decay, cancer, heart disease and stroke, if eaten in excess.
c) How can a diet high in dietary fibre help keep you slim?

6.5 Food additives

Yum – additives!

When food is processed in factories, chemicals called **food additives** are often added.

Around 3500 additives are used by British food manufacturers Below are the main kinds.

Colourings make food look more appetizing. Tinned peas and strawberries owe their bright colours to added chemicals. Egg yolks are given a brighter colour by chemicals put in hen food.

Flavourings put back flavours lost when food is processed in factories. They are sometimes used instead of 'real' flavours. Fruit-flavoured puddings and drinks owe their taste to chemicals and not real fruit.

Preservatives help to keep food fresh. So food can be stored in shops for some time before it goes bad. Pork pies, sausages, bacon, and dried fruits have preservatives in them.

Are additives dangerous?

There is no proof that additives cause serious illness in humans. But here are a few reasons why some people avoid eating these chemicals:

1 Additives fed to animals have caused damage to their kidneys, liver, and digestive system, and some have caused cancers.

2 Some additives destroy vitamins in food. For example, the preservative sulphur dioxide destroys vitamin B_1.

3 Some may cause asthma, itchy skin, headaches, and **hyperactivity** (uncontrollable behaviour) in young children.

If you want food free of additives, eat fresh vegetables and fresh fruit, fresh meat, 'real' fruit juices and yoghurt, and muesli.

Black Forest Gateau

INGREDIENTS

Sponge: Wheat flour, sugar, liquid egg, water, cocoa powder, dried skimmed milk, glucose, emulsifiers E471 E475, starch, salt, flavouring, colours E102 E122 E124 E142 155. Filling and Decoration: Cream (cream, dextrose, stabiliser E407), cherry topping (cherries, sugar, glucose, modified starch E1422, citric acid E330, flavouring, colours E123 E142 E422, thickener E412, preservative E211), chocolate flavoured coating (sugar, vegetable fat, fat reduced cocoa powder.

How many additives are there in this gateau? An additive is shown by the letter E, followed by a number.

Questions

1 Why are additives put in food?

2 Why do some people avoid food that contains additives?

3 Have a competition to see who can find the food containing the most additives. (Look on the labels of food packets and tins.)

6.6 Teeth

Teeth grow out of holes called **sockets** in your jawbone. Before digestion starts, teeth are used for biting and chewing food.

Enamel forms the hard, biting surface of a tooth.

Dentine is similar to bone.

The **pulp cavity** contains nerve endings and blood vessels.

Cement holds the tooth in its socket.

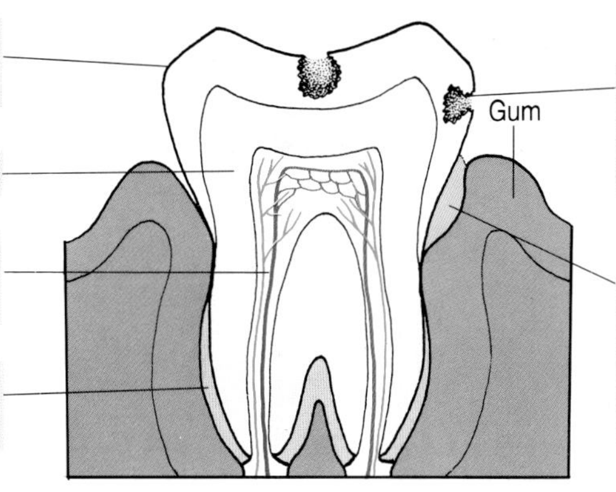

Tooth decay starts on the top of large back teeth, and where one tooth touches another.

Gum disease starts between the gum and teeth. It can destroy tooth cement and cause teeth to fall out.

Why teeth decay

Tooth decay happens when the **bacteria** in your mouth turn the **sugar** in your food into **acid**. The acid eats a hole in tooth enamel and dentine. When the hole reaches a nerve you get toothache.

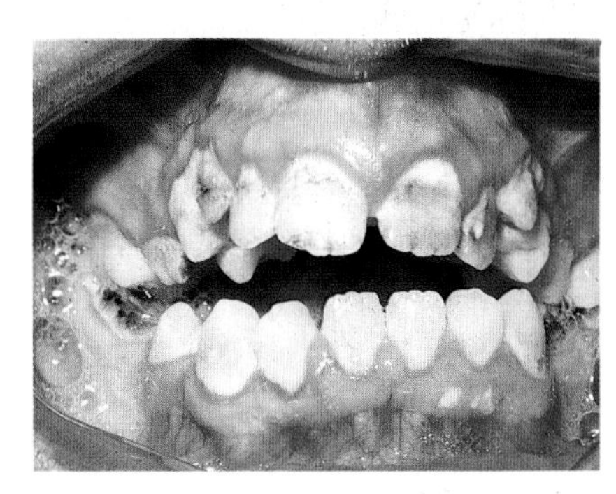

. . . so don't forget to brush your teeth!

Gum disease

If you don't clean your teeth regularly they become covered with a layer of food and bacteria called **plaque**. Gum disease starts if the plaque gets between the gum and a tooth. The plaque rots and causes the gum to swell and bleed.

Preventing tooth decay and gum disease

1 Eat less sugary foods, because **sugar** + **bacteria** = **acid** which rots teeth.

2 Brush your teeth at least once a day to remove plaque.

3 Use fluoride toothpaste. Fluoride strengthens tooth enamel.

4 Visit a dentist every six months.

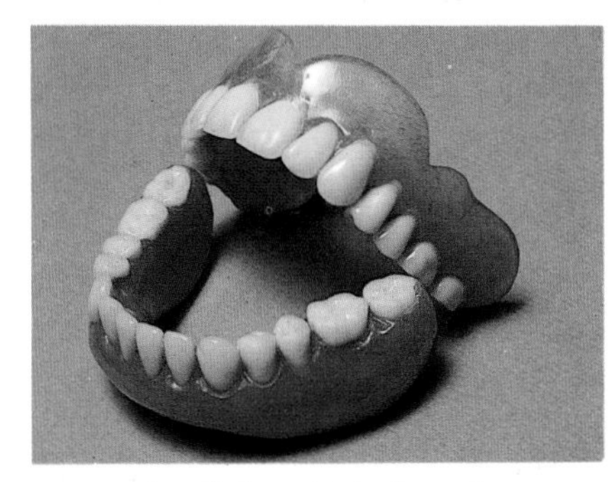

. . . so don't forget to brush your teeth!!

Questions

1 What is plaque?

2 How does plaque cause tooth decay and gum disease?

3 Why should you use fluoride toothpaste?

4 Name three foods which can rot your teeth. Explain why in each case.

6.7 Digestion

No matter what you eat, it is turned into liquid in your gut.

Bread and bacon and rice and all the other things you eat must be changed into a liquid inside your body, before your body can use them. The breaking down of food into a liquid is called **digestion**.

Food is digested inside a tube in your body called the gut, or **alimentary canal**. Your alimentary canal is over seven metres long and coiled inside your body.

Where your food is digested

Digestion starts when food is taken into the mouth. This is called **ingestion**. The teeth are used to break the food into small pieces. The pieces are mixed with a liquid called **saliva** and swallowed.

The walls of the gut produce chemicals called **digestive enzymes**. Enzymes break down food into liquids.

Liquid food passes through the gut wall into the bloodstream. This is called **absorption**. Blood carries food to all parts of the body.

Food is taken in by cells, and used for energy, growth, and repair. This is called **assimilation**.

Dietary fibre and other things in food which cannot be digested pass out through the anus when you visit the toilet. This is called **defecation**.

Substances which cannot be digested are called **faeces**.

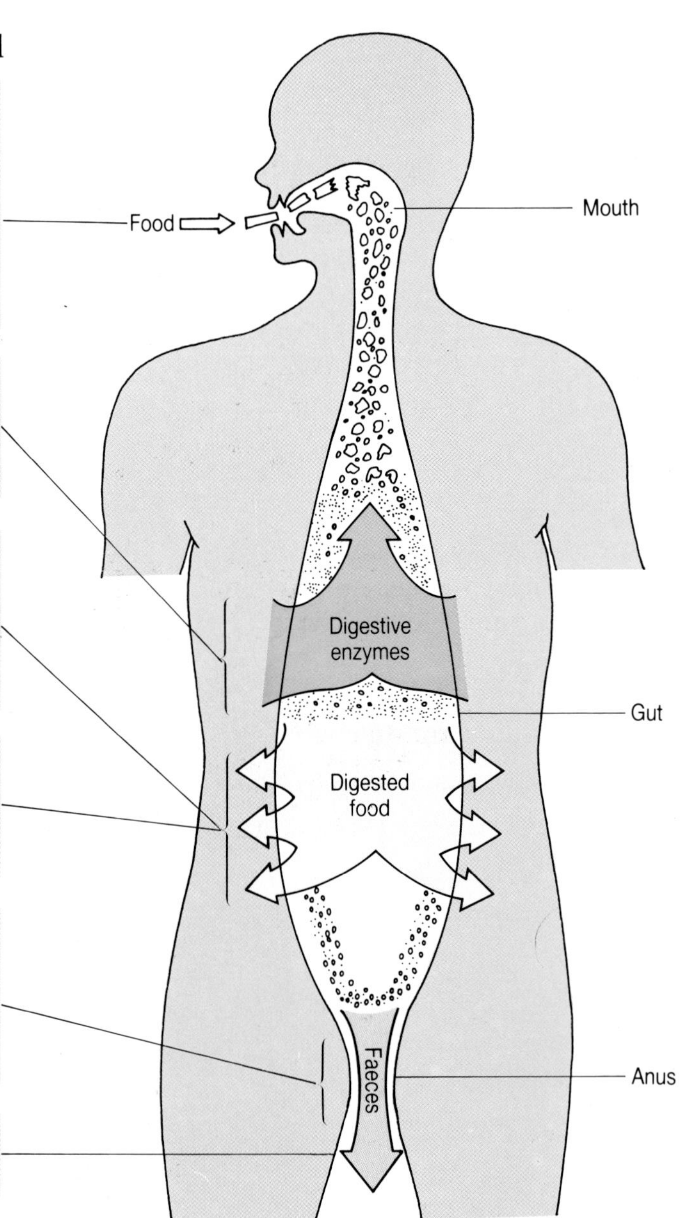

More about digestive enzymes

Digestive enzymes change the food you eat by breaking down large food molecules into smaller ones which can be absorbed into the blood. Each type of food needs a different enzyme to break it down.

Amylase enzymes break down carbohydrates like starch into glucose.

Lipase enzymes break down fats and oils into fatty acids and glycerol.

Protease enzymes break down proteins into amino acids.

This is what happens when starch is broken down by amylase enzyme and absorbed.

Starch is a carbohydrate found in rice, bread, and potatoes. Starch molecules are made up of many glucose molecules joined together, like a string of beads.

Amylase enzymes break down the starch by cutting it up into separate glucose molecules, like you might cut up a string of beads with scissors.

Glucose molecules are so small that they can pass through the cells which form the gut wall, and then through blood vessel walls into the bloodstream.

Blood carries the glucose to the cells of the body for assimilation.

Other kinds of food are broken down by other enzymes in the same way.

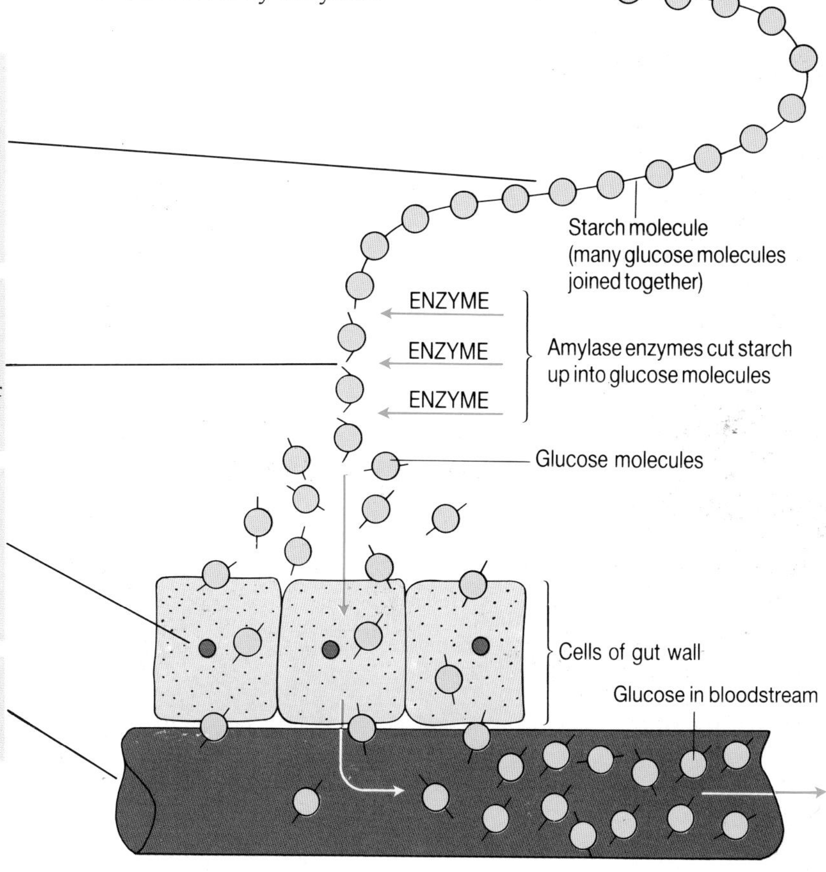

Questions

1 What is your alimentary canal?

2 What is the scientific word for:
  a) taking food into your mouth?
  b) breaking down food into small molecules?
  c) movement of food molecules into the blood?
  d) getting rid of undigested food?

3 Name the enzymes which digest: protein, carbohydrates, fats.

4 Which of the foods in question 3 are digested into: amino acids? glucose? fatty acids and glycerol?

6.8 A closer look at digestion

Digestion in the mouth

Digestion starts in your mouth. When you chew, food is ground into a pulp. At the same time, it is mixed with **saliva** from salivary glands. Saliva does two main things:

1 It wets the food, so that it slips easily down your throat.

2 It contains an **amylase enzyme** which starts to digest the starch in your food into sugar.

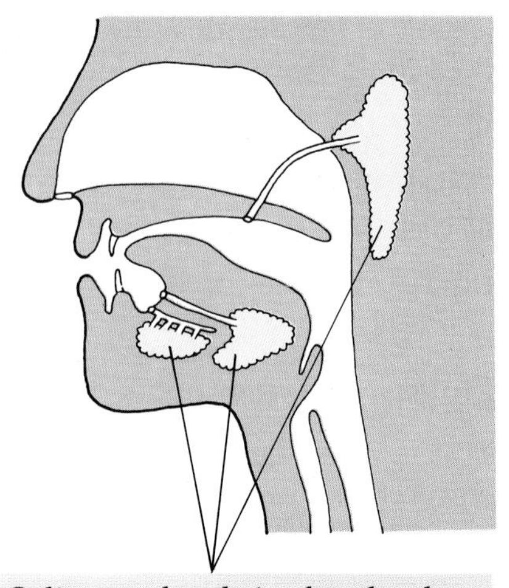

Salivary glands in the cheeks and under the tongue.

Swallowing

Before you swallow food, your tongue shapes it into a round lump called a **bolus**, and squeezes it to the back of your mouth.

The bolus pushes the **soft palate** upwards. This stops food getting into the space behind your nose.

A flap of skin called the **epiglottis** drops over the top of the windpipe. This stops food getting into your lungs.

The bolus is squeezed past the epiglottis, into your gullet or **oesophagus**.

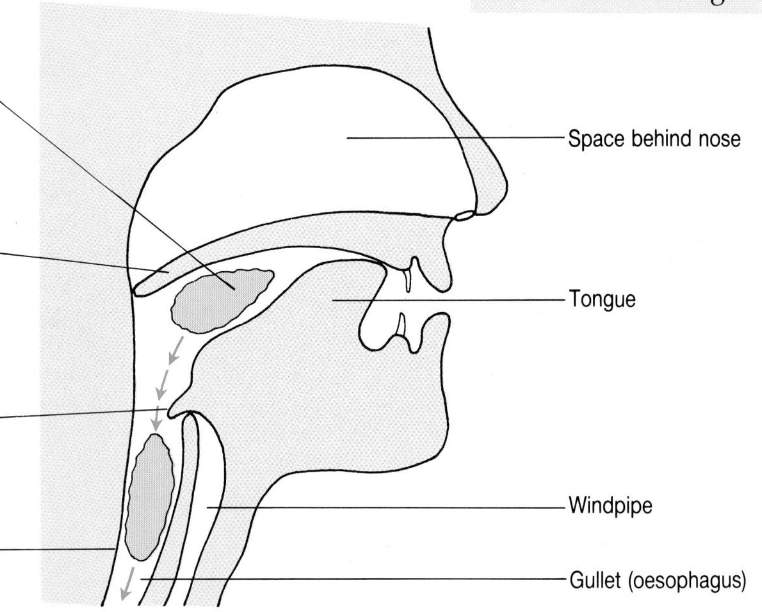

How food moves along the alimentary canal

Your gullet, and the rest of your alimentary canal, has **circular muscles** in its walls.

These muscles contract behind a bolus, and relax in front of it. So the bolus gets pushed along.

This contraction and relaxation of circular muscles is called **peristalsis**. It takes place all the way along your alimentary canal, and keeps food moving.

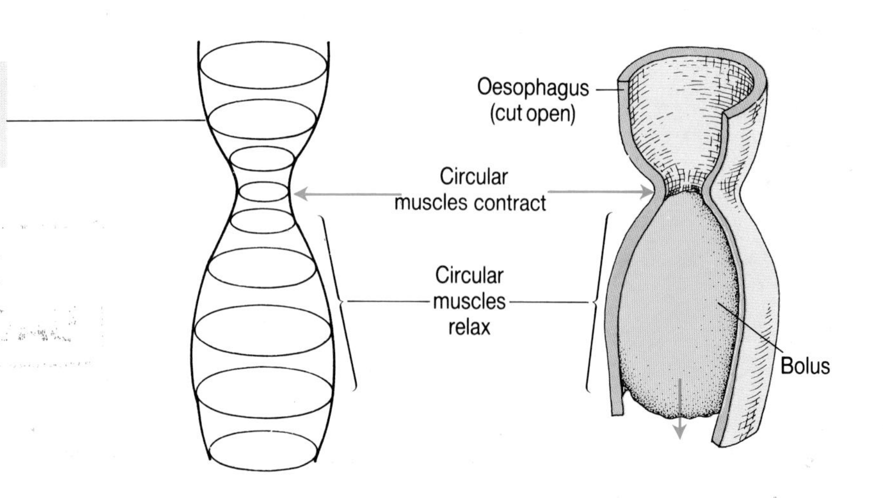

Digestion in the stomach and intestine

This is a simplified diagram of the alimentary canal. The drawing on page 128 shows what it really looks like, and its position in the body.

1 From the gullet, food passes to the **stomach**. Stomach muscles squeeze and relax mixing the food with **gastric juice** and **acid** made in the stomach wall. Gastric juice contains **protease enzymes** to digest protein. The acid kills germs in food and helps the enzyme work.

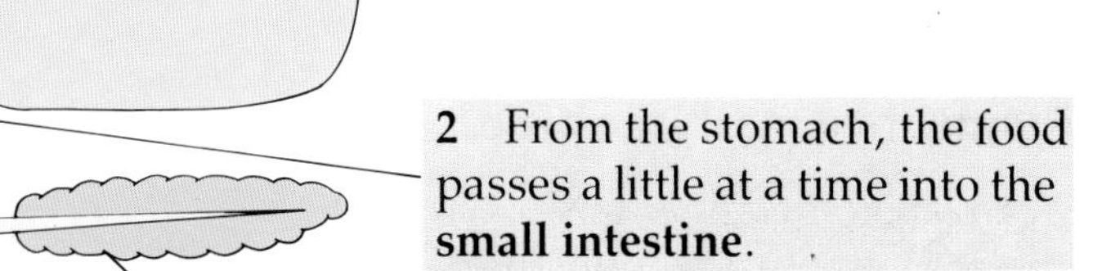

2 From the stomach, the food passes a little at a time into the **small intestine**.

In the small intestine three liquids are mixed with the food:

i Pancreatic juice from the pancreas contains **protease** to continue protein digestion, **amylase** to digest starch, and **lipase** to digest fats and oils.

ii Bile from the liver breaks up oil droplets into an emulsion to make them easier to digest. It also neutralises stomach acid to get food ready for pancreatic and intestinal enzymes.

iii Intestinal juice from the lining of the intestine contains more protease, amylase, and lipase, which complete digestion. These turn proteins into amino acids, carbohydrates into glucose, and fats and oils into fatty acids and glycerol.

3 Digested food is absorbed into the blood through the wall of the **small intestine**.

4 Undigested food passes into the **colon** where water is absorbed from it. It becomes a nearly solid waste called **faeces**.

5 Faeces are stored in the **rectum**.

6 Faeces pass out of the body through the **anus** when you go to the toilet.

Questions

1 Where in your body does digestion begin?
2 What are salivary glands for?
3 What might happen if you had no epiglottis?
4 What is the scientific name for your gut?
5 What substances are mixed with food in your stomach? How does each help digestion?

6.9 Absorption, and the liver

Digestion breaks down food so it can be absorbed into your blood and carried to all parts of your body. Carbohydrates are digested into glucose, proteins into amino acids, and fats and oils into fatty acids and glycerol. But vitamins, minerals, and water in food don't have to be digested since they are already easily absorbed.

Digested foods are absorbed through your gut wall into the blood and carried away. Most absorption goes on in your small intestine.

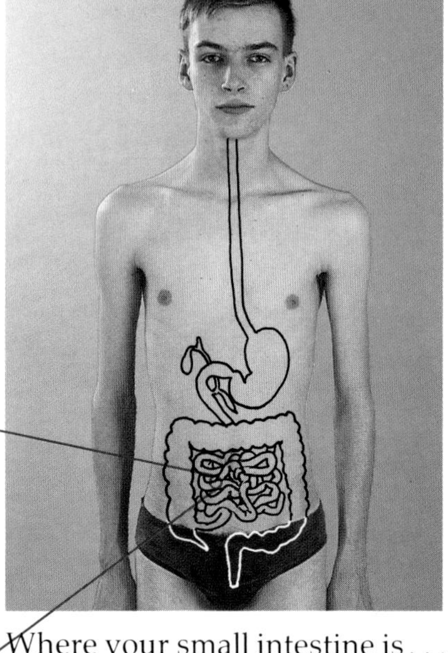

Where your small intestine is . . . It extends down from your waist.

Inside the small intestine

Your small intestine is over six metres long. Inside, millions of tiny fingers called **villi** stick out of its walls. One villus is about a millimetre long.

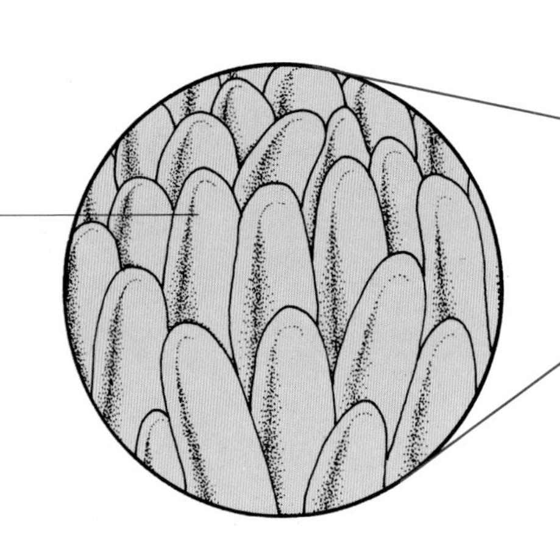

Absorption takes place through the villi. They provide a large surface for absorbing digested food.

Inside a villus

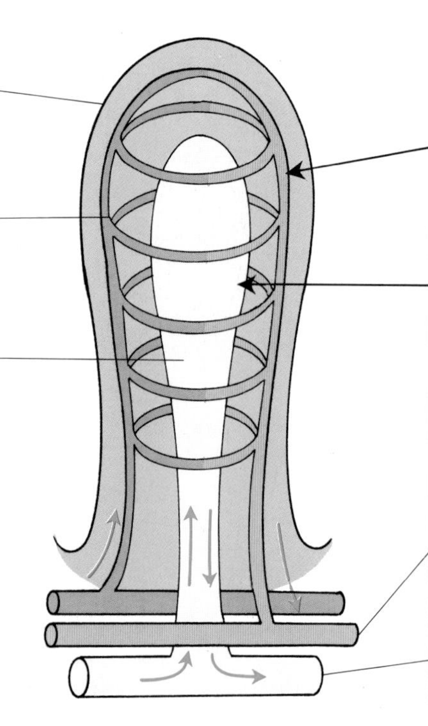

A villus has a surface layer only one cell thick. This lets digested food pass through easily.

It has a network of **capillaries** which carry blood.

It has a lymph vessel called a **lacteal** which carries lymph.

Glucose, amino acids, and some fatty acids and glycerol pass into the blood in the capillaries. Blood carries them to the liver, which is described on the next page.

Most of the fatty acids and glycerol pass into the lacteal. They are carried in the lymph to a vein in the neck. From there they flow into the blood and are carried to all parts of the body.

Blood vessel carrying digested food to the liver.

Lymph vessel carrying digested food to the bloodstream.

What happens to digested food in the liver

Your liver is the largest organ in your body. It weighs over one kilogram and stretches all the way across your body, just above the waist.

The liver is a chemical factory, a food store, and a central heating system. Here are just a few of the jobs it does:

1 It stores glucose as **glycogen**. It changes this back to glucose when the body needs it.

2 It stores the minerals copper and potassium, as well as iron needed to make red blood cells.

3 It stores vitamins A, B, and D.

4 It takes the goodness out of unwanted amino acids, and changes what is left into a waste called **urea**. Urea is removed from your body by your kidneys.

5 It takes some poisons from the blood and makes them harmless. These poisons come from germs, alcohol, and drugs.

6 It makes **bile**, which is needed for digestion.

7 It makes **fibrinogen**, which is needed for blood to clot in wounds.

8 These and many other jobs done in the liver produce heat, which the blood carries around your body to keep it warm.

Glucose, amino acids, vitamins, minerals, and some fatty acids and glycerol are absorbed into the villi.

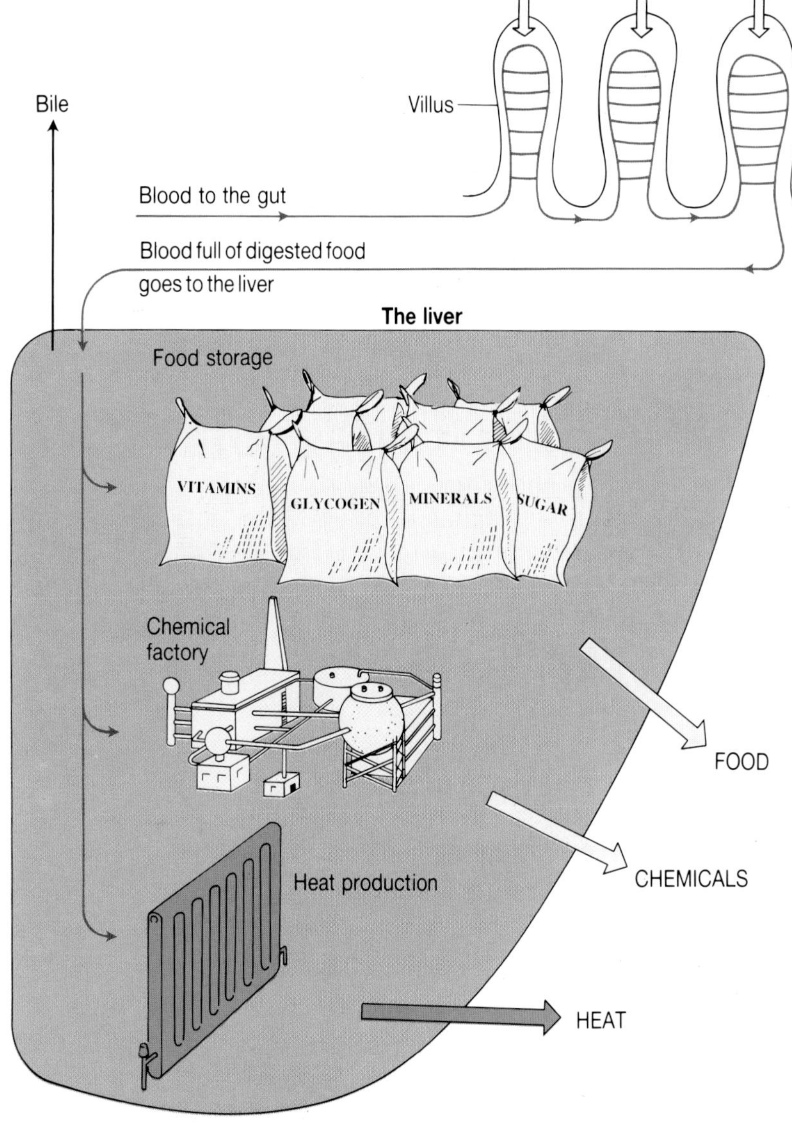

Questions

1 Digested food is absorbed inside your body. What does that mean?

2 Think of a reason why food has to be digested before it can be absorbed.

3 Where in your body are villi? What are they for?

4 Where in your body is your liver?

5 Name three things stored in your liver.

6 How does your liver help give you a steady supply of glucose?

7 Explain how:
- **a)** your liver helps digestion;
- **b)** your liver helps cuts to stop bleeding;
- **c)** your liver keeps you warm.

Questions on Section 6

1 *Recall knowledge*
Copy the following sentences into your exercise book. Then fill in the gaps using words from the list: **carbohydrates, proteins, fats and oils, vitamins and minerals, dietary fibre.**

a) .... are body-building foods.
b) Foods rich in .... are stored in your body to provide energy in the future.
c) You need .... to avoid constipation.
d) Food rich in .... provides energy quickly.
e) You need about twenty different .... to stay healthy.
f) Starch and sugar are examples of ....
g) Fresh fruit contains plenty of ....
h) Wholemeal bread contains plenty of ....
i) Eggs and fish are rich in ....
j) Butter is rich in ....

2 *Recall knowledge*
a) Write down three headings: *Rich in proteins, Rich in carbohydrates, Rich in fats.*
b) Now sort this list of food into three groups under these headings:
jam, lean ham, peanuts, toffee, lard, cake, chicken, egg, butter, potato, fish.

3 *Recall knowledge*
Copy these sentences, then fill in the gaps.

a) ..... and vitamin ..... are needed for strong bones and teeth.
b) Vitamin ..... is needed to heal wounds quickly.
c) ..... is needed for making red blood cells.
d) Lack of vitamin ..... causes scurvy.
e) Lack of vitamin ..... causes rickets.
f) Lack of ..... causes anaemia.

4 *Recall knowledge*
Instant mashed potato is an example of a processed food. To make it, potatoes are put through special processes in a factory.

a) What good things may be taken out of food when it is processed?
b) Name three types of food additives which may be added to food during processing.
c) Why are these chemicals added to food?
d) Think of four reasons why people buy processed foods.
e) Think of four reasons why it is better to buy fresh foods than processed foods.
f) What is the difference between:
wholemeal bread and white bread?
juice from a crushed orange and an orange-flavoured drink?
garden peas and tinned peas?
roast jacket potato and instant potato?

5 *Recall knowledge*
Below is a diagram of the digestive system.

a) Name the parts labelled A to J.
b) Match the labels on the diagram with the descriptions below:
where digested food passes into the bloodstream;
where food is mixed with gastric juice;
where faeces leave the body;
a gland which produces enzymes that digest protein, starch, and fat;
glands which produce saliva;
where water is absorbed from the undigested remains of food;
pumps bile into the small intestine;
stores glycogen, vitamins, and minerals.

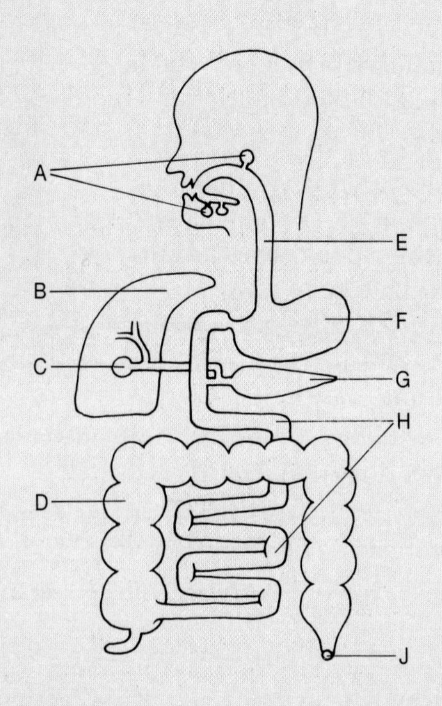

6 *Understanding/evaluation/design*
A group of students designed the experiment opposite to find the energy value of peanuts.

a) Their result was 1900 KJ per 100 g of peanuts.

b) How was the apparatus used to obtain this result?

c) Official tables give the energy value of peanuts as 2445 KJ per 100 g. Give two reasons why the students obtained a lower result.

d) Redesign the method to improve the accuracy of the result.

e) The students decided to use this method to find the energy value of bread. Which experimental conditions must they keep the same so that results can be compared with those for peanuts?

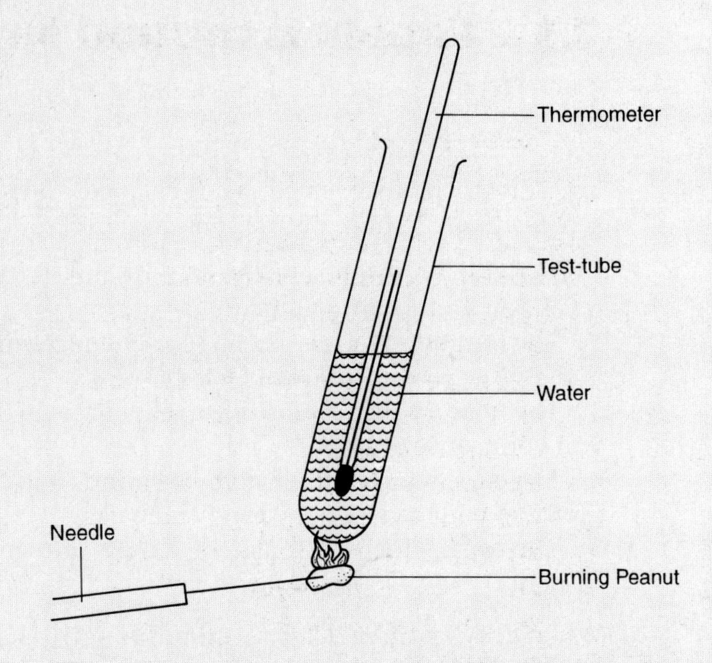

7 *Interpret data*
A dish was filled with a jelly (agar) containing starch. Five cavities were cut in the surface of the jelly, and each was filled with a different liquid (see diagram opposite). The apparatus was kept warm for an hour then flooded with iodine solution. The result is shown in the diagram opposite.

a) How does iodine affect starch?

b) Describe three things this result tells you about saliva.

c) Is is possible to say that wheat grains contain a substance similar to saliva? If they do, what use could it be during germination?

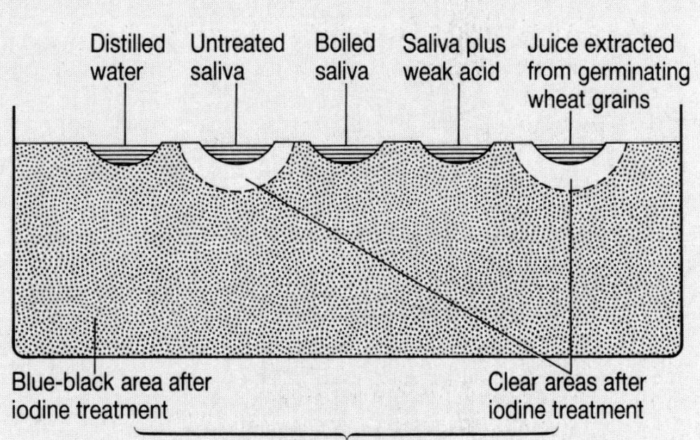

8 *Understanding/interpret data*
The chart opposite shows the average amount of energy which British and Japanese people obtain from various foods.

a) Cereals and meat are sources of protein. Which also contains the most fibre, and which contains the most fat?

b) How much more meat, fat and oil do the British eat than the Japanese?

c) A low-fibre, high-fat diet can cause heart disease. Would you expect heart disease to be less common in Japan or in Britain? Explain your answer.

d) Would you expect tooth decay to be less common in Britain or Japan? Explain your answer.

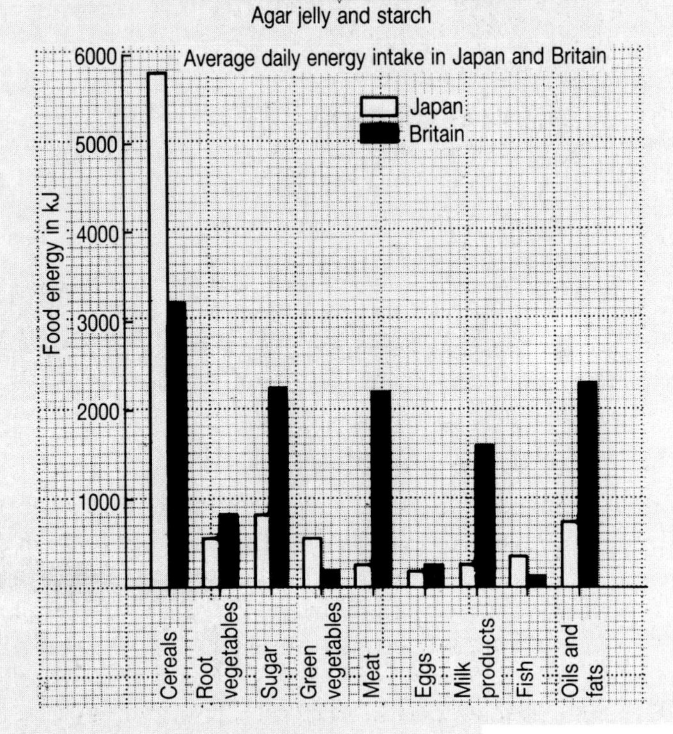

7.1 Touch, taste, and smell

You can touch, taste, smell, hear, balance, and see because of your sense organs. In humans the sense organs are:

1 The **skin**, which is sensitive to pressure, heat, and cold and gives you your sense of touch.
2 The **tongue**, which is sensitive to chemicals in food and drink, and gives you your sense of taste.
3 The **nose**, which is sensitive to chemicals in the air, and gives you your sense of smell.
4 The **ears**, which are sensitive to sounds and movement, and give you your senses of hearing and balance.
5 The **eyes**, which are sensitive to light, and give you your sense of sight.

When a sense organ detects a **stimulus** such as sound or light, it sends messages along nerves to the brain. The brain then gives you feelings or **sensations** such as hearing or sight.

A cat has a special set of sense organs – its whiskers. They are the same width as the cat's body. It uses them to judge whether it can squeeze through a gap.

The skin

Your sense of touch is produced by the ends of nerve cells in your skin. These are called **nerve endings**.

These nerve endings deep inside the skin are sensitive to heavy pressure. They warn you about pressure which could bruise you.

These nerve endings near the skin surface are sensitive to light pressure. They tell you about the texture of an object, for example whether it is rough or smooth.

These nerve endings very close to the skin surface make you feel pain if, for example, your skin is cut or burnt.

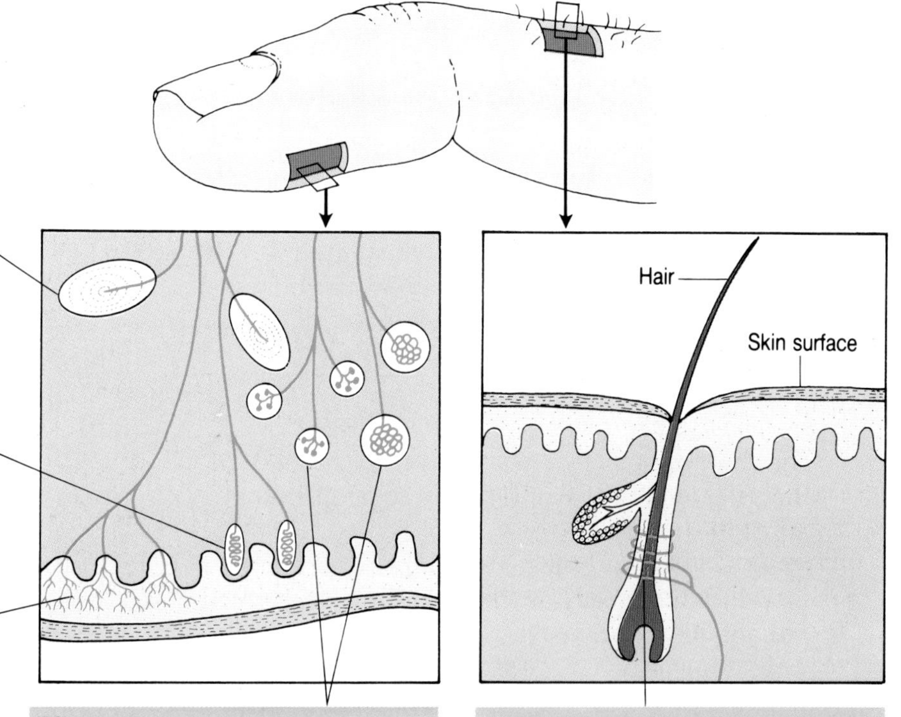

These nerve endings are sensitive to heat and cold. They detect changes in temperature. For example, when the weather changes, or when you touch cold or hot objects.

These nerve endings wrapped around the base of a hair detect if the hair is moved or pulled.

The tongue

Your tongue has little bumps on it. These bumps contain tiny sense organs called **taste buds**. Taste buds are sensitive to chemicals in food. These chemicals must dissolve in saliva before you can taste them. This is why dry food has no taste until you chew it to mix it with saliva.

Your sense of taste is useful.
1 It stimulates your stomach to produce gastric juice for digestion.
2 Many poisons and bad foods have a nasty taste. So you can spit them out before they harm you.

There are different taste buds for tasting bitter, sour, salty, and sweet foods.

A food can stimulate more than one kind of taste bud at the same time.

The taste buds for **bitter tastes** are at the back of your tongue.

The taste buds for **sour tastes** are at the sides of your tongue.

The taste buds for **salty and sweet** tastes are at the front of your tongue.

The nose

When a bad cold blocks your nose, food seems to lose its flavour. This is because many flavours are really smells!

Smells are chemicals in the air. The chemicals dissolve in moisture on the lining of your nose. This stimulates nerve endings in your nose to send messages to your brain which produces the sensation of smell.

Humans can detect about 3000 different smells. Smells help animals to hunt food and find their way. Smells can also warn of certain dangers.

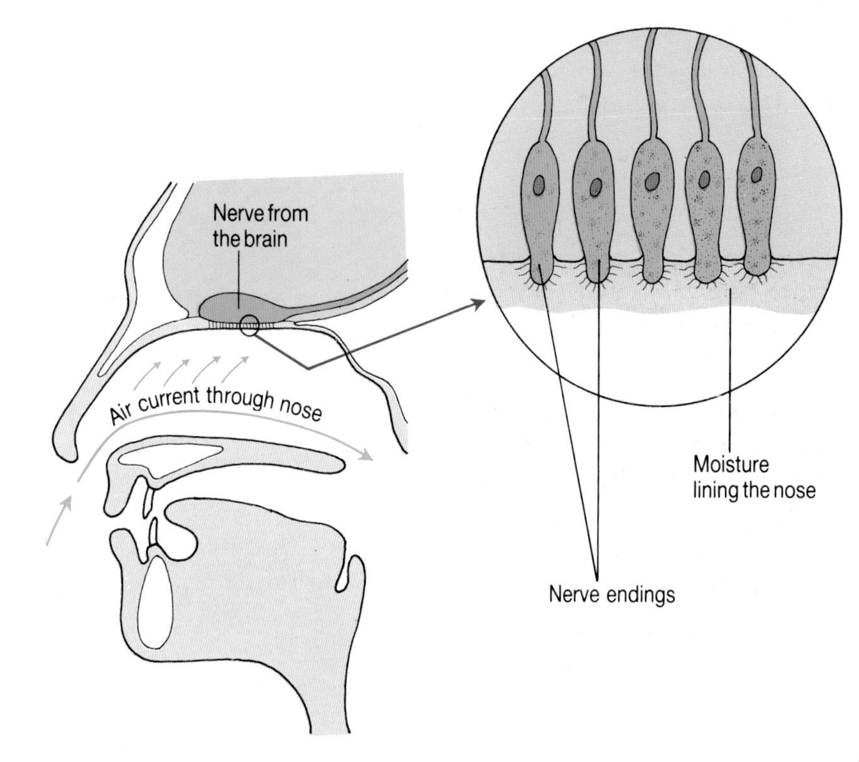

Questions

1 Name the five sense organs in humans.

2 If you gently stroke the hairs on your arm you feel a tingling sensation. Try to explain why.

3 Why do foods seem to have less flavour when you have a cold in the nose?

4 How do taste and smell protect you?

7.2 The eye

How you see things

1 Light goes from an object to your eye.

2 Light is bent as it passes through your eye.

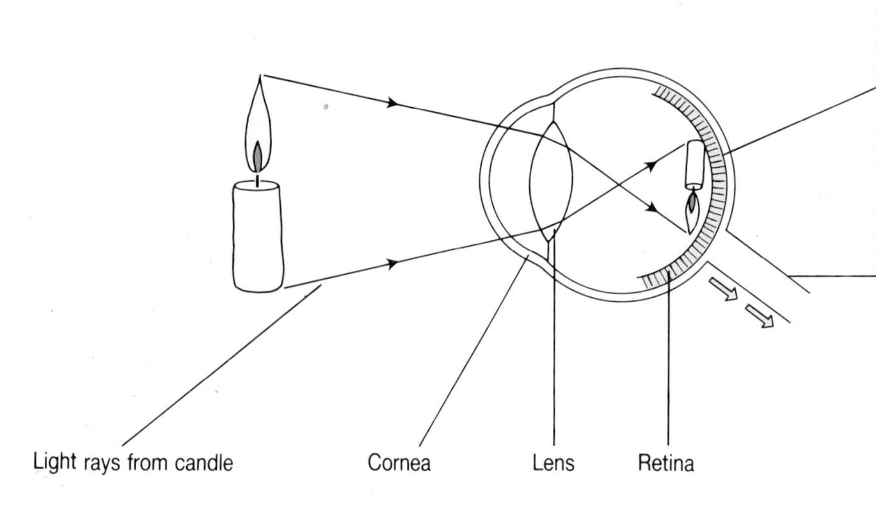

3 An upside-down picture of the object is focused on a layer called the **retina** at the back of your eye. The retina is made of cells sensitive to light.

4 The retina sends messages along the **optic nerve** to your brain. Your brain allows you to see a picture of the object which is the right way up.

How eyes are protected

Your eyes are set in holes called **orbits** in your skull. So all except the front of each eye is protected by bone.

There are **tear glands** behind the top eyelid. They make tears that wash your eye clean when you blink. Tears are produced faster if dust or smoke get into your eyes.

The **iris** is the coloured part of the eye. It is a ring of muscle with a hole called the pupil in the middle. It protects the eyes from bright light.

Eyelashes form a net in front of the eye which protects it from dust.

The white of the eye is a tough protective layer called the **sclerotic.**

The **pupil** lets light into the eye. If the light is too bright the iris muscle makes the pupil smaller. In dim light the iris muscle make the pupil bigger.

The parts of the eye

The **conjunctiva** is a thin clear skin which covers the front of the eye.

The **cornea** is a clear window in the sclerotic in front of the iris. It lets light into the eye.

The **iris** controls the amount of light entering the eye.

The front part of the eye is filled with a watery liquid called **aqueous humour**.

The **lens** helps focus a picture on the retina. The lens is clear and can change shape.

The **suspensory ligaments** hold the lens in place.

The **sclerotic layer** is the tough, white protective layer of the eye.

The back of the eye is filled with a jelly called **vitreous humour**.

The **choroid** is a black layer that stops light being reflected round the inside of the eye.

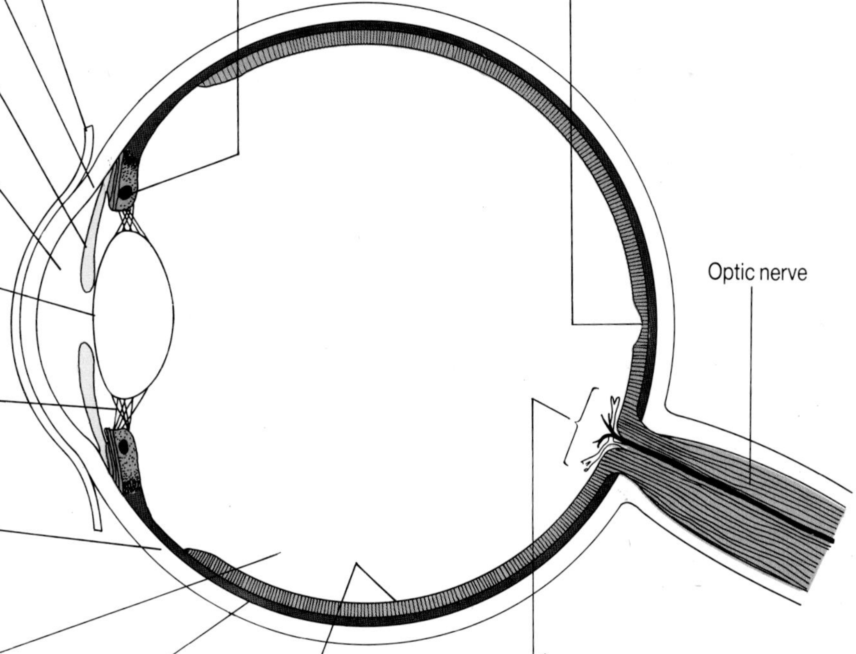

Ciliary muscles change the shape of the lens during focusing.

The **yellow spot** is the most sensitive part of the retina. It lets you see colour.

Optic nerve

The **retina** is a layer of cells which are sensitive to light. They send messages to the brain.

The **blind spot** is where blood vessels and nerves join the eyeball. It has no light-sensitive cells, so it sends no messages to the brain.

Questions

1 Give the scientific name for each of these:

a) Carries messages from an eye to the brain.
b) Muscles which change the shape of a lens.
c) A layer of light-sensitive cells.
d) It controls the amount of light entering the eye.
e) It prevents light being reflected around the eye.
f) They make liquid which washes the eyes.
g) A hole in the middle of the iris.
h) A clear window at the front of the eye.
i) Changes shape to focus a picture on the retina.

2 If you walk from a dark room into sunlight and back again how would your pupils alter in size? Why does this happen?

7.3 Vision

The light that goes into your eye has to be bent or **focused** on to the retina to let you see clearly.

Most of the bending of light takes place as it passes through the **cornea** and **aqueous humour**. The **lens** bends it a little more to make a perfectly clear picture on the retina. The **ciliary muscles** change the shape of the lens to bend light. A fat lens bends light more than a thin lens.

If the lenses in your eyes don't focus properly, wearing glasses will help.

To see a near object

Light from a near object needs to be bent more than light from a distant object. So the lens is made much fatter.

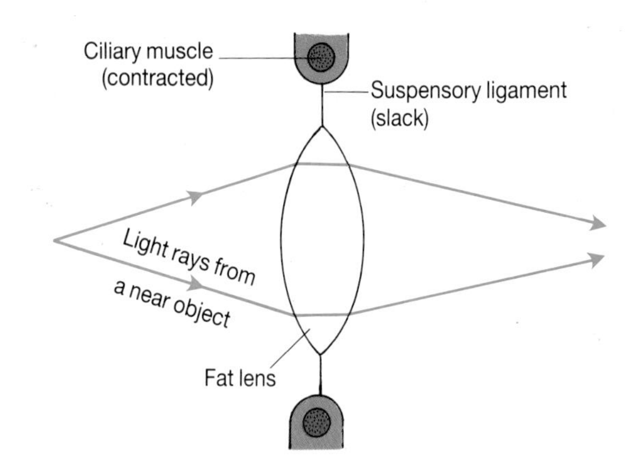

1 The ciliary muscles are in a circle round the lens. When they contract, the suspensory ligaments become slack, which allows the lens to become fatter in shape.

2 The lens can now bend light enough to make a clear picture on the retina of a near object.

To see a distant object

Light from a distant object needs to be bent very little. So the lens is stretched to make it thin.

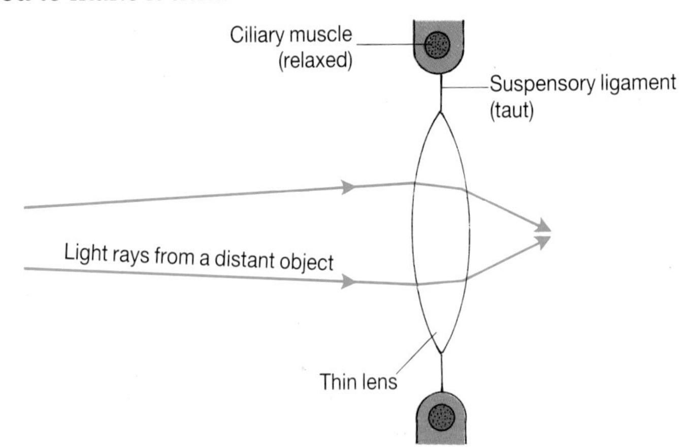

1 When the ciliary muscles relax, pressure inside the eyeball stretches the lens into a thin shape.

2 The thin, flat lens bends light just a little, to make a clear picture on the retina of a distant object.

Three dimensional vision

Each of your eyes gets a slightly different view of an object.

Your brain puts these two views together, so that you see the object as three-dimensional rather than flat.

Three-dimensional vision helps you judge how far away an object is.

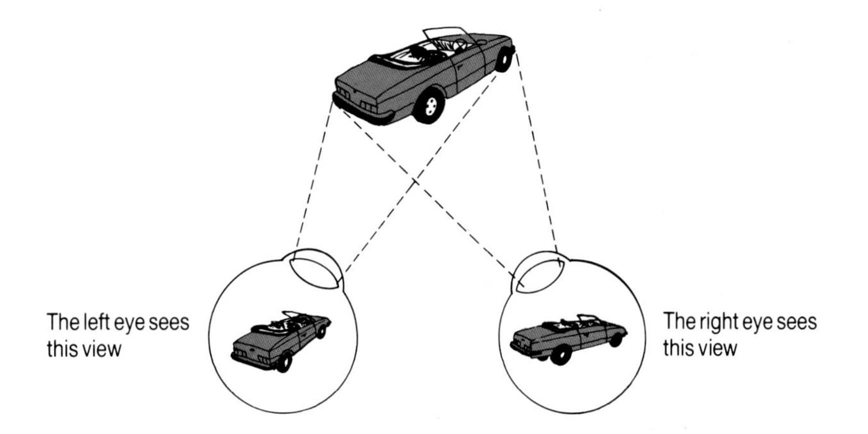

Two eyes

Rabbits, chickens, fish, and many other animals have eyes which look sideways, and not forwards like yours. Each eye sees a different view. They can even see what is happening behind them. This is useful if other animals hunt you for food!

Rabbits can see what's going on at each side...

...but owls look straight ahead.

More about the retina

There are two kinds of light-sensitive cells in the retina. They are called **rods** and **cones**.

Cones only work in bright light, but give a very clear picture and are sensitive to colour. The **yellow spot** in the middle of the retina is made entirely of cones.

If you want to see something very clearly you look straight at it, so that its picture falls on the yellow spot.

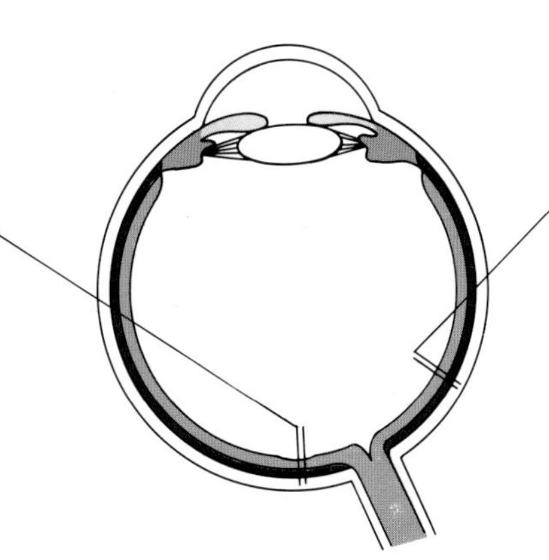

The rest of the retina is mostly rods, with a few cones. Rods do not give as clear a picture as cones and are not sensitive to colour. But rods work in dim light.

This explains why you don't see colours clearly in dim light.

Questions

1 Which parts of an eye bend light?

2 What shape are your eye lenses when you look at your hand? What shape are they when you look at a tree in the distance?

3 **a)** Where in the retina are rods and cones found?

b) Which detect colour, and which work in dim light?

7.4 Ears, hearing, and balance

The air around you is full of vibrations called **sound waves**. You do not hear sound waves with your ears. You 'hear' them inside your brain. This is what happens.

Sound waves move through the air at 332 metres per second from bells, radios, birds, cars and so on. When they reach your ear they are changed into 'messages' called **nerve impulses**.

Nerve impulses travel along a nerve from each ear to your brain. Your brain transforms them into a sensation of sound.

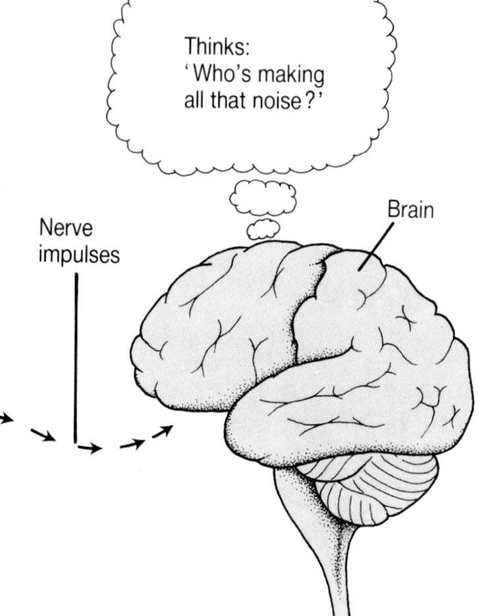

Sound waves

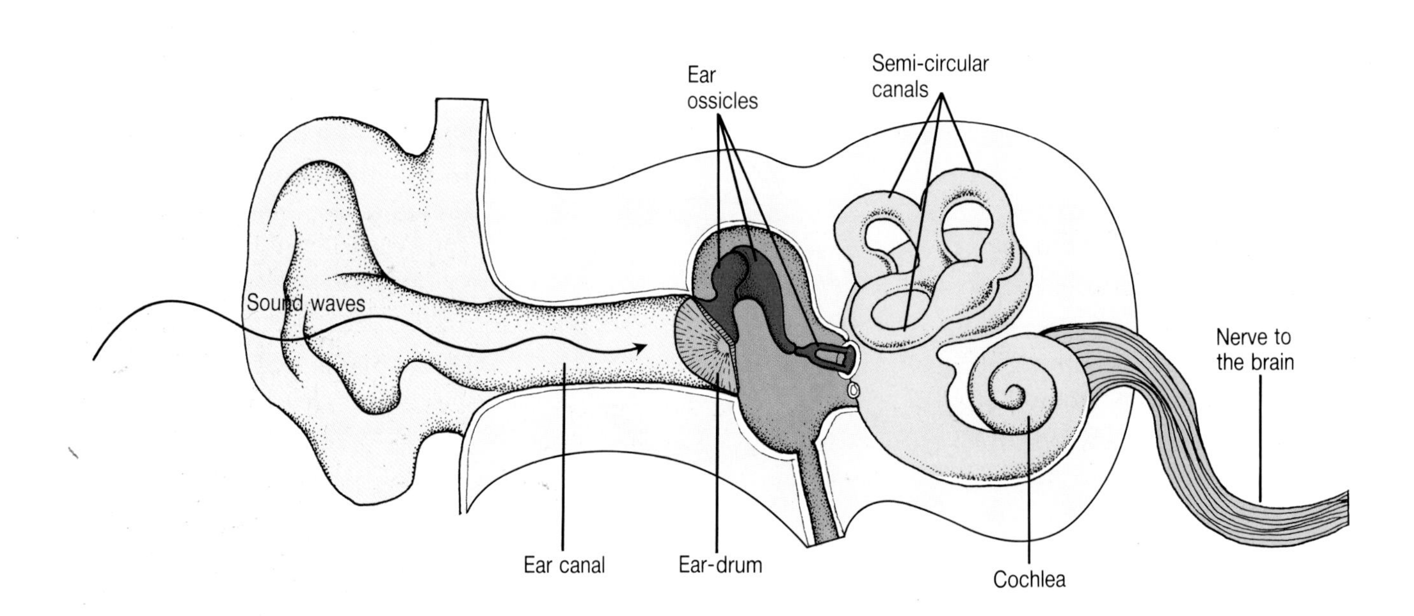

The part which you call your ear is a flap of skin in the shape of a funnel. This leads to a tunnel, the **ear canal** into your head. Sound waves are collected by the funnel. They move down the ear canal to a thin sheet of skin called the **ear-drum**.

The ear-drum is connected to the inner ear by three tiny bones called **ear ossicles**. The part of the inner ear concerned with hearing is called the **cochlea**. It is coiled like the shell of a snail.

How the ear works

This diagram shows the cochlea straightened out.

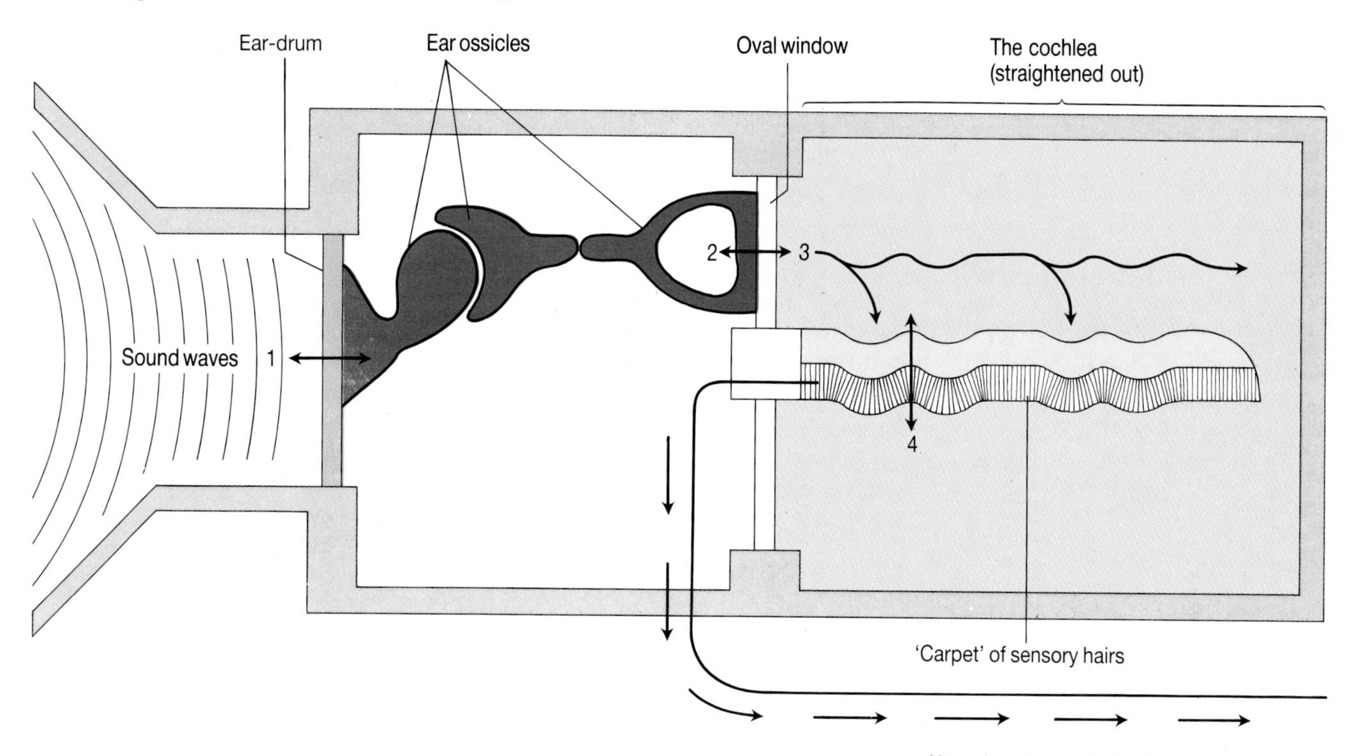

1 Sound waves make the ear-drum vibrate in and out. This makes the ossicles vibrate.

2 The ossicles are connected to another, smaller sheet of skin called the **oval window**. When the ossicles vibrate, they make the oval window vibrate in and out, but with a force which is 30 times greater than vibrations of the ear-drum. So, the ossicles magnify sound by this much.

3 When the oval window moves in and out it causes vibrations to move through the cochlea.

4 The cochlea contains a carpet of tiny hairs which are connected to nerves. Vibrations in the cochlea cause these hairs to vibrate up and down and send impulses to the brain where we hear the sounds.

Keeping your balance

You keep your balance in two ways. Your eyes tell you if you are upright or tilted. The semi-circular canals of your inner ears tell you if you are moving off balance.

The semi-circular canals are shown in the diagram on the opposite page. They are three tubes full of liquid. When you move, the liquid moves. Sensitive hairs inside the tubes detect this movement and send impulses along nerves to your brain. Your brain detects when you are losing your balance and sends impulses to muscles to keep you upright.

Questions

1 Which part of your ear:
  a) changes vibrations into nerve impulses?
  b) keeps you from falling over?

2 Explain how the man in the photograph above keeps his balance?

7.5 More about senses

An adult's body contains about 40 litres of water. If 5 per cent is lost we feel thirsty; a loss of 10 per cent makes us very ill; but if 20 per cent is lost we die, even though at least 30 litres of water remain in the body.

How many senses do you have?

If asked this question, most people would name the five **special senses**: touch, taste, smell, hearing, and sight. In fact we have more.

Senses inside your body

Hunger and thirst You cannot live without food and water, and there are senses which detect when more of these are needed. Hunger comes partly from sensory nerve endings in your stomach wall which are stimulated when your stomach contracts as it empties. When your body is short of water, body fluids become more concentrated. This is detected by receptors in your brain which produce the sensation of thirst, so that you drink to replace lost fluids.

Muscle and joint senses Sense organs called **stretch receptors** inside muscles and joint ligaments sense the amount of tension and stretch in these tissues. Your brain uses information from these receptors to tell you how much you are pushing or pulling something, to control the speed of your movements, and to judge the position of your limbs without looking at them.

Try this test. Shut your eyes. Then raise one arm and bend the elbow, wrist, and fingers to any angle you wish (try to bend each joint to a different angle). *Without looking*, try to bend your other arm so that its joints are at exactly the same angles. How do you do it?

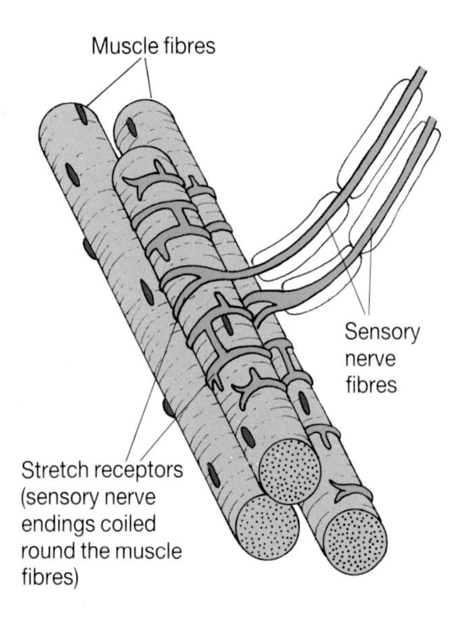

'Touch' is more than one sense

Using touch alone, you can feel if objects are soft or hard, rough or smooth, hot or cold, wet or dry, and guess their weight and shape.

This is done using skin senses described in Unit 7.1. Texture and shape are judged using skin pressure senses. Weight is also judged using pressure senses and muscle stretch receptors. You cannot judge actual temperature, only differences in temperature. Try the test opposite to investigate temperature sense.

1 Half-fill one beaker with hot (not boiling) water, a second with ice-cold water, and a third with warm water. Put one hand in the hot water and the other in the ice-cold water.

2 After a minute, put both hands in the warm water. How does this feel to each hand? How does this show that skin is only sensitive to *differences* in temperature?

More about hearing

One of our most astonishing skills is the ability to separate sounds. When you listen to an orchestra made up of strings, woodwind, brass, percussion etc., the many different sounds are reduced to a single, long string of vibrations. Your ear and brain can process these so that each sound is separated out.

The loudness of a sound is measured in **decibels (dB).** A whisper is about 20 dB, and a jet taking off about 140 dB. Loudness becomes painful, and harmful to the inner ear between 120 – 130 dB. Sudden loud bursts of sound, such as explosions, are more harmful than continuous loud sound such as factory machinery and loud music.

It takes at least 36 hours to regain normal hearing sensitivity after only two hours' listening to loud music.

How clear is your vision?

Try this test. Rest this book on a desk. Lower your head until it is about 20 cm from the page. Close one eye. With the other eye, look at the 'A' in the diagram opposite. *Without moving your eye from the 'A'*, try to read the other letters. How many are clearly visible?

Most people can only see two or three letters clearly. In fact only a small part of what we see is clearly visible, and this is the part which is focused on the **yellow spot** of the retina (Unit 7.2). But you are not normally aware of this because your eyes are rarely still. They constantly scan a scene so that most of it is seen by the yellow spot.

H
G
F
E
D
C
B
H G F E D C B A B C D E F G H
B
C
D
E
F
G
H

Part of each eye is blind

Try this test. Hold this page at arm's length. Close your *left* eye and fix the other on the black circle opposite. *While keeping your eye on the circle*, bring the book towards your face. Notice, out of the 'corner of your eye', what happens to the plus sign.

The plus sign disappears when its image falls on the **blind spot** of your right eye (see diagram on p. 135). It is impossible to make the plus disappear with both eyes open because its image can never fall on the blind spot of both eyes at once.

● +

Questions

1 How do you know how hard you are pushing or pulling against something?

2 How do you know about the position of your arms and legs without looking?

3 How many things can you find out about objects using sense of touch alone?

7.6 The nervous system

Your nervous system is like a manager inside your body. Its job is to **control** and **co-ordinate** the parts of your body so that they work together, doing their jobs at the right time.

Your nervous system co-ordinates muscles so that you can do things which need thought, like cycling, dancing, or reading.

It also co-ordinates things which you don't need to think about, like heartbeat and breathing.

The nervous system consists of the **brain**, the **spinal cord**, and millions of **nerves**. Together the brain and spinal cord are called the **central nervous system**.

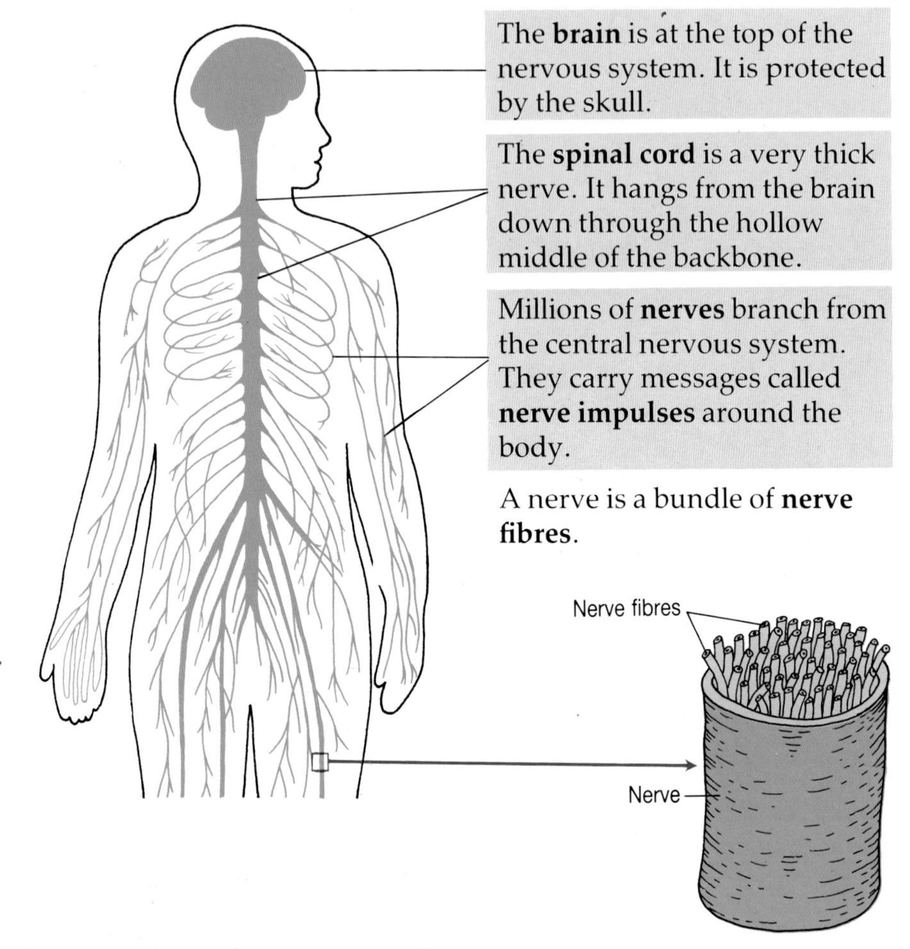

The **brain** is at the top of the nervous system. It is protected by the skull.

The **spinal cord** is a very thick nerve. It hangs from the brain down through the hollow middle of the backbone.

Millions of **nerves** branch from the central nervous system. They carry messages called **nerve impulses** around the body.

A nerve is a bundle of **nerve fibres**.

Nerve cells

The nervous system is made up of **nerve cells**. Most cells are small and rounded. But nerve cells are not. They are stretched out into long thin **nerve fibres** that can be over a metre long. Nerve impulses travel along nerve fibres in *only one direction*.

Sensory nerve cells carry impulses from sense organs into the central nervous system.

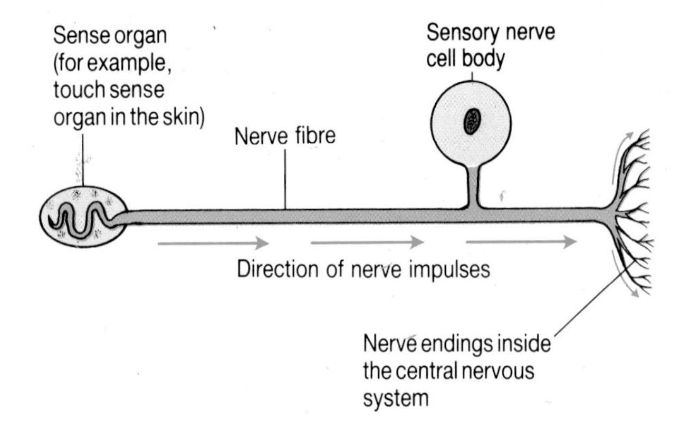

Motor nerve cells carry impulses from the central nervous system to muscles and glands.

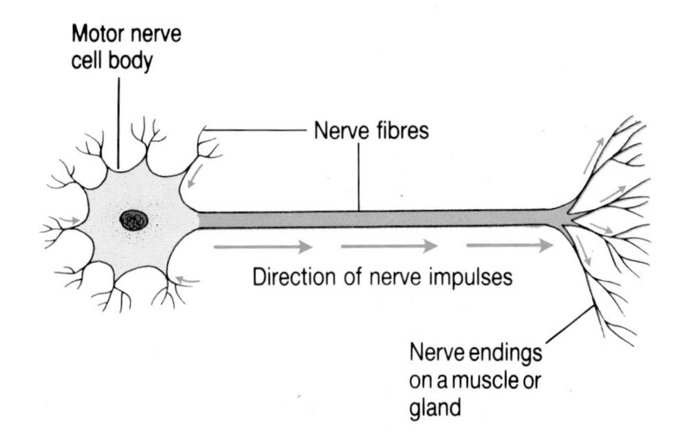

How the nervous system works

If you sat on a drawing pin you would jump up yelling with pain. This is an example of a **stimulus** and a **response**. The stimulus is pain. The response is jumping and yelling. Your nervous system controls the response.

This diagram of the nervous system shows how it works.

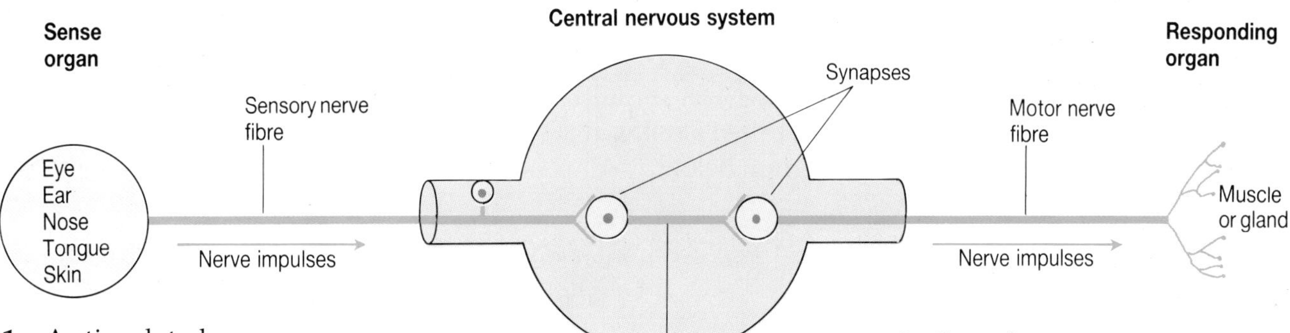

1 A stimulated sense organ sends nerve impulses along **sensory nerve fibres** to the central nervous system. This works out the best response to the stimulus, then sends impulses to muscles and glands that carry out the response.

2 There is a tiny gap called a **synapse** where one nerve cell meets another. Only impulses from a strong stimulus can cross a synapse; weak stimuli do not affect a responding organ.

3 Impulses must cross many synapses as they pass from sensory nerve fibres through the central nervous system. If the stimulus is strong enough, impulses eventually travel along **motor nerve fibres** to make a response in muscles or glands.

Reflex and voluntary actions

Reflex actions are actions you do without thinking, to protect yourself: coughing clears your windpipe, and shivering keeps you warm. The pupils get smaller in bright light to protect the retina, and larger in dull light to help you see.

Voluntary actions are actions which need thought, like speaking to a friend or writing a letter.

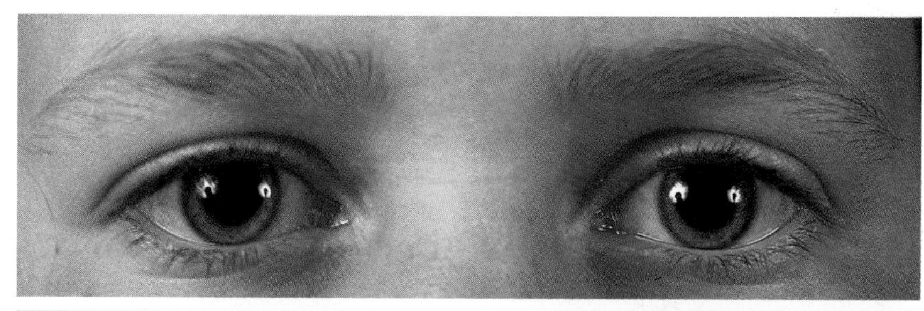

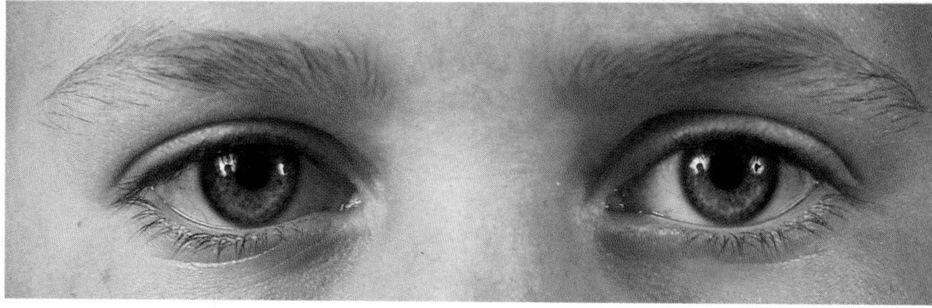

Questions

1 Name the parts of the nervous system.
2 Name two kinds of nerve cell.
3 How are nerve cells different from other cells?
4 What is a nerve impulse?
5 What is a reflex action?
6 What is a voluntary action?
7 Sort these into reflex and voluntary actions: coughing, reading, sneezing, sweating, writing.

7.7 More about reflexes

Why is sneezing an example of a reflex action? Why do we sneeze?

What is the difference between pulling your hand away from something hot, and reaching out to pick something up? The first is a **reflex response** – behaviour which you do automatically without thinking, and which you do not have to learn. The second is a **voluntary response.** It has to be learned and requires thought.

Reflex responses are extremely fast because nerve impulses travel through a simple arrangement of nerves called a **reflex arc.** Trace the path of nerve impulses through the diagram below.

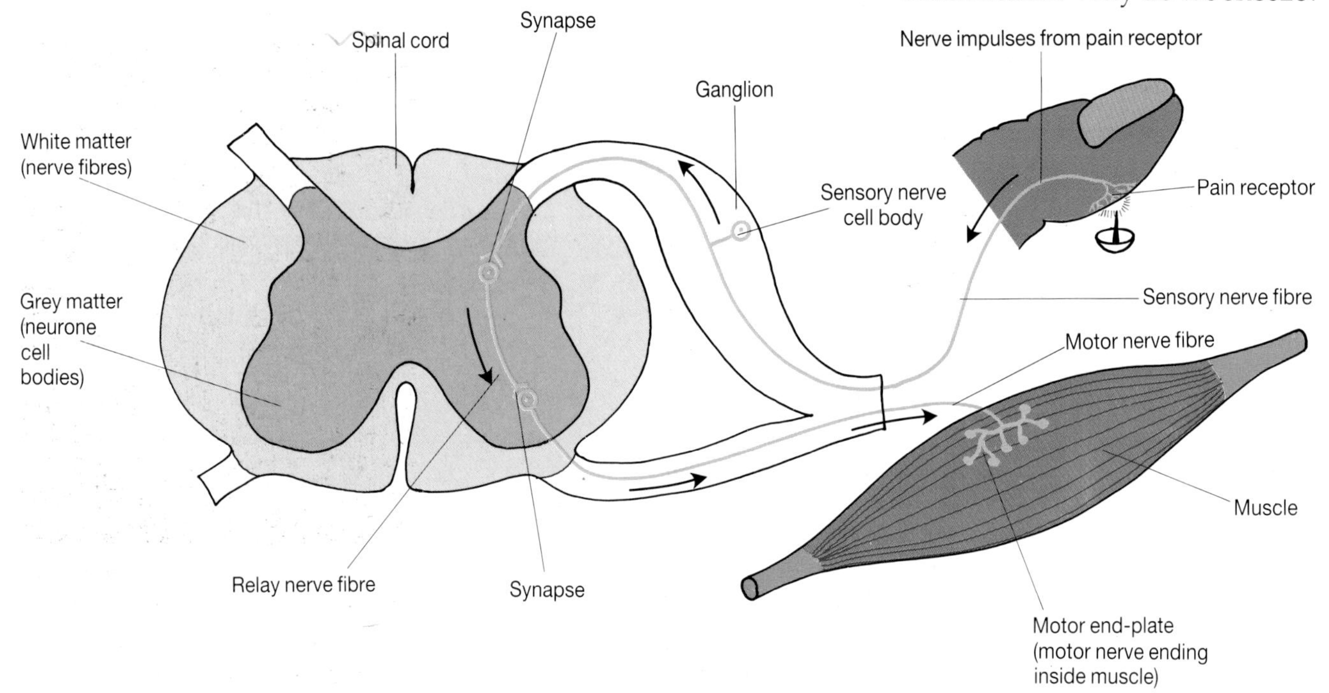

The stimulus in this example is a pricked finger. This causes impulses to travel from a **pain receptor** in the skin along a **sensory nerve fibre** to the **spinal cord.**

Here impulses must cross a synapse to reach a **relay nerve fibre** which carries them through the nerve cord to a synapse with a **motor nerve fibre.** This fibre carries impulses to muscles which respond by pulling the finger away.

Questions

1 What reflex actions occur when:
 a) dust blows into your eyes (two examples);
 b) when a bright light suddenly shines in your eyes (two examples);
 c) when you run fast in hot weather (three examples);
 d) if you move suddenly from a warm room to a cold, windy day outside;
 e) when you are hungry and smell cooking food;
 f) when food accidentally enters your windpipe?

2 Which of these reflexes protect you from harm and which help conserve energy and body materials?

7.8 The brain

There are about 10 000 million nerve cells in your brain. Each cell is linked with thousands more. This linking allows your brain to do many different jobs all at the same time.

Nerve cells from the cerebellum of the brain. Their job is to control muscles and balance.

The parts of the brain

Your brain is at the top of your spinal cord and is protected by your skull. It has three main parts: the cerebrum, the cerebellum, and the medulla.

The **cerebrum** is the dome-shaped roof of the brain.

Its **sensory areas** (blue) receive impulses from your eyes, ears, tongue, nose, and skin, and give you sensations or feelings.

Its **motor areas** (red) control your muscles during movement.

Its **association areas** (yellow) control memory and thinking.

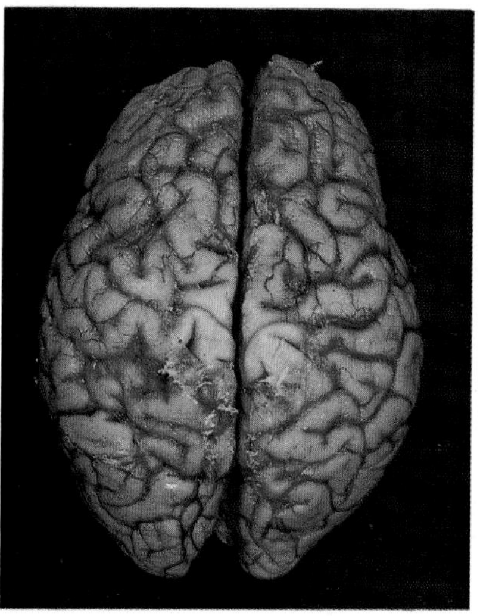

Your brain looks like this from above. The cerebrum is divided into two halves, called the cerebral hemispheres.

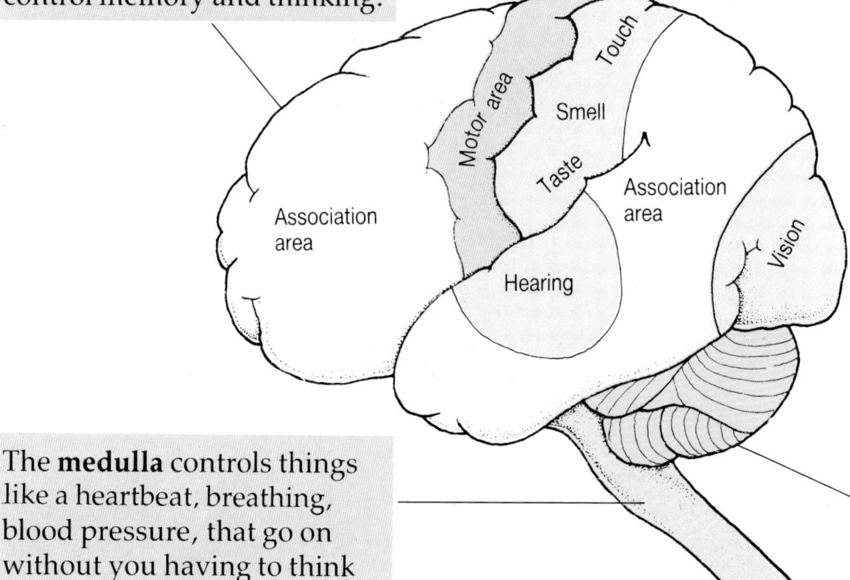

The **medulla** controls things like a heartbeat, breathing, blood pressure, that go on without you having to think about them.

The **cerebellum** helps control your muscles and balance during walking, running, cycling, and so on.

Question

1 Which part of the brain:
 a) do you use to think?
 b) controls your breathing?
 c) helps you balance while cycling?
 d) gives you the sensation of touch?

7.9 Hormones

Hormones are made by a set of glands called the **endocrine system.** Like the nervous system, hormones co-ordinate the body but they do this in different ways.

A nerve impulse travels from the brain or spinal cord along a nerve fibre directly to one particular muscle or gland. Hormones are produced by endocrine glands and released in tiny amounts into the bloodstream, which carries them all over the body. But only certain parts called **target organs** respond to them. Responses to hormones may last a few minutes, or may go on for years.

The diagram below shows the main endocrine glands, the hormones they produce, and summarizes the jobs which the hormones do.

How the endocrine system works

Stimulus
↓
Endocrine organ produces a hormone
↓
Hormone carried all around the body by the blood
↓
Hormone-sensitive tissues respond (e.g. heart, blood vessels, liver, sex organs)

The **thyroid gland** is attached to the windpipe. It makes a hormone called **thyroxine**.

Thyroxine controls the speed of chemical reactions in cells. In children, too little thyroxine leads to slow growth and mental development.

The **adrenal glands** are on top of the kidneys. When you are angry or frightened they make a hormone called **adrenalin**.

Adrenalin prepares your body for action. It speeds up heart-beat and breathing, raises blood pressure, and allows more glucose to go into the blood, to give you energy.

In females the **ovaries** make a sex hormone called **oestrogen**. Oestrogen gives girls their female features such as breasts, soft skin, and a feminine voice. It also gets the womb ready for a baby.

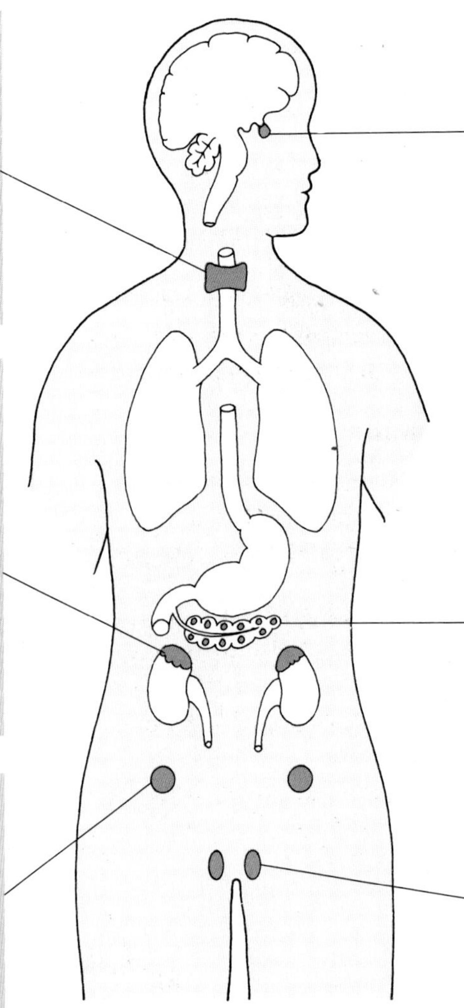

The **pituitary gland** is under the brain. It makes many hormones.

1 One of these hormones controls growth. A person with too little growth hormone becomes a dwarf. A person with too much becomes a giant.

2 In females, it makes hormones which control the release of eggs from ovaries, and the birth of a baby.

The **pancreas** is below the stomach. The pancreas produces a hormone called **insulin**. Insulin controls the amount of glucose in the blood. People with an illness called **diabetes** make too little insulin. So their liver releases harmful amounts of glucose into the blood.

In males the **testes** make a sex hormone called **testosterone**. This gives boys their male features such as deeper voices and more body hair than females.

Diabetes

If the pancreas produces too little insulin, this sets off a complex chain of events in the body. First there is a massive increase in glucose level in the blood because glucose from digested carbohydrate is no longer turned into glycogen (Unit 5.7). Also, without insulin both fat and protein are broken down, yielding more glucose. Fat also breaks down into poisons such as acetone and acetic acid.

Some of the excess glucose is excreted in urine, but this uses large amounts of water, so diabetics develop an insatiable thirst. Their respiration rate also slows down because insulin is needed for cells to absorb glucose. Without insulin treatment, either as injections or tablets, diabetics may become unconscious because of dehydration, acetic acid poisoning, and slow respiration.

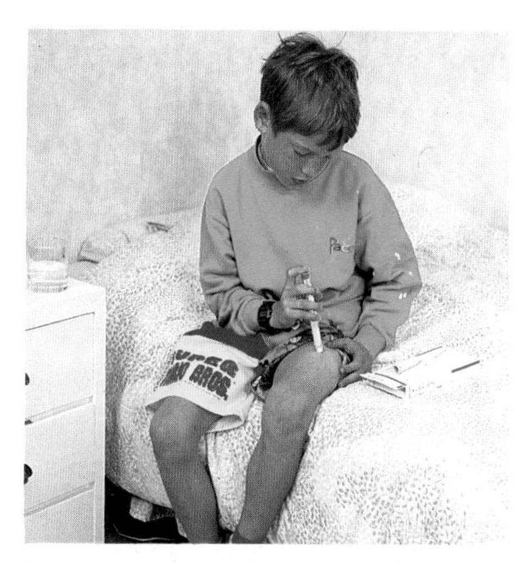

Using a special pen-like dispenser to inject a measured dose of insulin.

Another job for the pituitary gland

Control over the amount of water in the body is a vitally important job for the pituitary gland.

Too much water, and cells swell and may even burst. Too little, and the chemical reactions of life slow down and you may die of dehydration. The diagrams opposite show how the brain, pituitary gland and kidneys stop this happening.

Too little water
in the blood and tissue fluid
(the body becomes dehydrated)

↓

Detected by the **hypothalamus** inside the brain

↓

The hypothalamus makes the pituitary gland produce **anti-diuretic hormone (ADH)**

↓

ADH causes the kidneys to reabsorb **more** water, so **less** urine is produced

Too much water
in the blood and tissue fluid

↓

Detected by the hypothalamus

↓

The hypothalamus stops the pituitary gland producing ADH

↓

The kidneys stop reabsorbing water, so **more**, very dilute, urine is produced

Adrenalin – the fight or flight hormone

In dangerous, frightening, or exciting situations, the effects of adrenalin are almost instantaneous, to prepare you for action:

- Your heart beats faster, pumping more blood and oxygen to your muscles.
- Blood vessels in your muscles enlarge to deliver more blood.
- Lung bronchioles enlarge to let a greater volume of air into your lungs, and breathing rate increases.
- Blood with extra oxygen, and extra glucose released from the liver, flows to your muscles and brain.
- Together, these changes boost the speed and power of your reactions to events.

The familiar symptoms of adrenalin's effects are a dry mouth, a pounding heart, and a 'sinking' feeling in the stomach.

Question

1 Which hormone:
  a) prepares the body for action?
  b) controls the amount of glucose in blood?
  c) gives boys a deep voice?
  d) gives girls soft skin?
  e) controls chemical reactions in cells?

Questions on Section 7

1 *Recall knowledge/understanding*
Three students were asked to sit in a row with their tongues out.
At the same moment crystals of sugar were placed on the front of the first tongue, on the sides of the next tongue, and at the back of the third one.

a) Which student was the first to taste the sugar? Explain why.
b) What will the results be if the test is repeated using salt? Using lemon juice? Using bitter-tasting instant coffee?
c) To taste food and drink properly, you should spread it all over your tongue. Explain why.
d) How do taste and smell protect us from harm?

2 *Recall knowledge*
The diagram below shows the inside of an eye.

a) Name the parts labelled A to L.
b) How does the shape of part J change: when the eye is focused on a near object? when it is focused on a distant object?
c) Describe the main differences between the parts of the eye labelled C, D, and E.
d) What happens to part I: in dim light? in bright light?
e) Describe two functions of part B.
f) What passes from the eye along part F?

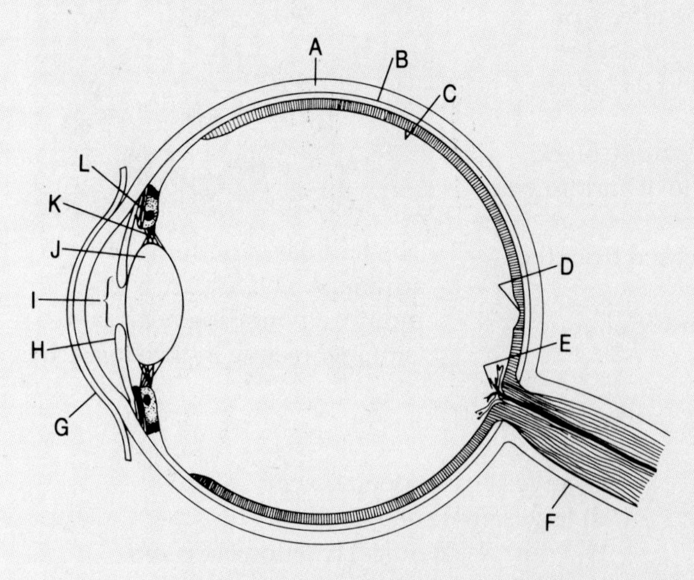

3 *Understanding*

a) What is a reflex action?
b) What reflex action happens:
when dust blows into your eyes?
when a bright light shines in your eyes?
when a very cold wind suddenly blows over you?
when food 'goes the wrong way' and gets into your windpipe?
c) Explain how each of these reflex actions protects you from harm.

4 *Understanding/interpret data*
A group of students were asked to test the sensitivity of skin to touch. They were given pieces of cork through which two needles had been passed. In some corks the needle points were 0.5 cm apart, in others they were 1.5 cm apart.

While one student sat with his eyes closed another touched his skin with either one needle point or two and the student was asked if one or two points were touching his skin.

A record was made only of answers given when *two points* were used.

The students used this method on skin of the arm, palm and fingertips. The table below shows the results of 15 pairs of students. It gives the number of times each student was *correct* in judging that two needle points were touching his skin.

a) What do these results tell you about the difference in sensitivity of the arm, palms and fingertips?
b) What does it tell you about the distance apart of nerve endings in the skin which are sensitive to touch?

Distance apart at needle points	Arm	Palm	Fingertips
1.5 cm	104	155	173
0.5 cm	95	122	158

5 *Recall knowledge*

Name the glands labelled A to F on the diagram opposite, then match each label with one of these descriptions.

a) Its hormones control the speed of cell chemistry.

b) Its hormones change boys into men.

c) Its hormones control the female reproductive system.

d) Its hormones prepare the body for action.

e) Its hormones change girls into women.

f) Its hormones control the amount of glucose in blood.

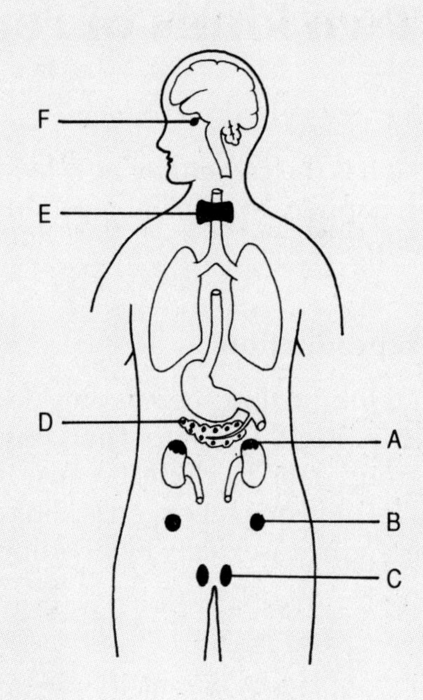

6 *Recall knowledge/interpret data/understanding*

A student with one eye covered used her other eye to look at the following objects within a period of eight seconds.

A – a stationary object nearby
B – a stationary object far away
C – an object moving towards her
D – an object moving away from her

An instrument was used to measure the curvature of her lens (i.e. whether it became fatter or thinner – Unit 7.3). The graph opposite shows changes in lens curvature during the eight seconds of the test.

a) Use the letters A, B, C or D to indicate which object she was looking at during the following periods

0–2 seconds, 2–4 seconds
4–6 seconds, 6–8 seconds

b) Which set of ciliary muscles was contracted to look at object A and to look at object B?

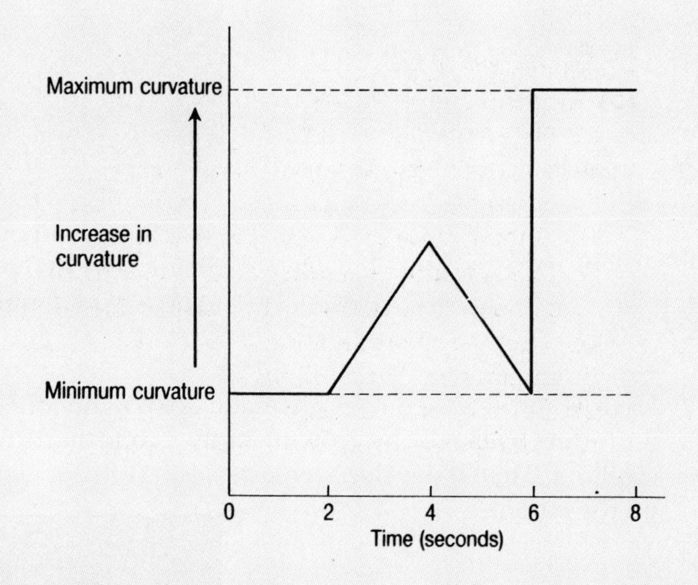

7 *Recall knowledge/understanding*

The diagram opposite shows connections between the ear and the brain.

a) A fire alarm rings. Use letters in the diagram to indicate all the parts, in their correct sequence, involved in hearing this sound.

b) Which letters point to the part(s) that:
are a set of three small bones;
changes sound vibrations into nerve impulses;
interpret nerve impulses as sounds.

c) What part do B, H and G play when you run for safety?

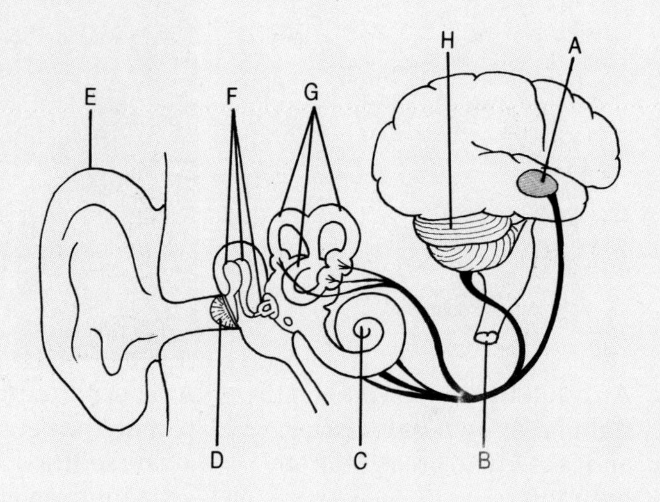

8.1 Two kinds of reproduction

Reproduction is the creation of new living things. There are two ways it can happen: by **asexual reproduction**, or by **sexual reproduction**.

Asexual reproduction

In asexual reproduction, there is only one parent, and its young are exact copies of itself. The parent and its young are a genetically identical group, called a **clone,** because their development is controlled by only one set of genes. Single-celled creatures, like *Amoeba*, reproduce asexually by dividing to form two cells.

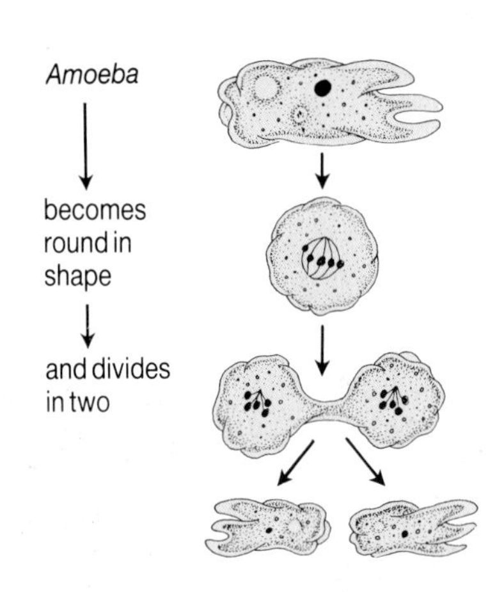

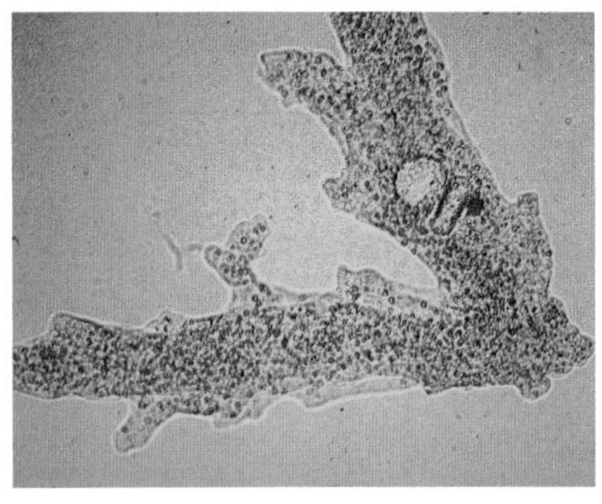

Amoeba takes about an hour to divide. The two parts separate to give two identical organisms

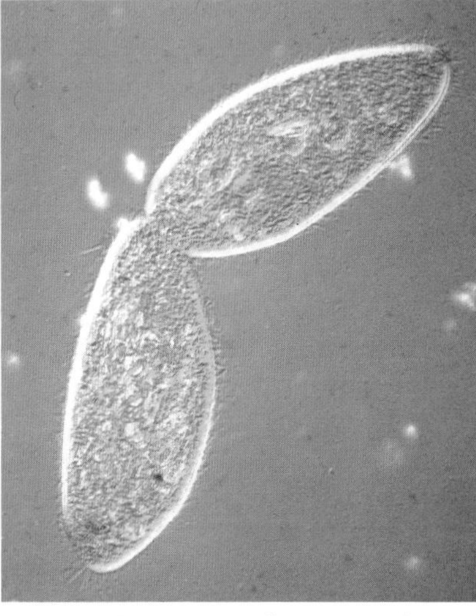

Paramecium reproduces in the same way. Notice the fine hairs or cilia round both parts.

Hydras and sea anemones are made up of many cells. They reproduce asexually by growing buds. A bud starts off as a swelling. Then it develops tentacles, a mouth and a gut, and splits off the parent.

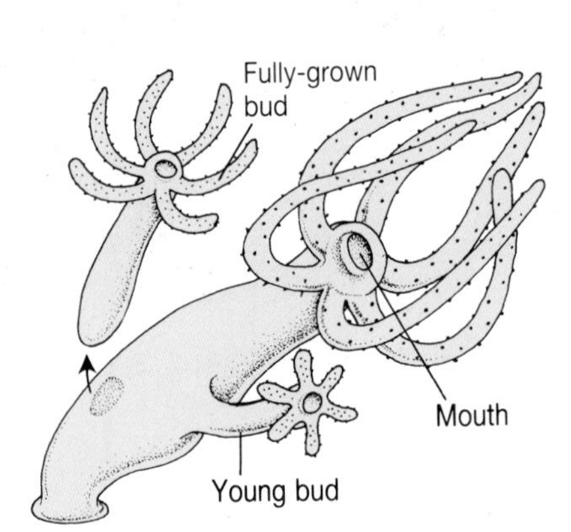

An adult hydra. The bud on the right has grown its tentacles and is ready to break off from the adult.

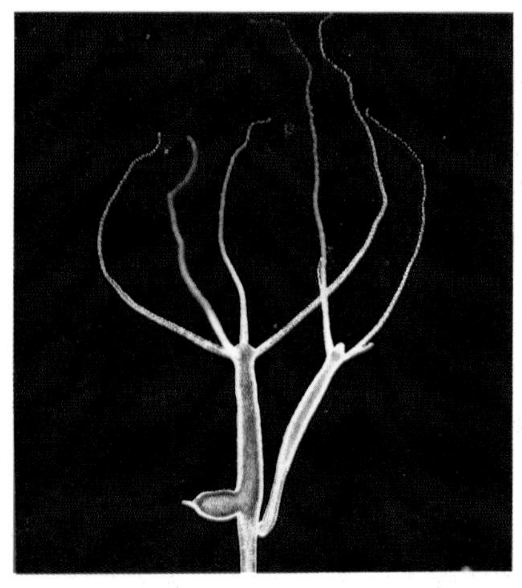

After separating from the parent, the new bud floats away, settles down to feed and grows into an adult.

The beadlet sea anemone grows pink buds inside its body, then squirts them out of its mouth.

Sexual reproduction

In sexual reproduction there are two parents, and each has **sex organs** which produce **sex cells**. In animals, male sex cells are **sperms** produced by sex organs called **testes.** Female sex cells are **eggs** or **ova** (one is an ovum), produced by sex organs called **ovaries**. Young produced by sexual reproduction differ from their parents because their development is controlled by a mixture of genes from both parents. Sticklebacks reproduce sexually.

Male sticklebacks produce sperms.

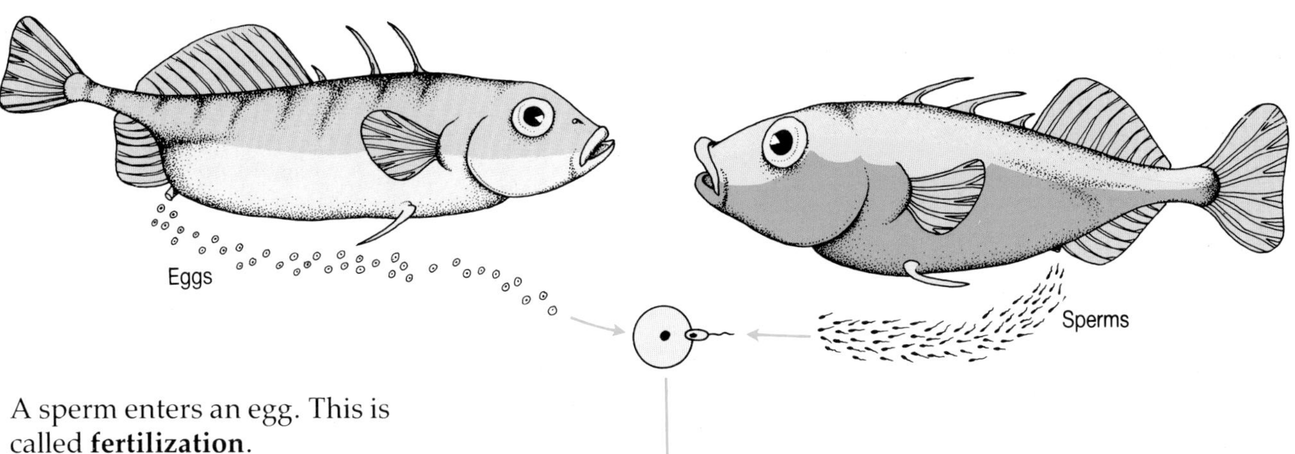

A sperm enters an egg. This is called **fertilization**.

A fertilized egg divides many times to form a ball of cells. The cells develop into a baby stickleback. A partly-formed baby is called an **embryo**.

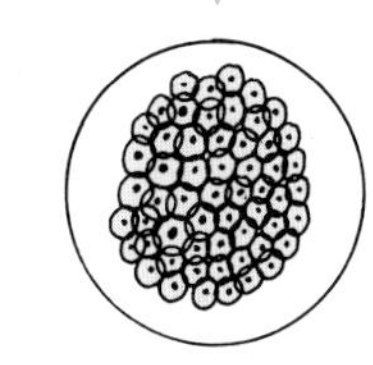

External and internal fertilization

A female stickleback's eggs are fertilized outside her body. This is an example of **external fertilization**.

In insects, reptiles, birds, and mammals (including humans) the male puts his sperm into the female's body. So the eggs are fertilized inside her. This is called **internal fertilization**.

Plants reproduce sexually. You can read about them on page 70.

These stickleback embryos are nine days old.

Questions

1 How is sexual reproduction different from asexual reproduction?

2 a) What are male sex cells and female sex cells called?

b) Name the organ which produces male sex cells.

c) Name the organ which produces female sex cells.

8.2 Human reproduction I

Sexual development in boys

Sometime between 11 and 16, a boy goes through these changes:

1. His voice becomes deeper.
2. Hair starts to grow on his face and body.
3. His muscles develop.
4. His testes start producing **sperms**.

These changes are caused by a hormone called **testosterone** produced by the boy's testes.

The changes that take place in young people during sexual development are called puberty.

Male sex organs

Male sex organs make male sex cells called **sperms**. They pass them into a woman's body during sexual intercourse.

A **testis** makes sperms and the male hormone testosterone. A man has two testes.

The **penis** passes sperms from the man's body into the woman's body during sexual intercourse.

These spaces are called **erectile tissue**. Before intercourse they fill with blood, making the penis stiff and erect.

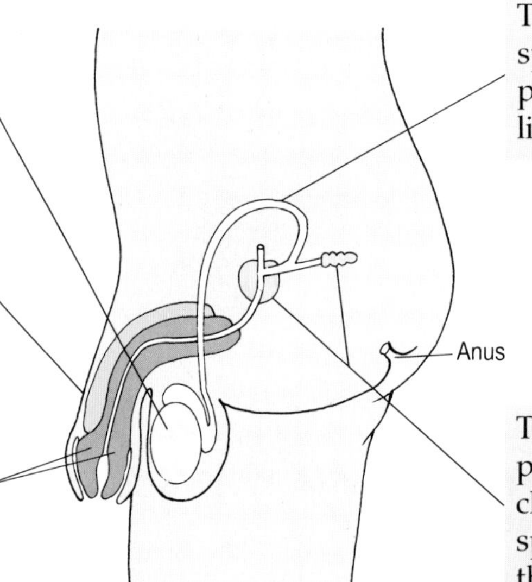

These **sperm tubes** carry sperms from the testes to the penis. Sperms are carried in a liquid called **semen**.

These **glands** make the liquid part of semen. It contains chemicals which cause the sperms to swim after they enter the woman's body.

Sexual development in girls

Sometime between 8 and 15, a girl goes through these changes:

1. Her breasts get bigger.
2. Her hips get more rounded.
3. Hair starts to grow on parts of her body.
4. Her ovaries start releasing eggs (ova).

These changes are caused by a hormone called **oestrogen** produced by the girl's ovaries.

Girls develop sexually earlier than boys do. But of course it varies from person to person.

Female sex organs

Female sex organs make ova, and protect and feed an ovum if it is fertilized and develops into a baby.

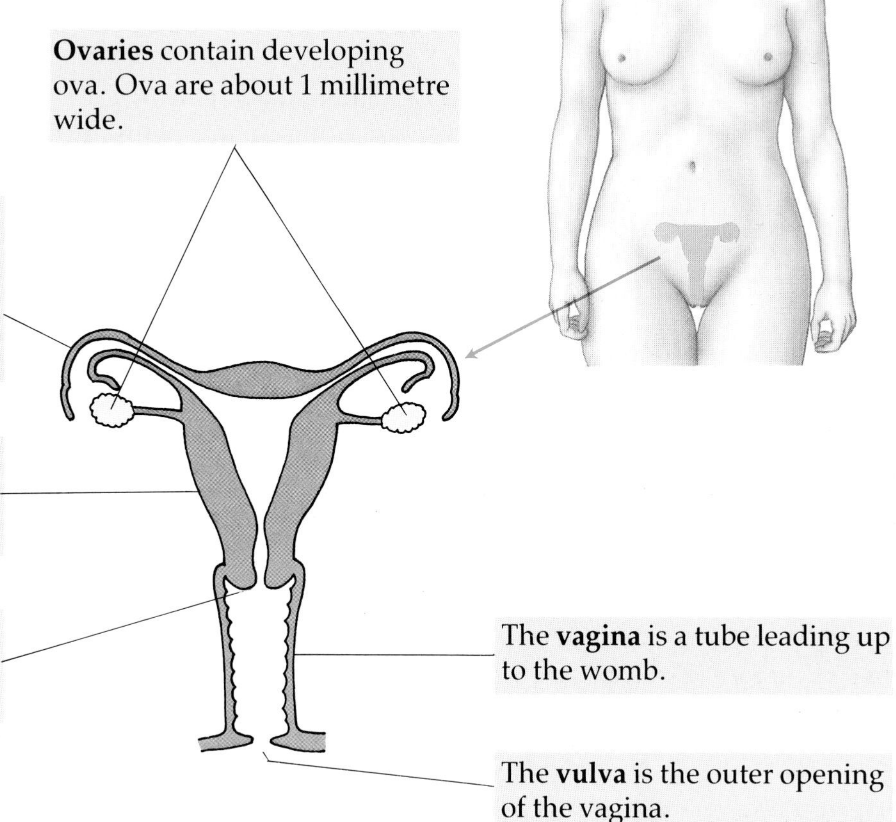

Ovaries contain developing ova. Ova are about 1 millimetre wide.

Fallopian tubes have funnel-shaped openings which catch ova as they come out of the ovaries. Ova move down these tubes to the womb.

The **womb** or **uterus** is a bag in which a fertilized ovum develops into a baby.

The **cervix** is a ring of muscle which closes the lower end of the womb.

The **vagina** is a tube leading up to the womb.

The **vulva** is the outer opening of the vagina.

Sexual intercourse

During sexual intercourse the man's penis becomes stiff and erect. He moves it backwards and forwards inside the woman's vagina. This causes semen to pump from his testes, through his penis, into the woman's body.

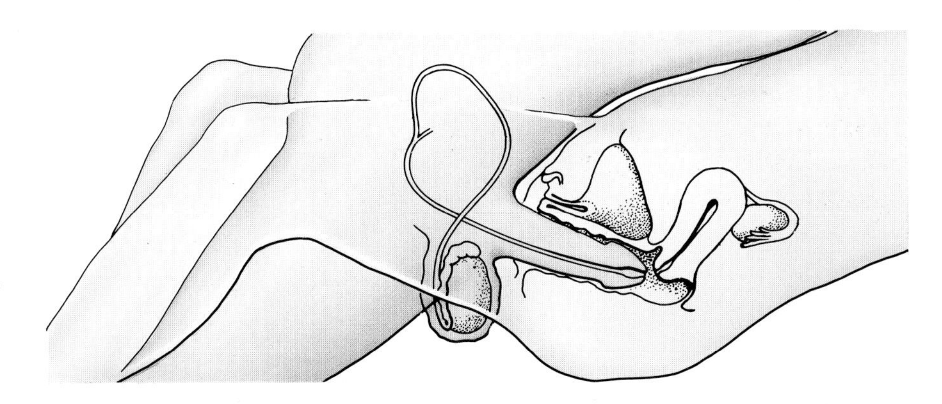

Questions

1 Which part of the body:
   a) produces sperms?
   b) produces ova?
   c) passes sperms from a man to a woman?

2 Name the liquid that contains sperms.

3 How does oestrogen affect a girl's body?

4 How does testosterone affect a boy's body?

5 What are female sex cells called?

6 How do sperms get into a woman's body to fertilize an ovum?

7 Where does a fertilized ovum develop into a baby?

8.3 Human reproduction II

The start of a new life

Human reproduction begins with sexual intercourse. During sexual intercourse millions of sperms pass from a man's penis into a woman's vagina. One of them may fertilize an ovum. This diagram shows how it all happens.

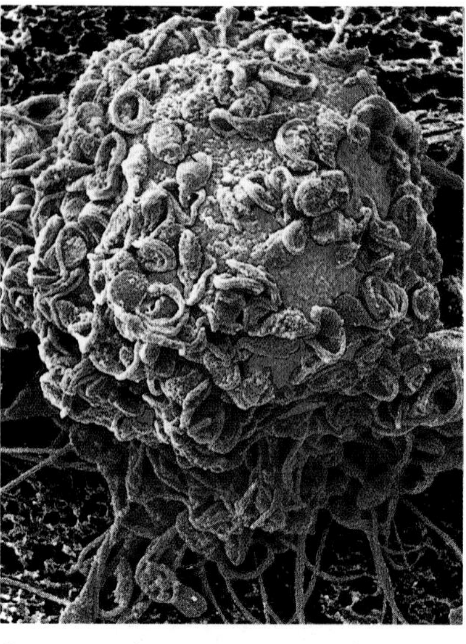

Sperms clustering round an ovum. Only one sperm will succeed in getting through the wall of the egg, to fertilize it.

2 The follicle bursts, squirting its ovum out of the ovary. This is called **ovulation.** The ovum is sucked into a Fallopian tube.

3 Sperms in the Fallopian tube are attracted to the ovum. One sperm burrows into the ovum. The sperm nucleus and the ovum nucleus join. This is called **fertilization**. A skin forms round the ovum to keep out the other sperms.

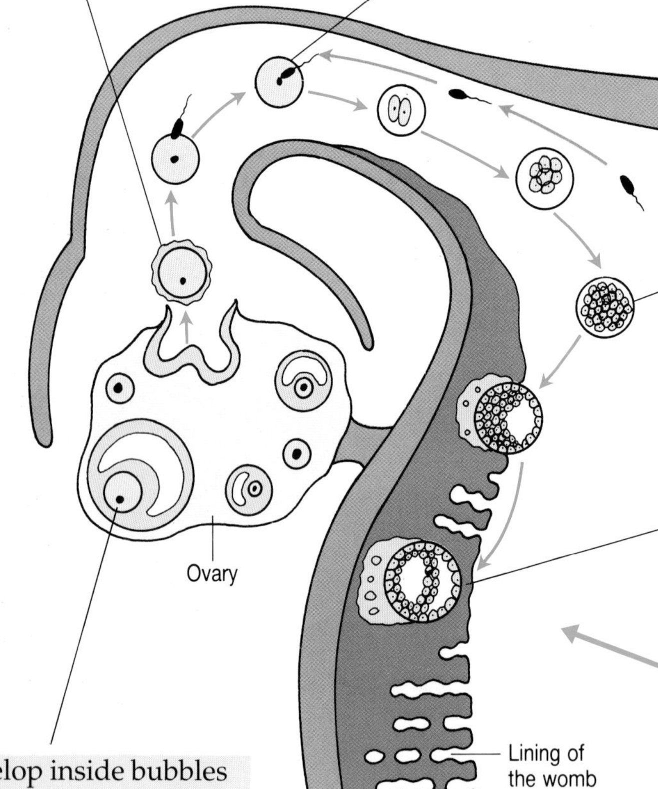

4 The fertilized ovum divides to make a ball of about a hundred cells. This is called an **embryo**. The embryo moves down the Fallopian tube to the womb.

5 The embryo sinks into the thick lining of the womb. This is called **implantation**. The embryo gets food and oxygen from the blood vessels in the lining of the womb. This allows it to grow into a baby.

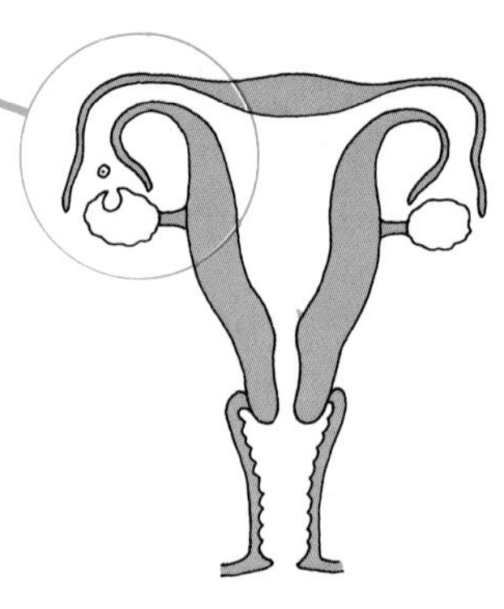

1 **Ova** develop inside bubbles called **follicles** inside an ovary. The follicle grows until it bulges from the ovary.

If an ovum is *not* fertilized after ovulation, it dies. But another ovum is released about a month later, as part of the **menstrual cycle**.

The menstrual cycle

Every month changes take place in a woman's body. These changes are called the **menstrual cycle**, or **monthly period**. They usually begin when a girl is between 8 and 15 years old.

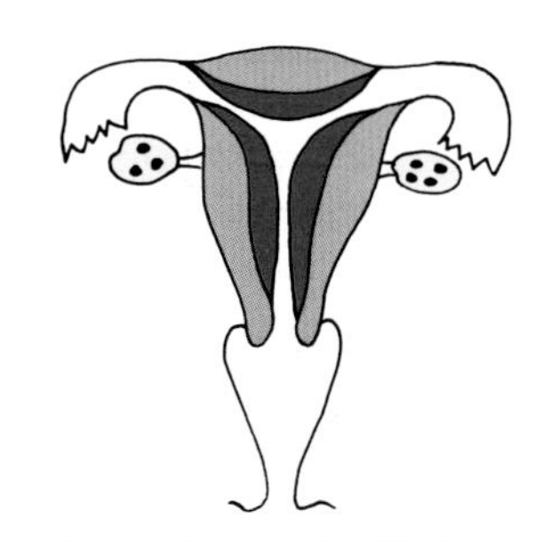

2 During the week after ovulation, the part of the follicle left behind in the ovary produces the hormone **progesterone.** This makes the womb grow a thick lining of glands and blood vessels, and stimulates the breasts to begin developing glands which make milk. The womb is now ready to protect and feed a fertilized ovum.

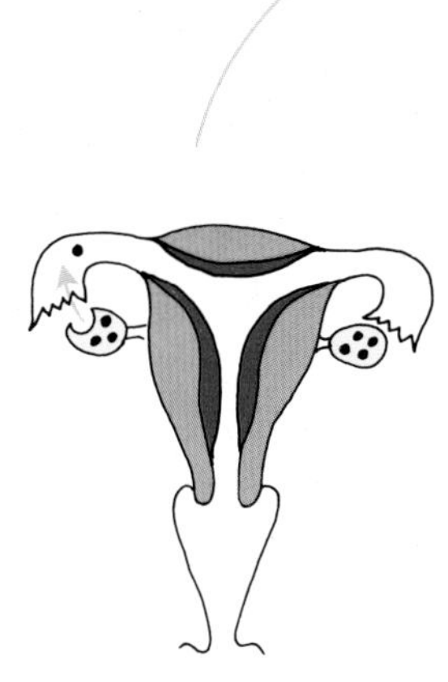

1 Ovulation produces a new ovum about every 28 days. The ovum is sucked into a Fallopian tube. The woman is now **fertile**. This means that, if she has sexual intercourse at this time, the ovum could be fertilized and she could become **pregnant.** But if the ovum is not fertilized, it dies within a few days.

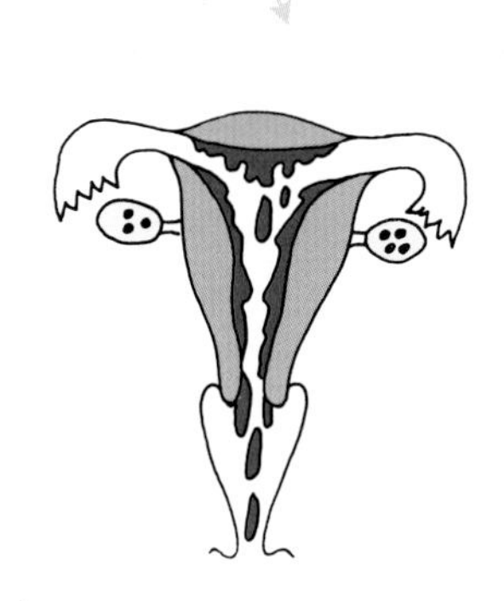

3 During the 14 days after ovulation, if the woman is not pregnant, progesterone production slows, then stops. This causes the thick lining of the womb to break down. This is called **menstruation**, or a **period**. When bleeding stops, the pituitary gland releases **follicle-stimulating hormone** which starts a new cycle by making the ovary develop a new follicle. The follicle produces the hormone **oestrogen** which starts the womb growing a new thick lining.

When a woman is between 45 and 55 years old, her periods stop. This change is called the **menopause**, or **change of life**.

Questions

1 What is ovulation? How often does it happen?

2 Where in the body does an ovum get fertilized?

3 How do sperms reach the ovum?

4 How does a fertilized ovum become an embryo?

5 Explain what implantation means.

6 During the menstrual cycle:

a) why does the womb grow a thick lining?

b) what happens when this lining breaks down?

8.4 Life before birth

A baby feeds, breathes, and gets rid of waste inside a mother's body. The mother's blood brings it food and oxygen and carries the wastes away. But the mother's blood does not mix with the baby's blood. Food, oxygen, and wastes are exchanged in the **placenta**.

The placenta is shaped like a pancake, and is attached to the wall of the womb. There are blood vessels and spaces inside it. The baby's blood flows into the blood vessels. The mother's blood flows into the spaces. Substances are exchanged through the baby's blood vessel walls.

It is important for a pregnant woman to eat healthy food, to give the baby a good start in life.

1 The baby's blood carries carbon dioxide and other wastes to the placenta, along an **artery** in the **umbilical cord**.

2 The mother's blood flows into the spaces in the **placenta**, carrying food and oxygen.

3 The food and oxygen pass into the baby's blood through the thin walls of the blood vessels. At the same time the **wastes** pass out of the baby's blood.

4 The mother's blood carries the wastes away.

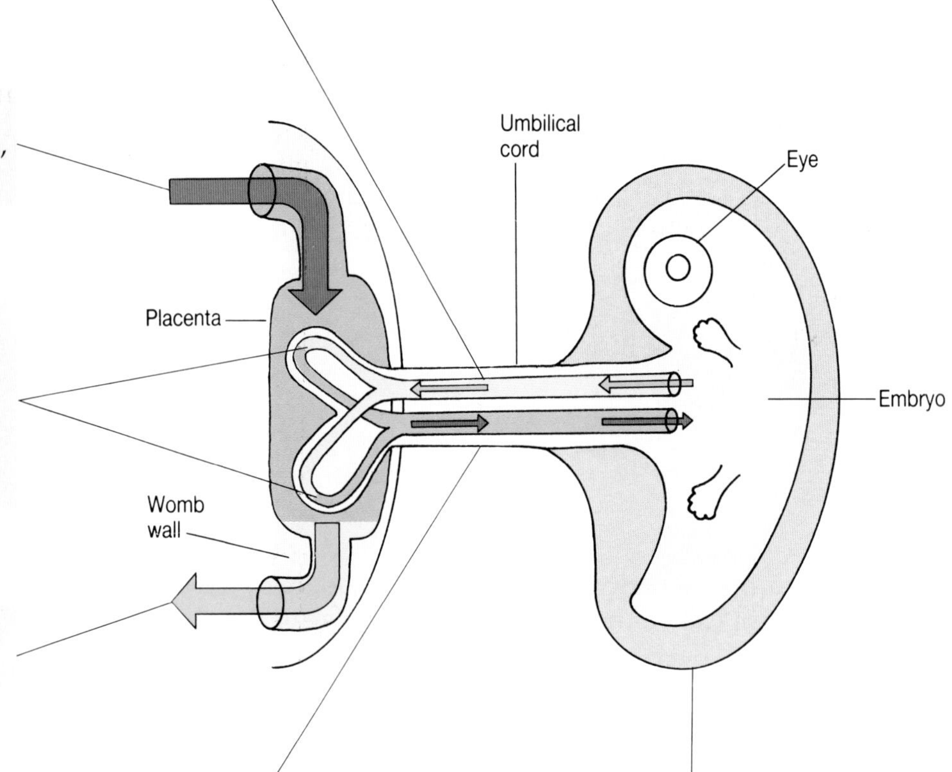

5 The baby's blood carries food and oxygen back to the baby along a **vein** in the **umbilical cord**.

The baby is in a bag of liquid called the **amnion**. This cushions it against knocks and bumps as the mother moves about.

How the baby develops

A fertilized ovum is smaller than a full-stop. In nine months, it will have grown into a baby weighing around three kilograms, and measuring about half a metre.

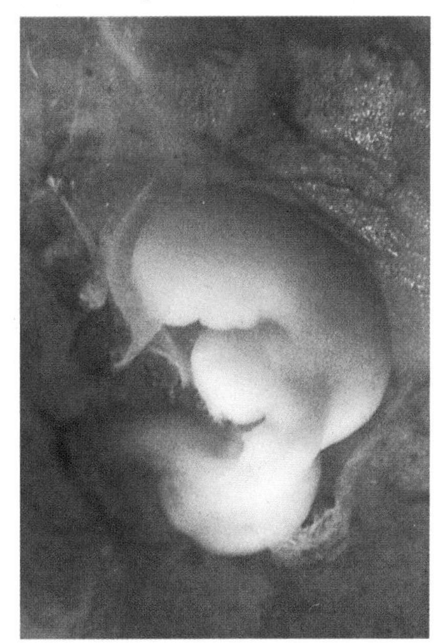

At 4–5 weeks

The embryo is about half a centimetre long. It has a bulge which will be its head, containing a developing brain, eyes, and ears. It has a tail, and lumps which will turn into arms and legs. A small heart pumps blood along an umbilical cord to the placenta.

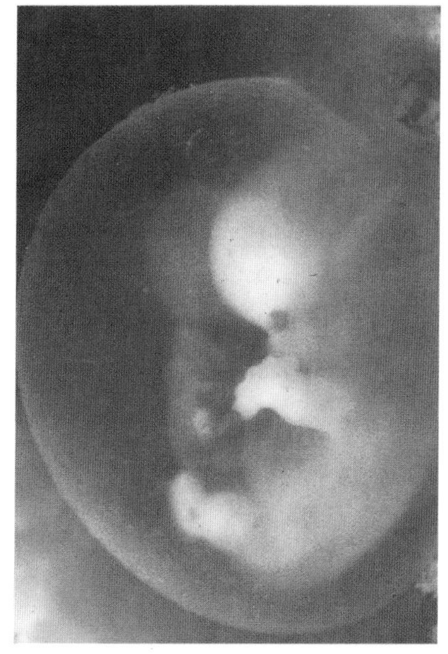

At 8–10 weeks

It is now 4 centimetres long. It has a face, and limbs with fingers and toes. Most of its organs are fully developed. From now until birth it is called a **foetus**.

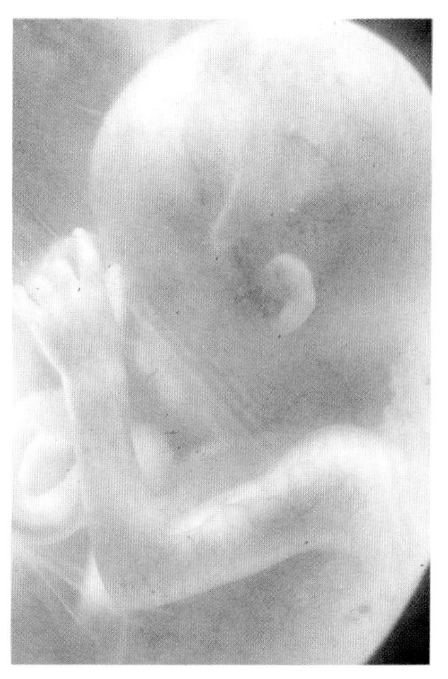

At 16 weeks

It is about 16 centimetres long. It is turning into a boy or a girl. The mother will soon feel it moving inside her.

Use the drawing on the right to name the parts of the baby in each photograph.

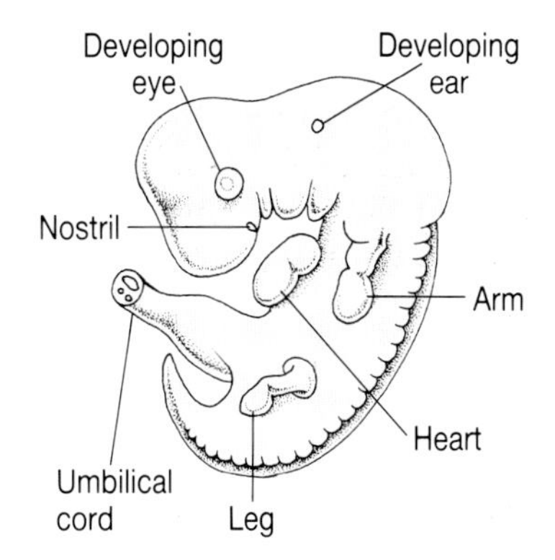

Questions

1 How does a baby get food inside the womb?

2 How does a baby get oxygen inside the womb?

3 How does a baby get rid of its wastes while inside the womb?

4 Think of a reason why the mother's blood is not allowed to mix with the baby's blood.

5 The umbilical cord has two tubes inside it. What are these tubes? What do they do?

6 What is the amnion? How does it help the baby?

7 What is a foetus?

8.5 Birth, and birth control

During pregnancy, the wall of the womb develops thick, strong muscles ready for the baby's birth.

Birth begins with **labour**. The womb muscles start to contract. After some time the contractions push the baby head-first out of the womb.

At first the womb muscles contract about every half hour. Then the contractions get faster and stronger. They push the baby's head through the cervix into the vagina.

The contractions become very strong. The baby moves to a face-down position and is pushed out of the mother's body. It breathes air for the first time.

A few minutes later more contractions push the placenta and umbilical cord out of the womb. These are called the **afterbirth.**

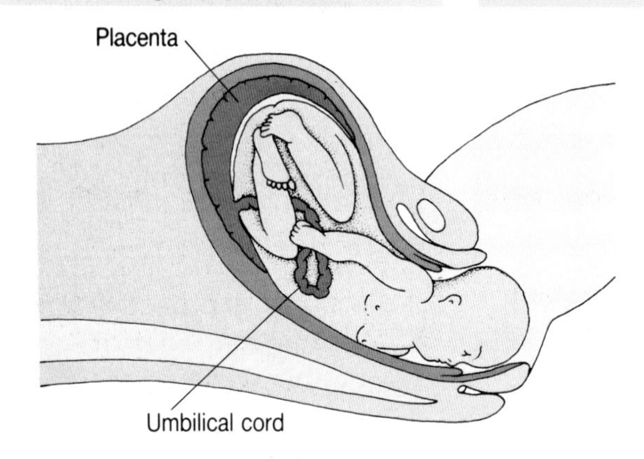

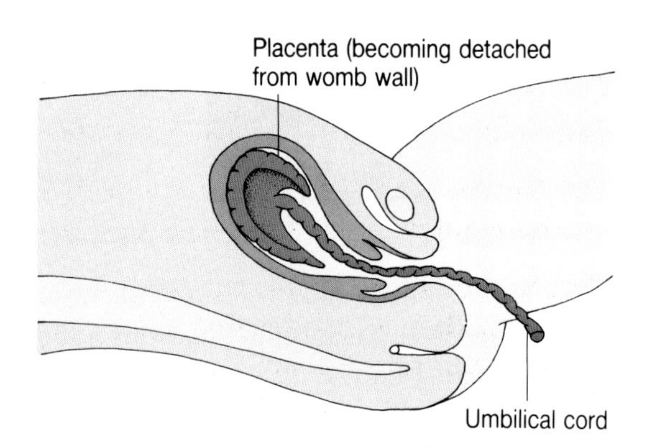

Health of the mother

When a women is pregnant, she must take care of her health.

German measles (rubella) If a mother gets rubella during the first 12 weeks of pregnancy the germs can damage her baby, causing deafness, blindness, and heart disease. Pregnant women should avoid people who have rubella. Young girls should be innoculated against the disease.

Alcohol If a pregnant woman drinks alcohol, it can slow her baby's development, damage its brain, and cause it to be born early.

Smoking Pregnant women who smoke have smaller babies than non-smokers. Their babies are more likely to be born dead.

Health of the baby

When the baby is born it is put on the mother's breast or stomach immediately. This helps love develop between mother and child. Mother's milk is the perfect food for babies. It is free, always ready, and pure. Mother's milk protects babies from germs. If a mother is unable to breast-feed, special powdered cow's milk may be used.

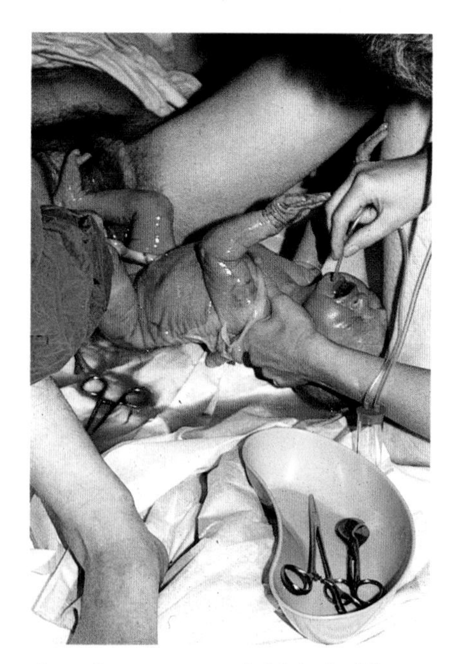

Another successful birth. The scissors is for cutting the umbilical cord.

Birth control

Birth control, or **contraception**, lets couples decide how many children they want. These are some of the methods used.

The condom, or sheath, is a thin rubber cover for the penis. It is put on before intercourse. Sperms get trapped inside it.

Condoms are quite reliable. But they are even safer if the woman uses a **spermicide**. This is a chemical which kills sperms. The woman puts a spermicide cream or jelly into her vagina before intercourse.

The diaphragm, or cap, is a circle of rubber with a metal spring round it. Before intercourse, the woman smears it with spermicide. Then she puts it into her vagina, to cover the cervix. It stops sperms getting into the womb. For safety, it is left in place for at least eight hours.

Contraceptive pills, or 'the Pill', contain female hormones. They do three things:

1. They prevent the ovaries from releasing ova.
2. They stop an ovum moving along a Fallopian tube.
3. They fill the entrance to the womb with a sticky substance so that sperms cannot swim through.

Contraceptive pills are very reliable. But they should not be used by women with liver disease or diabetes. Also some doctors think that women who smoke may develop blood clots, migraine headaches, or heart conditions, if they take the Pill. Some may also develop breast cancer.

The rhythm method Some people feel it is wrong to use any kind of birth control except the **rhythm method**. In this method the couple avoids sexual intercourse when there is an ovum ready to be fertilized. Knowing when the ovum can be fertilized is difficult to work out. This method is not very reliable.

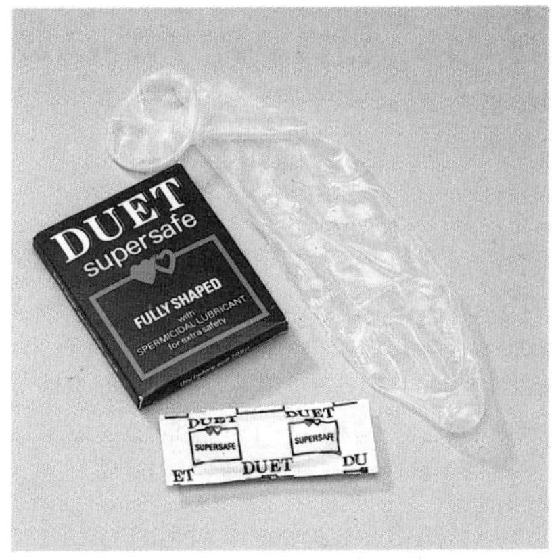

As well as being a reliable contraceptive, condoms also protect against infection.

Some women choose the diaphragm or cap as a contraceptive.

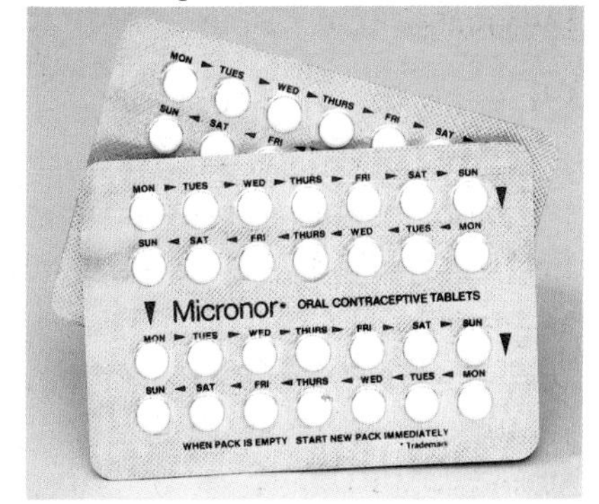

It is essential to talk to the doctor before using the Pill.

Questions

1. What causes a baby to be pushed out of the womb?
2. Which part of the baby normally comes out first? What is the afterbirth?
3. Why should young girls be innoculated against rubella?
4. Why should pregnant women avoid alcohol and smoking?
5. How does a condom prevent pregnancy?
6. How does a spermicide work?
7. Which women might find contraceptive pills harmful?

8.6 Embryos, babies, and children

Research involving human and animal embryos has aroused strong argument and extreme emotions. This Unit helps you understand why this is happening, so that you can form your own opinions.

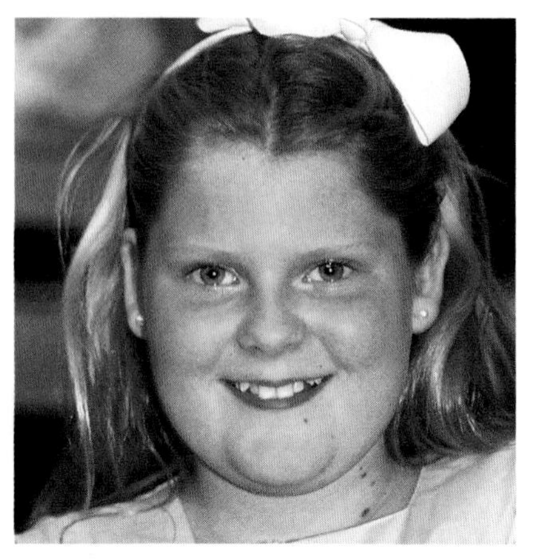

Louise Brown, the world's first test-tube baby, on her tenth birthday .

Human embryo research

It is now possible to take ova from a woman and fertilize them to produce embryos needed for the **test-tube baby** technique, described below. This enables a woman who cannot have a baby by normal means to have a growing embryo implanted in her womb.

Some ethical problems to think about Is it wrong to take embryos produced by one woman and implant them into another – called a **surrogate mother**? Think of arguments for and against women who are past normal child-bearing age using this technique to become pregnant. Since these embryos are partly developed human beings, it is unethical to throw them away if they are not wanted for implantation? Is it wrong to use human embryos for medical research even though this may improve test-tube baby techniques and help prevent disease? Should embryos have the same legal rights as human beings?

How test-tube babies are produced Human embryos can be obtained by giving a woman doses of a fertility drug containing **follicle-stimulating hormone** to stimulate her ovaries to produce many follicles and ova. The ova are sucked from her ovaries with a special syringe inserted through her abdomen, and placed in a glass dish, with sperms, for about 60 hours. Sperms fertilize the ova, which develop into embryos (see diagram opposite). Embryos do not grow into babies inside test-tubes as the name of this technique suggests. Embryos are placed inside the womb of a woman, where there is a 10 per cent chance they will develop in the normal way.

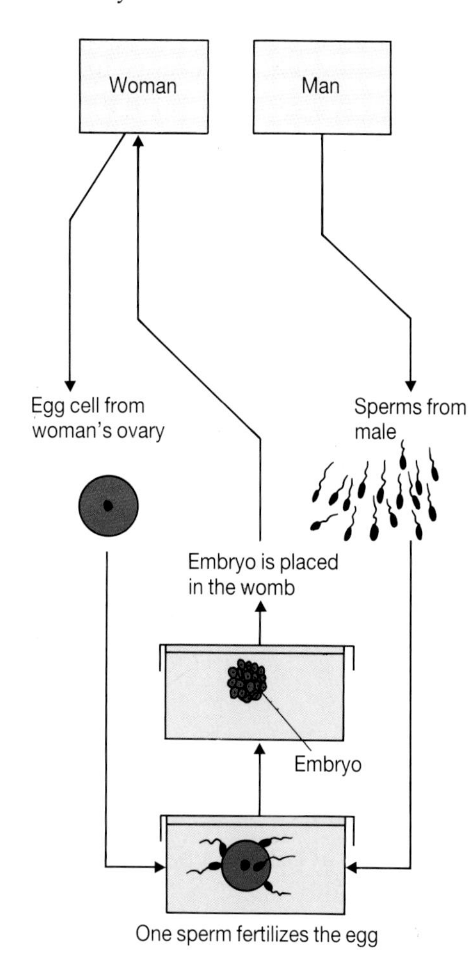

Preventing genetic diseases About 7000 babies with genetic diseases, such as Down's syndrome (Unit 2.8) and muscular dystrophy, are born each year in the UK. It is now possible to take one cell from an embryo, without damaging the rest, and study it for certain genetic defects. If these are absent, the rest of the embryo can be implanted in the mother and allowed to develop normally. If a disease is identified, parents must decide if they want the embryo destroyed. Can you think of arguments for and against this research?

Animal embryo research

Techniques using animal embryos are used extensively in many branches of agriculture, and also wildlife conservation.

Clones The cells of an embryo up to six days old can be separated into single cells which then reform into whole embryos. When implanted into a female, they grow into genetically identical animals, called **clones**. This method is used to produce many identical copies of prize bulls, cows, horses, and other valuable animals. Cloning can also be used to increase populations of rare wild animals. The female receiving an embryo need not be the same species as the donor: a horse, for example, can give birth to a zebra. So cloned embryos of endangered species can be implanted into domestic animals for development under safe, controlled conditions. Later the young can be released back into the wild.

This animal was produced by mixing embryo cells from *two different species*: a sheep and a goat. The cells of a four-cell sheep embryo were mixed with cells from an eight-cell goat embryo. They developed into a **chimera**, an animal which is part sheep and part goat.

Growth and development

The way children grow and develop depends on the genes they inherit, their health, diet, and the environment in which they live.

Puberty, the changes which transform a child into a sexually mature adult, can occur earlier in girls than boys. Between 8 and 15 years, female sex hormones called **oestrogens** cause the development of a girl's reproductive organs, and her periods begin. At the same time, her **secondary sexual characteristics** appear, such as wide, rounded hips, and breasts. The testes of boys aged 10 to 14 years produce sperms for the first time and release the hormone **testosterone**. This stimulates the appearance of characteristics including deepening of the voice and growth of body and facial hair.

Growth Rapid growth in height occurs at puberty, followed by a 'filling out' of the body. Girls usually stop growing by the age of 16. Boys start later and often continue to grow after 18.

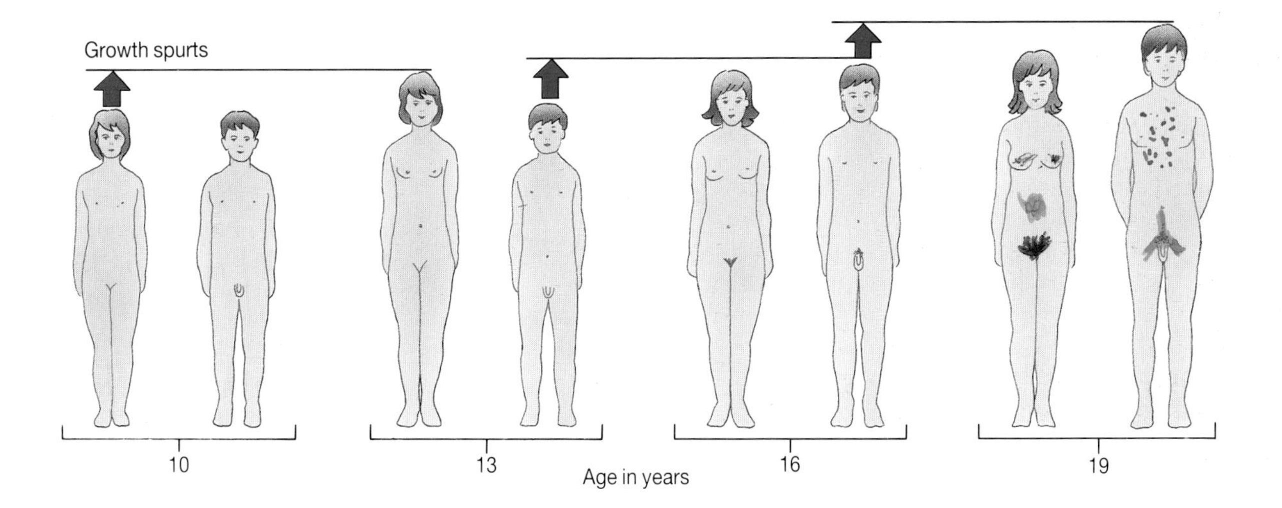

Questions

1 At least a million women in Britain are infertile. How has embryo research helped them have children?

2 Think of arguments for and against producing clones of humans.

3 What differences are there between the growth rates of girls and boys?

8.7 Human populations

Every day, the number of people in the world increases by 200 000. So every second there are two more mouths to feed.

The population explosion means that the world is getting crowded.

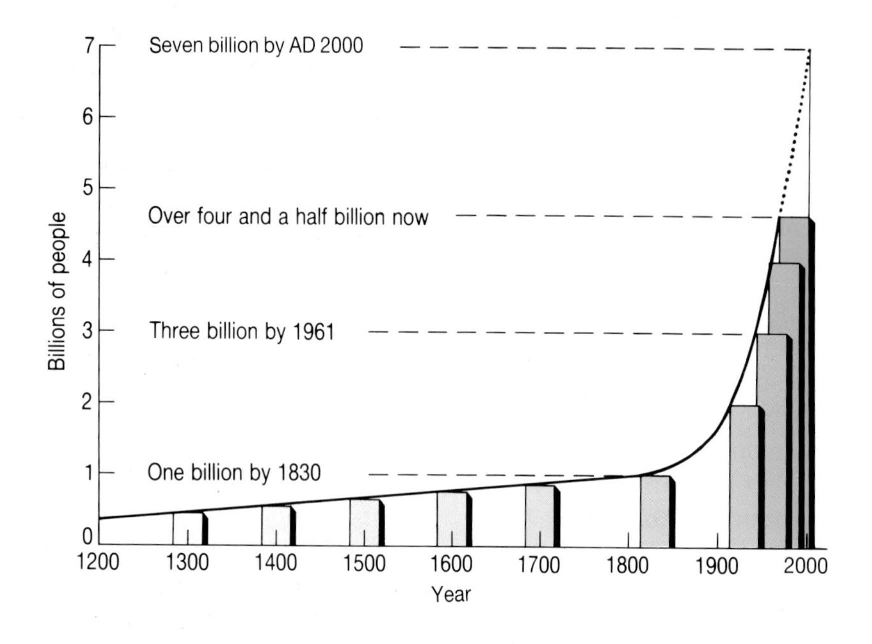

Why is the world's population increasing?

Until about 1800, world population was rising slowly. Then a population explosion started. Here are some of the reasons:

- Better medicines and health care help fight disease.
- Clean water supplies and better sewage disposal slow the spread of diseases.
- Modern farming methods in some countries produce more food.

This means fewer babies die, and adults live longer. The result is population growth at a rate which is causing the most severe problems for the human race. Failure to solve them will produce world-wide social, economic, and environmental catastrophe.

Health care improvements help control disease in poor countries and reduce the death-rate.

Compare birth-rates and death-rates in the table opposite. Population growth is much faster in poor, developing countries than rich, developed ones. This is because a fall in the death-rate of poor countries has not been followed by a fall in birth-rate. Consequently the problems of population growth are very different in poor and rich nations.

Area	**Pop. 1985** (in millions)	***Birth-rate**	***Death-rate**	**% incr. yearly**	**Doubling time**	**Pop. 2000** (in millions)
Africa	551	45	16	+2.9	24 yrs	869
Asia	2829	28	10	+1.8	39 yrs	3562
Europe	492	13	10	+0.3	240 yrs	509
Latin America	406	31	8	+2.3	30 yrs	554
North America	264	15	8	+0.7	99 yrs	297
World	4845	27	11	+1.7	41 yrs	6135

*Births and deaths per 1000 of the population per year

Population growth in poor countries

Decreasing death-rate in poor countries is due mainly to imported drugs and improved health care against diseases such as malaria and yellow fever. But birth-rates are still high for many reasons:

- Children are needed to support the family and most are working by at least 10 years of age.
- Poor countries cannot afford to buy machines, so most work is done by hand. Therefore a large labour force is essential.
- Death-rates are still high enough to mean that a poor family must have 10 children to be sure of having a son who will live to 40 years.
- In some societies, even if a woman would like to have fewer children, her low social status, lack of education and shortage of contraceptives make this difficult to achieve.

In poor countries, many people are needed to run farms, as most jobs are done by hand.

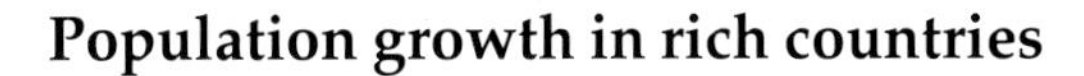

Population growth in rich countries

Population growth in rich countries is slowing, and may soon stop. The main reason is that machines have replaced people in industry and farming so a large work-force is not needed. Another reason is the high cost of supporting children.

In rich countries, there is a machine for almost every job on the farm.

The problems of population growth

Reduced birth- and death-rates in rich countries result in a population made up of more old people than young. Fewer people earn money and pay taxes, and more want pensions and extra health care: this can cause economic and social problems.

Rapid population growth in poor countries makes the task of maintaining food supplies, clean water, electricity, and transport, impossible to achieve. One solution is the use of contraceptives, though many people have religious objections. Contraceptives will not become widespread until better living conditions make large families unnecessary, and education standards are improved.

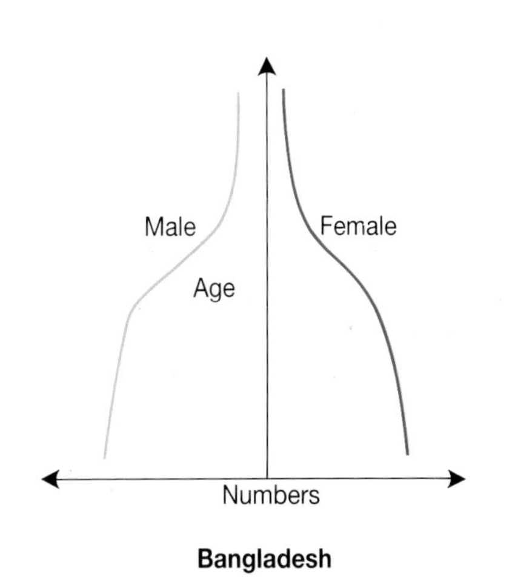

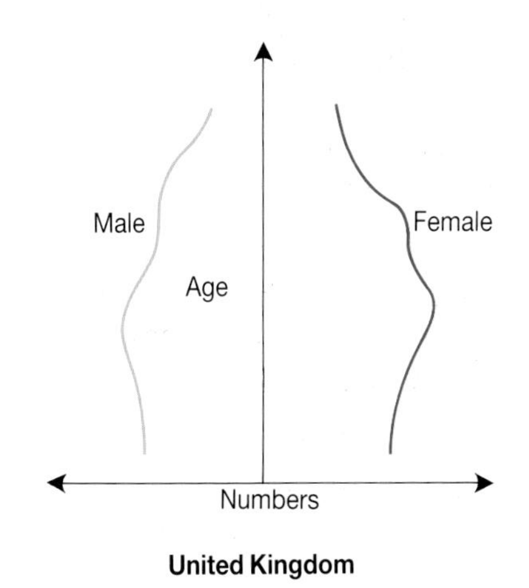

Questions

1 Study the graphs above. How do the proportions of young to old people in Bangladesh and the UK differ? What are the reasons for this difference?

2 The population in some African countries could double in the next 25 years. Describe the problems this will cause, other than those mentioned in this Unit.

8.8 Feeding the world's billions

Why do millions starve?

Right now, there is enough food in the world for everyone. Many scientists believe that the world is capable of producing enough food to satisfy our needs at least until the year 2000. So why do millions of people starve?

1 Unequal sharing People in the rich industrialized countries can afford to build up huge stores of food, and eat far more than they need. The average European gets through enough food and other materials for 40 people in a poor country.

As living standards rise even further in rich countries, the demand for food will rise and the 'food gap' between rich and poor will widen. Because of this, the number of underfed people in poor nations could rise almost 1.5 billion by the year 2000.

This drawing shows the amount of food eaten in rich and poor countries each year.

= 25 kg of food

2 Avoidable waste of food Up to 50 per cent of food is wasted between the field and the consumer. Poor harvesting causes grain to be spilled and broken open and fruit to be bruised. Damp food goes mouldy during storage, and badly stored food can be ruined by insects and rats. Solving these problems would be expensive, but worthwhile.

3 Growing the wrong crops Poor countries often use their best land for **cash crops** such as tobacco, tea, coffee, and other crops they can sell to rich countries. This provides much-needed income but prevents them growing enough food to feed their people. So food is imported. If the price paid for their cash crops falls, people can no longer afford food imports and have to go hungry.

Some countries, including several in South-East Asia, export food, although some of their own people are underfed.

Land is used to grow crops like coffee for sale instead of for feeding the local people.

4 Creating deserts As more land is used for cash crops, farmers are forced on to less fertile land or must cut down forests. When tropical forests are cleared to make space for crops , or for timber and firewood, their thin soils are exposed to the tropical Sun. Unless carefully tended, the soil is soon washed away by rainstorms, leaving a hard infertile subsoil. This is not only useless to farmers but cannot be re-colonized by forest plants. So, deforestation leads to desert formation.

A tropical forest being cut down in Brazil, to make way for crops and cattle ranches. The exposed topsoil will soon get worn away by wind and rain.

5 Over-grazing If herds of cattle are grazed for long periods on one area, they eat most of the plants and expose bare soil. Their feet compress the soil and it is baked by the Sun into a **hardpan**, which does not allow water to soak in. During rainstorms water pours off hillsides, stripping away surface layers and cutting deep channels.

Both forest clearance and over-grazing expose soil, which is washed away during rainstorms. Channels (gullies) may then be eroded.

Action needed

1 More of the best farmland should be used for crops. Forests must be preserved and managed properly so they provide wood, food from plants and animals, medicines and many other valuable products, without destroying them for ever.

2 Rich countries can help, but not just with food and money. They can help to improve education, farming methods, and irrigation.

3 Wild animals, such as the hippopotamus and oryx, shown right, can be a source of meat. Africa has at least 30 wild hoofed animals which, after millions of years of evolution, can live off wild plants and resist tropical diseases. Some can even live in deserts. Wildlife could become an endless supply of food and other materials, *but only if managed with care* so that population numbers are maintained.

Questions

1 It takes 30 times more land to produce 1 kg of animal protein as it does to produce the same amount of plant protein. Use this fact, and other facts from this Unit, to produce arguments against clearing wild habitats for pasture.

2 What is meant by the 'food gap' between rich and poor nations? What could be done to stop it widening?

3 In 1991 Africa lost $5.6 billion when cash crop prices fell. Why did this increase hunger?

8.9 Wild populations

Growth of wild populations

Wild populations, such as frogs in a pond or starfish in a seaside rock pool, cannot protect themselves from bad weather, predators, diseases and many other factors which slow, or stop, their population growth. As a result, the growth rates of human and wild populations can be quite different.

If a healthy population of wild animals enters an unoccupied area where there is plenty of food, space, shelter, and no predators, its numbers will increase rapidly. This happens because the extra food makes them more fertile and so birth-rate exceeds death-rate.

But this cannot continue for long. Available space fills up, competition for food begins, predators soon show up, and many other factors which limit population growth appear. These reduce birth-rate and increase death-rate. This produces changes in population numbers which, if plotted on a graph, give a **growth curve**, like a flattened letter 'S'. You can see this above right.

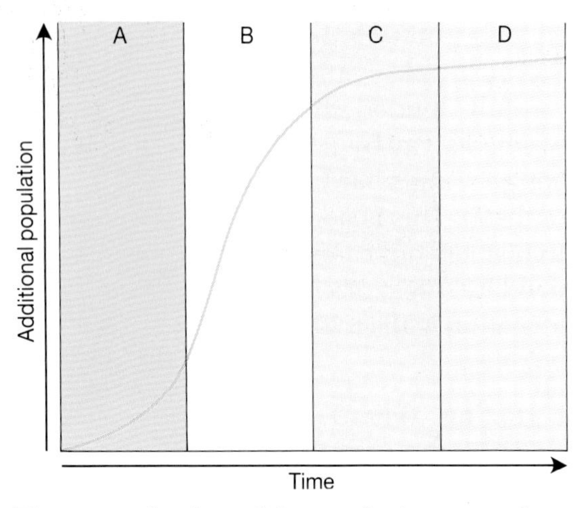

The growth of a wild population goes through four distinct phases: **A** – a slow start when there are few fertile members; **B** – maximum growth when birth-rate greatly exceeds death-rate; **C** – decelerating growth as food shortages and other problems appear; and **D** – fairly constant population numbers when birth- and death-rates are about equal.

Factors, such as food shortages, which limit population growth are called **environmental resistance.** As these build up, birth-rate decreases and death-rate increases until they are equal. Then population growth stops and numbers stay fairly constant. At this point the population has reached the maximum number that the area can support. This is called the **carrying capacity** of that area, for that particular species.

Factors which limit population growth

Shortage of food and other resources Wild environments have a limited supply of resources including food, nesting places, and shelter from the weather and predators. As numbers increase, competition for resources increases, and energy, at first available for reproduction, must be spent in the struggle for survival. Female fertility is reduced, which means they produce fewer eggs or babies, and many young die before maturity because there is not enough food for them. In addition, adult animals die younger as the strain of survival increases. Overcrowding may also result in competition between different species which share the same resources. These factors cause birth-rate to fall and death-rate to rise.

Certain insects, fish, reptiles, and birds such as robins take over and defend a breeding territory against intruders of the same species. The territory includes enough food to support a brood to maturity. But, as population increases, territories shrink and broods get smaller.

Climate Weather conditions and the change from summer to winter in temperate regions, and from dry to wet seasons in the tropics, can have dramatic effects on wild populations. Numbers fall when harsh weather conditions halt reproduction and increase death-rate.

Build-up of wastes In an enclosed environment such as a small pond, population growth can lead to increasing amounts of carbon dioxide and other, often toxic wastes. These limit further growth of some populations.

Diseases, parasites, and predators Overcrowding quickly leads to the build-up of disease organisms and parasites, since they can spread with ease from one organism to the next throughout the whole population. This, and the arrival of predators, are important reasons why the rapid growth phase of a population eventually comes to an end, and is often followed by a fall in numbers.

Lemmings are found in North America and Scandinavia. They are eaten by ....

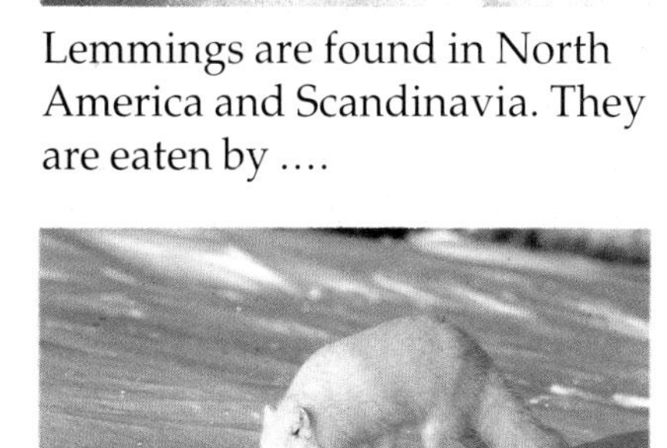

.... Arctic foxes. They have a predator-prey relationship in which changes in the lemming population are followed by changes in the fox population (see graph below).

Long-term population changes

When the growth curve of a population levels off, its numbers rarely stay constant for long. Frequently a population of animals undergoes a series of **explosions** and **crashes.** This is true of some herbivorous (plant-eating) mammals. As their numbers change, so do the numbers of carnivores which prey on them.

Predator-prey relationships If the growth curves of a herbivore population and its carnivore predator are plotted together on a graph (see opposite), it is clear that the high and low points of the herbivore are followed by high and low points of the predator's population.

Lemmings of North America and Scandinavia and the Arctic fox, one of their predators, are an example. A population explosion amongst the lemmings provides more food for the foxes which, in turn, increase in numbers. When the lemming population crashes due to over-grazing, the spread of disease, and increasing predation, the fox has less food and so its numbers fall.

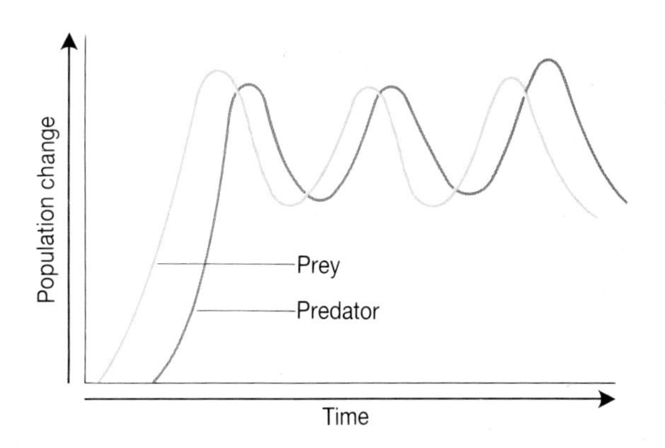

Questions

1 a) What is 'environmental resistance'?
b) What effect does it have on population growth?
c) What is 'carrying capacity', and what happens to a population when it is reached?

2 a) After a population crash amongst lemmings, what factors will eventually lead to another increase in their numbers?
b) Would removing Arctic foxes stop this cycle of boom-and-bust' population changes?

Questions on Section 8

1 *Recall knowledge*

Copy the following sentences and fill in the gaps.

a) In . . . . . reproduction there is only one parent.

b) *Amoeba* reproduces asexually by . . . . .

c) Male sex cells of animals are called . . . . .

d) They are produced by organs called . . . . .

e) Female sex cells of animals are called . . . . .

f) They are produced by organs called . . . . .

g) . . . . . is the joining together of a male sex cell and a female sex cell.

h) A fertilized cell is called a . . . . . It develops into a baby.

i) A partly-developed baby is called an . . . . .

2 *Recall knowledge*

a) Name some animals that have external fertilization.

b) Name some animals that have internal fertilization.

c) Write down four advantages of having a baby develop inside its mother's body.

d) Write down four disadvantages of having a baby develop outside its mother's body.

3 *Recall knowledge*

a) How many sperms are needed to fertilize an ovum?

b) What happens to the nucleus of a sperm and ovum during fertilization?

c) Why does a skin form around an ovum immediately after fertilization?

d) How long does it take for a fertilized human ovum to develop into a baby?

e) Where does the implantation of an embryo take place?

f) What does implantation mean?

4 *Recall knowledge*

a) What is the scientific name for German measles?

b) Describe two ways in which German measles can damage a developing baby.

c) How can girls avoid catching this disease?

d) Give three reasons why breast-feeding is better for babies than bottle-feeding?

5 *Recall knowledge*

Below is a diagram of a woman's reproductive system.

a) Name the parts labelled B to G.

b) Now match each of these descriptions to one of the labels:

where fertilization occurs;
the space where a baby develops;
ovulation occurs in this organ;
the lining of this part is shed once a month;
a baby passes through these parts during delivery.

c) Write down another name for a woman's period.

d) What happens to the lining of the womb just after ovulation? Why does this change happen?

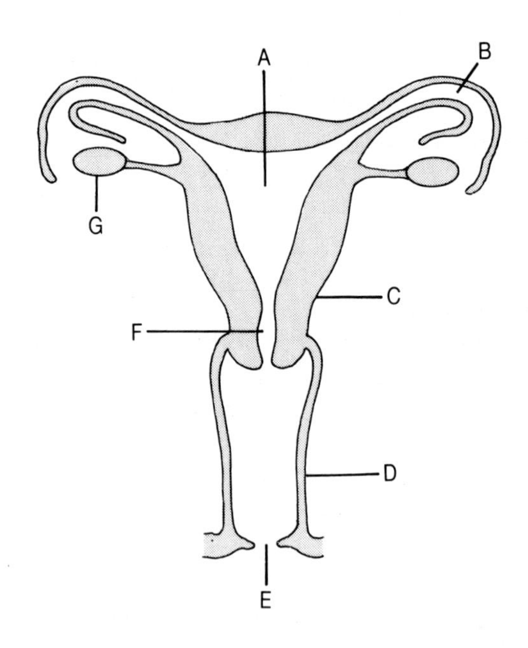

6 *Recall knowledge*

a) What is a condom?

b) What is a diaphragm?

c) Why should a woman put spermicide inside her vagina before sexual intercourse when using a diaphragm or condom?

d) What do contraceptive pills contain?

e) Describe three ways in which contraceptive pills can prevent pregnancy.

7 *Recall knowledge*

Opposite is a man's reproductive system.

a) Name the parts labelled A, B, C, D and E.

b) Match each of these descriptions to one of the labels:

passes sperms into the woman's reproductive organs during sexual intercourse;
produces the hormone testosterone;
makes the penis erect during intercourse;
carry sperms from the testes to the penis;
makes the liquid part of semen.

c) List the changes which the hormone testosterone produces in boys. At about what age do these changes begin?

d) List the changes which the hormone oestrogen produces in girls. At about what age do these changes begin?

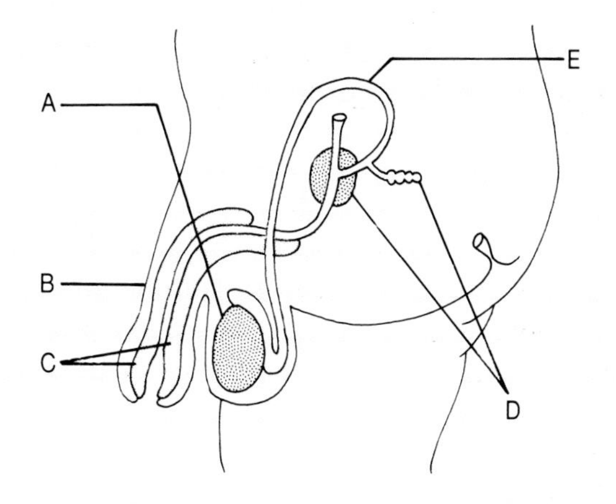

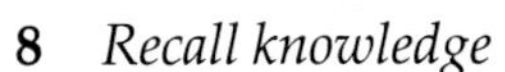

8 *Recall knowledge*

The drawing opposite shows a developing baby inside the womb.

a) Name the parts labelled A to D.

b) What does part C contain?

c) Name two substances which pass into the baby's blood from the mother's blood.

d) Name one substance which passes from the baby's blood into the mother's blood.

e) Where on the diagram do these substances pass between mother and baby?

f) What does part B do during birth?

g) A few minutes after a baby is born the afterbirth is pushed out of the womb. Name the parts of the afterbirth. Which letters on the diagram point to these parts?

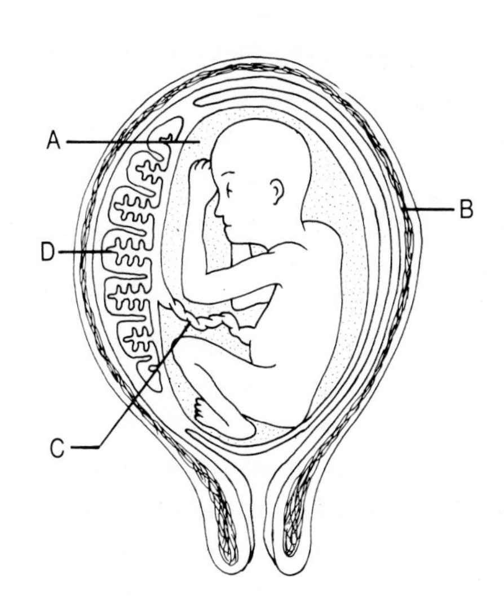

9 *Recall knowledge/interpret data*

a) From the graph opposite, find out the population of the world in 1900 and its population in 1970.

b) Describe how the population changed between those two years.

c) Write down as many reasons as you can to explain this change.

d) What problems has the growth in world population caused?

e) Give three reasons why millions of people are starving to death when there is enough food for everyone in the world?

f) Write down two actions that rich countries can take to help poor countries.

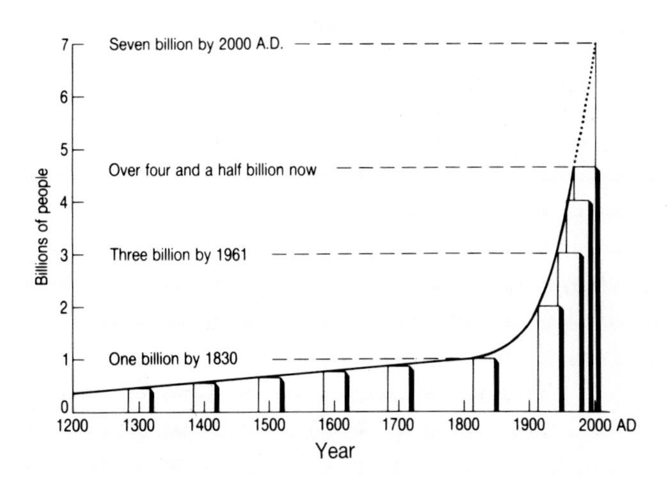

9.1 Depending on each other

Producers and consumers

All living things need energy. They get their energy from food. Plants are called **producers** because they make their own food. They use sunlight to make food by photosynthesis.

Without plants, humans and all other animals would starve to death. This is because they *cannot* make their own food. The only way animals can obtain energy is by eating, or consuming, plants or other animals. So animals are called **consumers**.

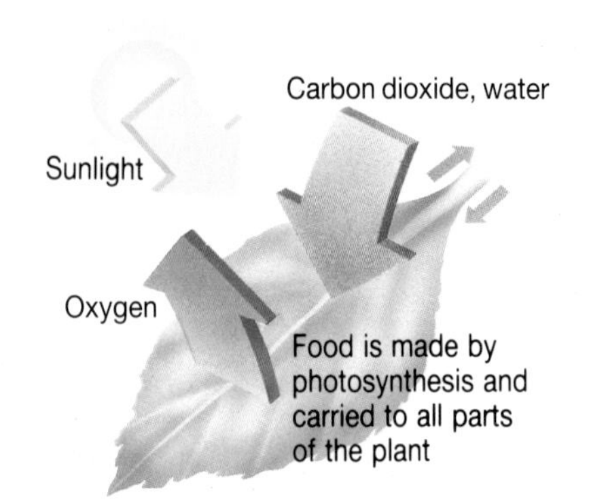

Food chains

1 Seaweed is a **producer**. It produces food by photosynthesis.

2 Periwinkles are called **first consumers** because they eat seaweed.

3 Herring gulls are called **second consumers**, because they eat periwinkles.

Seaweed ⟶ periwinkle ⟶ herring gull.

This is an example of a food chain.

Food chains show how one living thing is the food for another. Energy passes along a food chain from producers to consumers as one member of the chain eats the next.

Food pyramids

On the right is a **food pyramid**. It shows how consumers can get larger in size, but smaller in number, as you go along the food chain.

When a moth caterpillar eats a leaf it uses up some of the energy from the leaf for itself. So there is less to pass on to a robin. That means the robin needs to eat several moth caterpillars to get enough energy. So there are fewer robins than caterpillars.

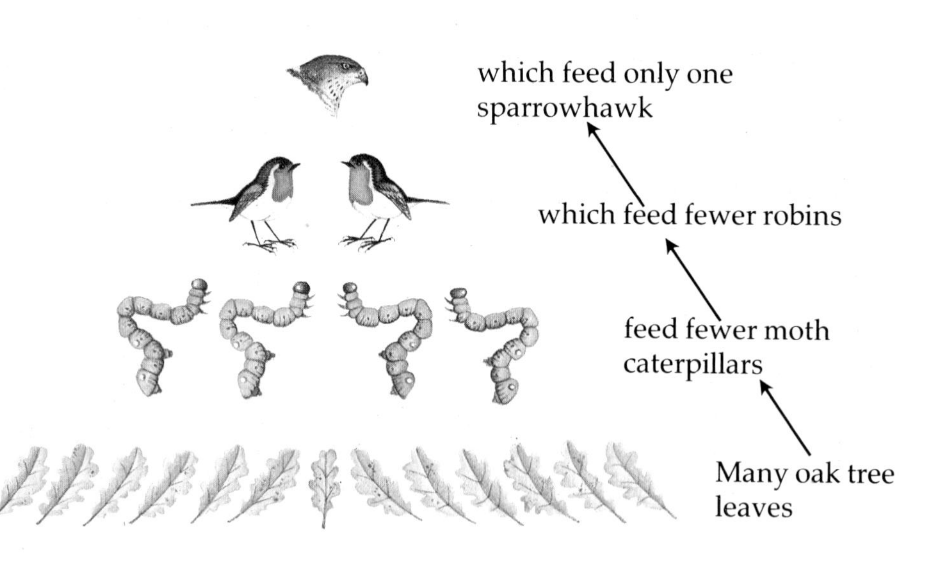

Food webs

A plant or animal usually belongs to several food chains. For example, seaweeds are eaten by limpets and winkles as well as by sea worms. In this way food chains are connected together to make **food webs**. This diagram shows a food web in the sea.

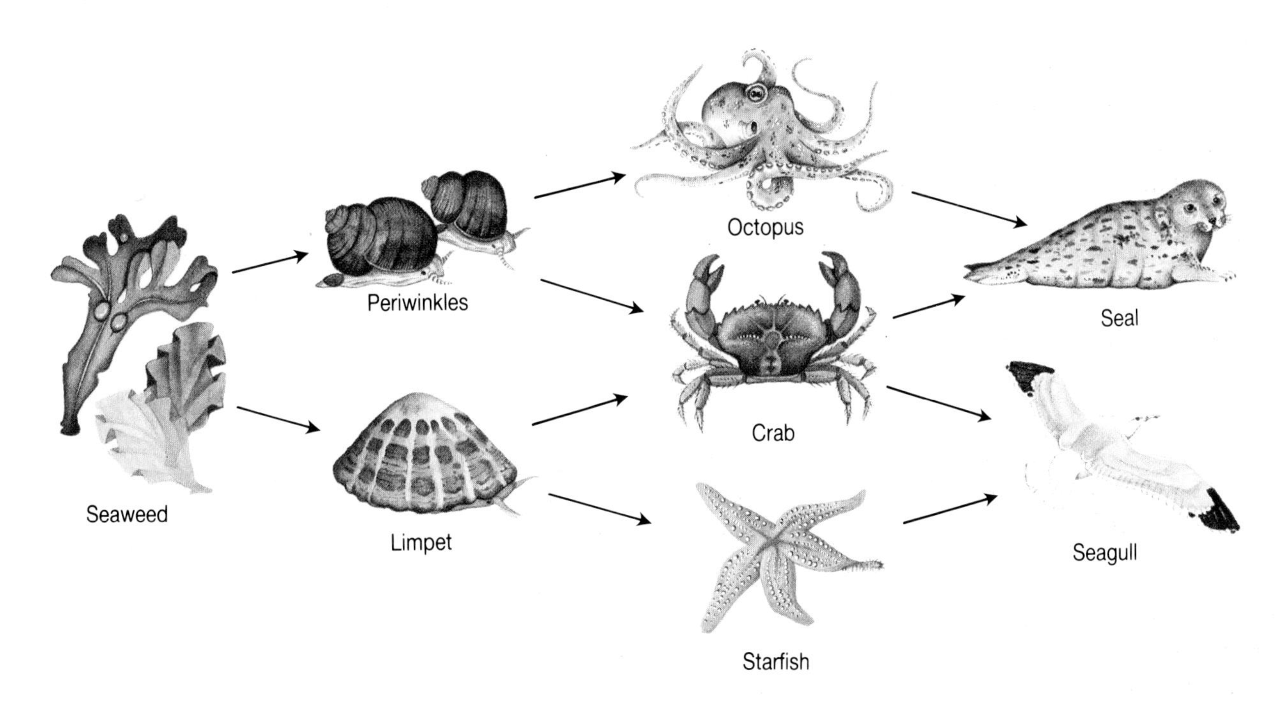

Decomposers – Nature's recyclers

Decomposers are bacteria and fungi which feed on dead animals and plants by making them decompose, or rot. They produce digestive juices which break down and dissolve dead tissues into a liquid, which they absorb.

Decomposers are vitally important because:
1 They remove dead animals and plants by decomposing them, releasing nitrates and other minerals from their bodies into the soil.
2 Minerals are essential for the healthy growth of plants, and without them soil would soon become infertile and all plant life would die.
3 Without plants, all food chains would collapse and animal life would die out.
4 Decomposers maintain life on Earth by recycling minerals taken from the soil by plants.

Fungi decomposing a dead tree and returning its mineral content to the soil.

Questions

1 Why are plants called producers?

2 Put this food chain in the correct order:
slug → fox → primrose leaf → frog

3 Explain how decomposers keep soil fertile.

4 In the food web above list:
a) the producers;
b) the second consumers.

5 Now write down two food chains which end with seals.

9.2 More about food chains

Trophic levels of food chains

The position of an organism on a food chain is called its **trophic level**. This position depends on whether it is a plant or an animal and, if an animal, on what it eats.

Plants occupy the first trophic level of food chains since they produce food, by photosynthesis, which supports consumers at all the other levels. The second trophic level is occupied by plant-eaters (herbivores), and third and fourth levels are occupied by flesh-eaters (carnivores).

Energy flow along food chains

Energy flows from one link to the next along food chains as producers are eaten by consumers, and consumers eat each other. But up to 90 per cent of this energy is lost at each level. So only 10 per cent is transferred to the next level.

Some energy is lost as heat, as plants and animals respire. Some is unused when parts of plants and animals are left uneaten. Some is lost as urine, and also when the indigestible parts of food pass through animal intestines and out as faeces.

Since there is less energy available at each trophic level in a food chain, fewer consumers can live at each level. There is rarely enough energy to support more than four levels (see diagram).

The food pyramid on p.170 illustrates this fact, but is misleading because it is a pyramid of numbers, and its base could just as well be shown as one tree rather than many leaves. Instead of looking at the *numbers* of organisms at each trophic level, more is learned by studying their total *mass*, and, since these are living things, this is called their **biomass**.

A pyramid of energy and biomass

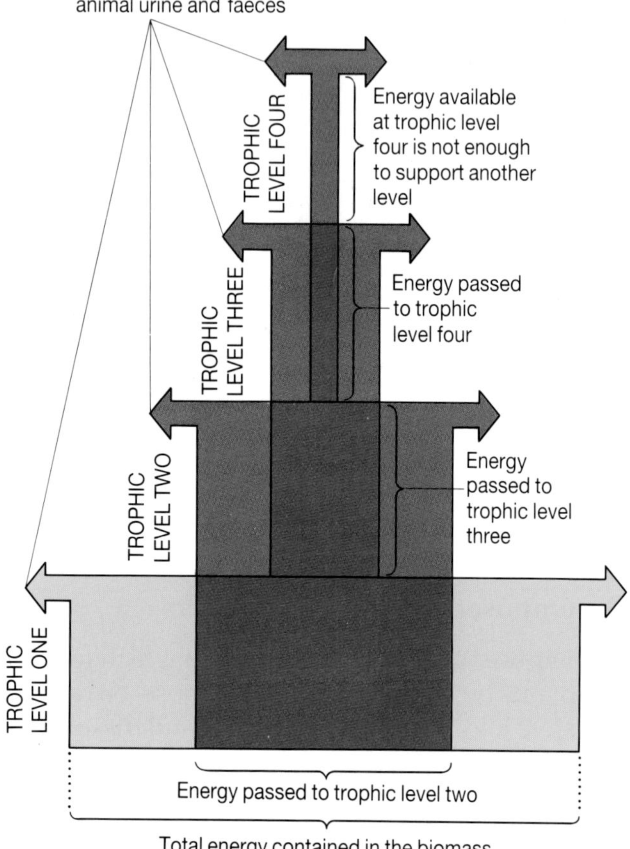

Since 90 per cent of the energy is lost at each level of a food chain, a biomass of plant material at level one with an energy value of 1000 kJ provides only 100 kJ of energy to level two, 10 kJ of energy to level three, and only 1 kJ to level four.

Question

Two fields are used to grow grain. The grain from field one is used to feed cattle which are turned into beefburgers. The grain from field two is fed directly to humans as bread and other grain products. Use information in this Unit to explain why one field is used in a way which wastes energy, while the other wastes less energy and so is a far more efficient use of the land.

9.3 The carbon cycle

All living things need carbon. It is needed for the proteins, fats, and other substances that make up living things. The carbon comes from carbon dioxide in the air.

Plants take in carbon dioxide from the air. They use it to make food by photosynthesis. Animals then get carbon by eating plants.

The amount of carbon dioxide in the air always stays the same, because it is returned to the air as fast as plants take it in.

1 Plants and animals give out carbon dioxide when they respire.

2 Bacteria and fungi respire and give out carbon dioxide while they are decomposing the bodies of dead animals and plants.

3 Wood, coal, gas, and petrol contain carbon. When they are burned the carbon combines with oxygen to form carbon dioxide. It goes off into the air.

Wood, coal, gas and oil started off as carbon dioxide. The gas is released again when they are burned.

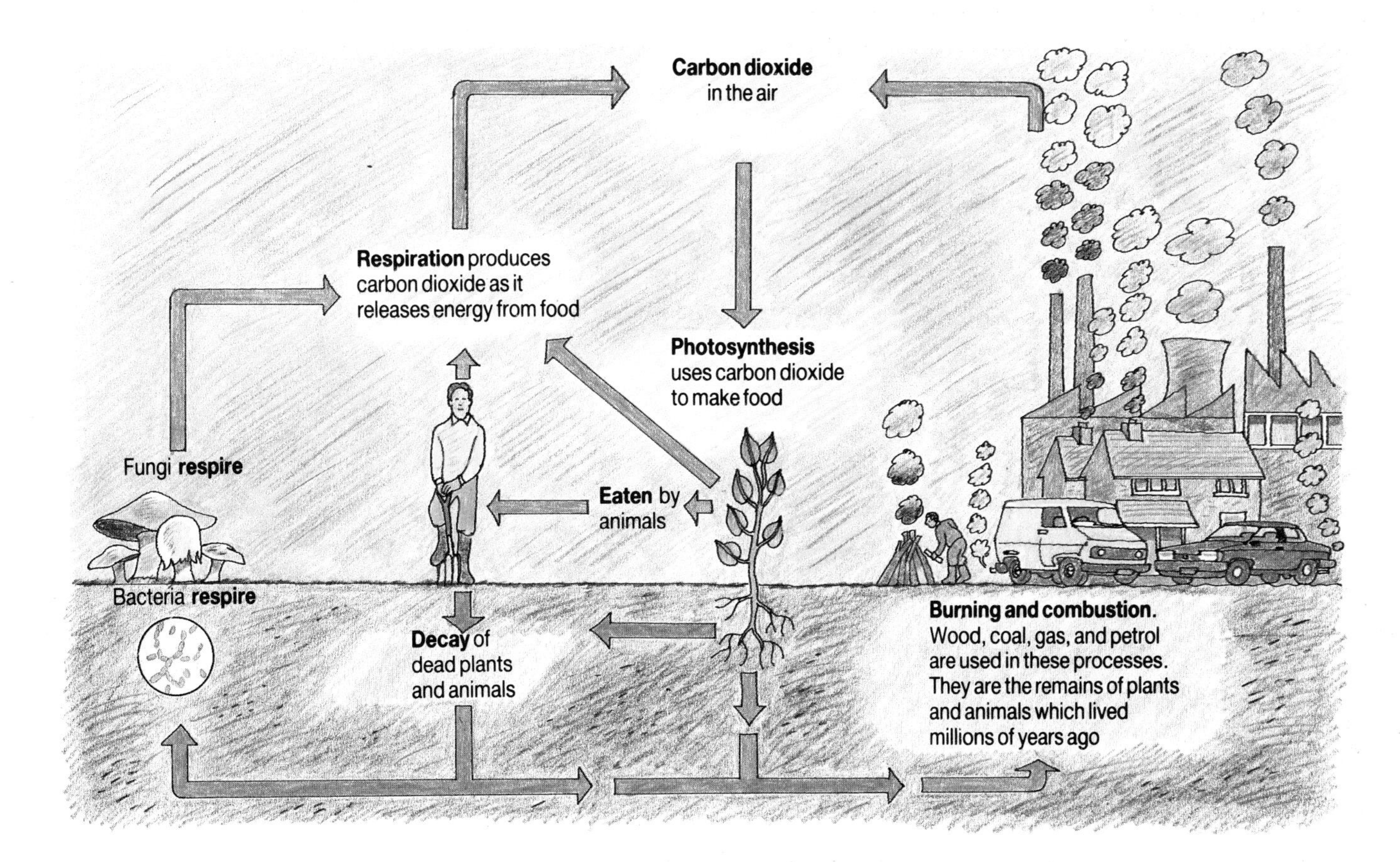

Questions

1 How does carbon get from the air into the bodies of plants and animals?

2 Explain why the amount of carbon dioxide in the air stays the same.

9.4 The nitrogen cycle

All living things need nitrogen to make proteins. Air is four-fifths nitrogen. But neither plants nor animals can take it in from the air. First it has to be changed into **ammonium** or **nitrate**.

Then plants obtain nitrogen by taking in nitrates from soil. Animals obtain nitrogen by eating plants, or other animals.

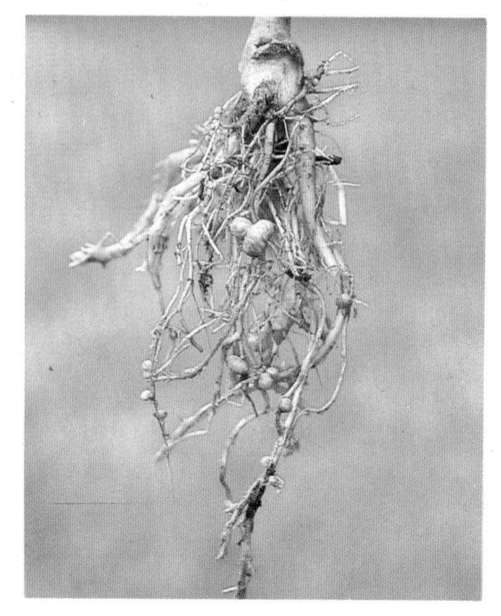

Root nodules on a pea plant, homes for nitrogen-fixing bacteria. Because the bacteria feed them well with ammonium, peas grow well even in poor soil.

How nitrogen is turned into ammonium and nitrate

1 **Lightning** makes air so hot that nitrogen and oxygen combine. They make chemicals which are washed into the soil, where they form nitrates. Only small amounts of nitrogen are formed this way.
2 **Nitrogen-fixing bacteria** live in soil, and in lumps called **root nodules** on the roots of clover, peas, and beans. They use nitrogen from the air to make ammonium, which is released into the plants or the soil.
3 **Nitrifying bacteria** make nitrates from ammonium in animal droppings, and from the bodies of dead animals and plants.

Some soil nitrates are changed back into nitrogen by **denitrifying bacteria**. These live in the waterlogged soil of ponds and marshes.

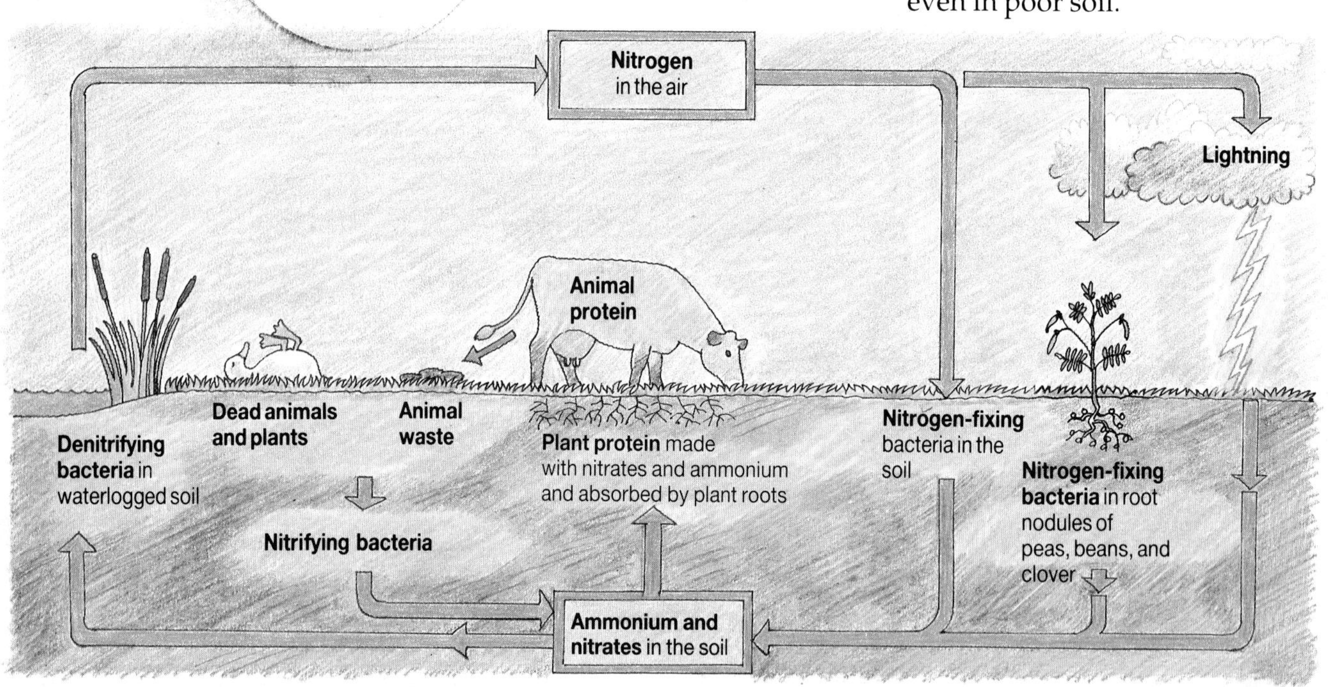

Questions

1 Why do living things need nitrogen?
2 How do plants obtain nitrogen?
3 How do animals obtain nitrogen?
4 Describe three ways in which nitrogen is changed into nitrates.
5 How are nitrates changed back into nitrogen?

9.5 The water cycle

Water is always on the move. It rises from seas, rivers, and lakes to the clouds and falls back again as rain. This is called the **water cycle.** Land plants and animals would die without the water cycle, and soil would turn to dust.

How the water cycle works

1 The sun warms seas, rivers, and lakes. The heat causes water to evaporate and turn into **water vapour**.
2 Plants and animals also give out water vapour.
3 As the water vapour rises it cools down and forms tiny water droplets. These gather as **clouds**.
4 When clouds cool down the water droplets join to form bigger and bigger drops. When the drops are heavy enough they fall as **rain**.
5 Streams and rivers take water back to the sea. Some water is taken up by plants and animals. The cycle starts again.

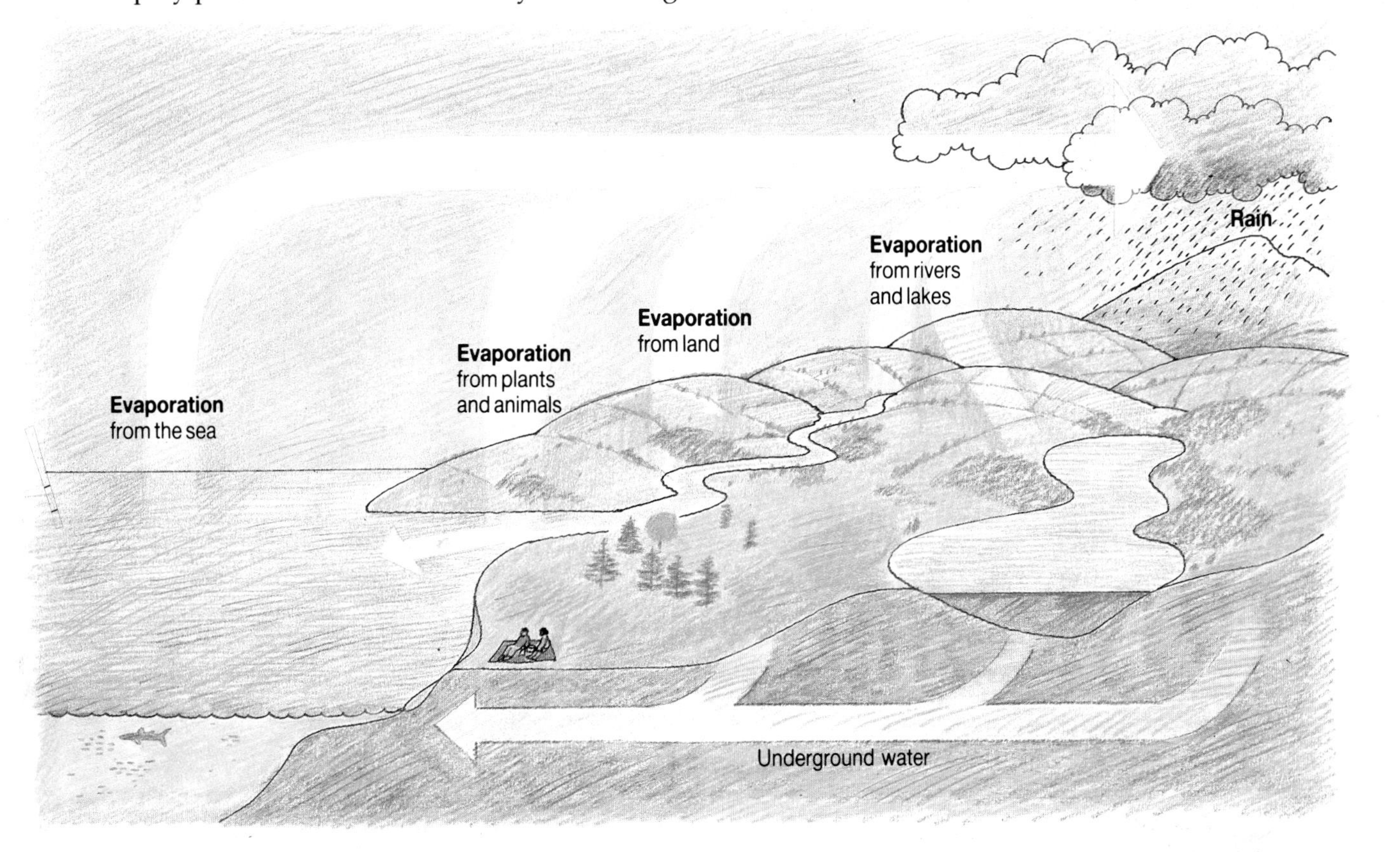

Questions

1 Why is the water cycle important?
2 What is the Sun's part in the water cycle?
3 How are clouds and rain formed?
4 How does water get back to the sea?

9.6 Planet Earth in danger

We need places like this to inspire us to conserve and repair the biosphere, rather than exploit it.

Never before in our history has planet Earth been in so much danger. The **biosphere**, the layers of air, soil, rocks and water which are the home of all living things, suffers continuous destruction at our hands.

If this destruction continues, wildlife will be reduced to a few tough species such as rats, flies, and sparrows. Large areas of land will become unproductive deserts. Lakes, rivers, and the sea will become sterile. And we will no longer have clean air to breathe, pure water to drink, and fertile land for our crops and animals.

Things we take from, and add to the biosphere

Deforestation Every year an area of tropical and other forests the size of Wales is cut down for timber or cleared for agriculture. Forests produce the oxygen we breathe, conserve soil, store water, and are the home to millions of different creatures.

Habitat loss Thousands of square kilometres of marsh and other wetlands are being drained throughout the world each year for agriculture. Coral reefs are destroyed by pollution and over-fishing. Britain has lost thousands of miles of hedgerows and 80 per cent of its heaths since 1945.

Soil erosion 25 billion tonnes of soil are blown away each year world-wide, mainly due to overgrazing, deforestation, and use of chemical fertilizers rather than manure. In European countries alone, 140 000 km^2 suffer a 'higher than acceptable' soil loss.

Hunting 50 million animals a year are killed for their skins. Elephants are killed for ivory, rhinoceroses for horn. Birds, fish, and reptiles are captured for pets. Millions of migrating birds are shot for fun.

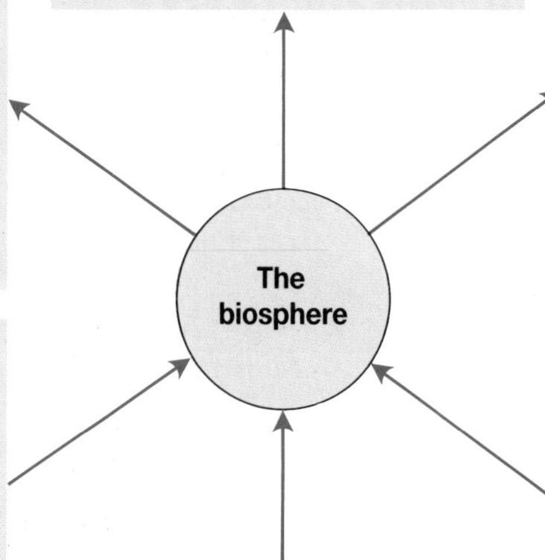

Over-fishing Fish provide 23 per cent of the world's protein. But many major ocean fisheries are now over-fished to the point of collapse. Drift nets up to 30 km long are used to catch tuna. They trap and drown dolphins as well.

Pesticides World pesticide production is nearly 3 million tonnes, and is increasing. Pesticides increase food production but can kill wildlife. They contaminate drinking water and food: most now contains pesticide residues.

Industrial waste These include phenols, dioxins, arsenic, cadmium, lead, mercury, and chlorinated hydrocarbons (PCBs). Many are toxic, long-lasting, and accumulate in living things. Radioactive waste can cause cancer and leukaemia.

Fertilizers 30 kg are used per person per year in the UK. They improve yields but pollute lakes, rivers, and drinking water. Nitrates damage red blood cells.

Sewage and refuse Millions of tonnes of untreated sewage are discharged into rivers and the sea. The bacteria that decompose it use up oxygen. This kills aquatic life. Refuse, much of which could be reused or recycled, is dumped in landfill sites. These can leak poisons which enter sources of drinking water.

Other pollutants Smoke and fumes from vehicles, homes, factories, and power-stations pollute air and cause acid rain. Oil spills kill sea birds and sea and shore wildlife.

This destruction occurs because people in industrialized nations do not always think of the world as a precious resource which could supply all our needs. Too often, we think of it as a source of riches to be exploited for profit and leisure. As developing nations become richer, they too want all the benefits which the world's resources can provide.

Many regret the price humans and the natural world have to pay for these benefits, but do not want to give up their standard of living and comfort. This is why people often try to escape to places where they can appreciate the beauties of nature. But such places are becoming increasingly difficult to find.

This situation will not change until we decide to limit our needs to those which can be satisfied without destroying the millions of other creatures which share our planet.

The last of Sweden's wolves was killed a century ago. Now, because public opinion is less hostile, they are returning to breed. It is their world too!

Questions

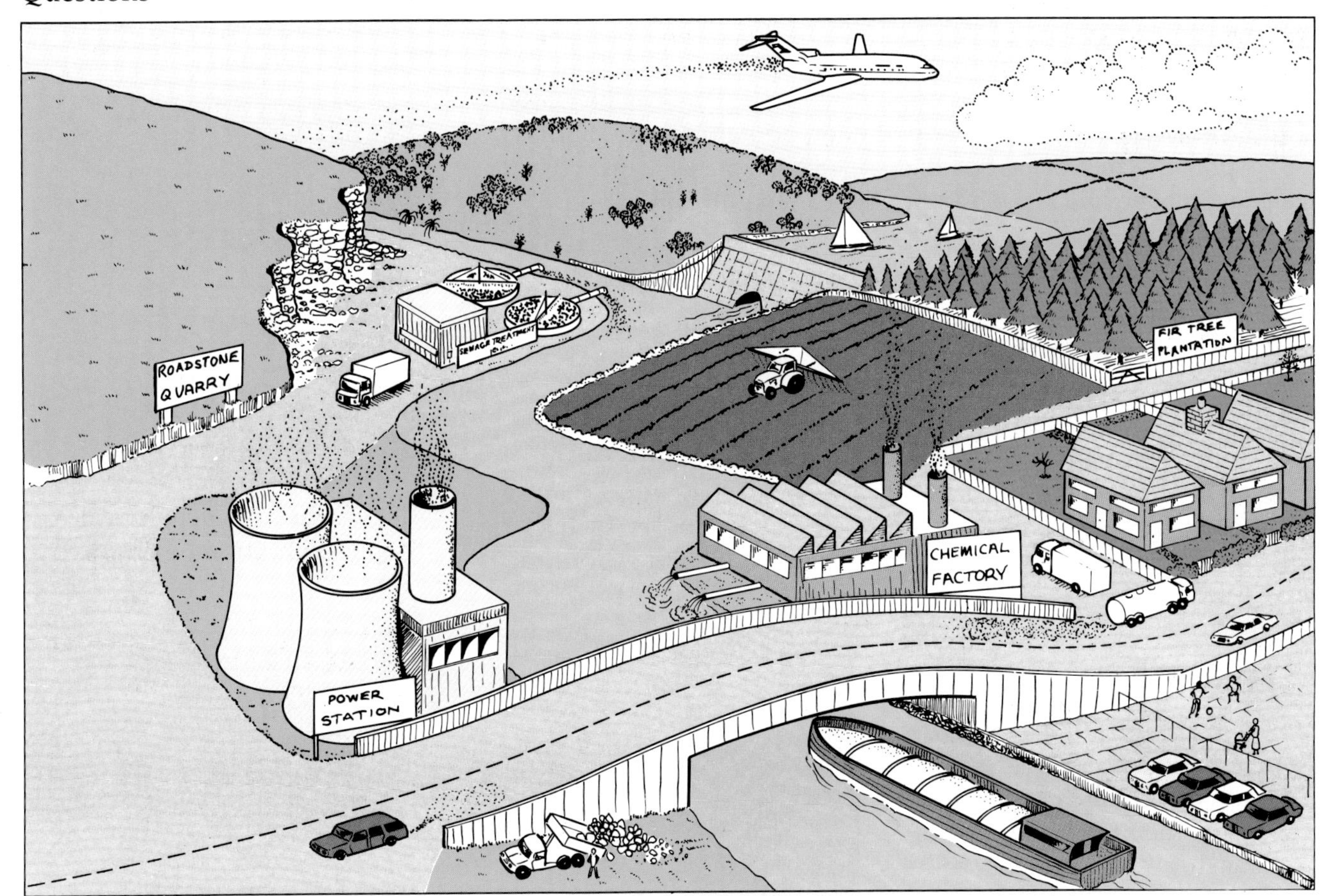

1 Describe places in this picture where:
  a) land is covered so nothing can grow;
  b) wild plants have been replaced with plants that people want;
  c) water and air are being polluted;
  d) land is being covered with chemicals which could harm wildlife;
  e) land is flooded, or removed altogether;
  f) wildlife can live relatively undisturbed.

2 Think of arguments for and against: forestry, quarries, pesticides, fertilizers, hunting, and cars.

9.7 Water pollution

What is pollution?

A substance is a **pollutant** when its presence harms living things. Pollutants can spread through air, water and soil, and through food chains. They harm and kill plants, animals, and humans.

The thousands of different plants and animals which live in water can be damaged by pollution. Also, human health depends on clean water supplies. Streams, rivers, and the sea are polluted by industrial and agricultural wastes, untreated sewage, and spilled oil. But the most easily-damaged aquatic environment is the still, or slow-moving, water of a lake.

Describe some of the ways in which drinking water could become polluted.

Eutrophication, or how pollution can kill a lake

Sewage and farm animal waste discharged into rivers accumulate in lakes, and chemical fertilizers spread on the land can be washed into lakes by rainwater. These pollutants contain nitrates and phosphates which act as fertilizers, causing a massive growth of algae. This can kill a lake by a process called **eutrophication**. This is what happens:

- Single-cell and filamentous algae (Unit 1.6) cloud the water and form a dense, green blanket on the surface which reduces light to submerged water plants. These die and are decomposed by bacteria.
- The bacteria use up oxygen dissolved in the water. This suffocates many fish and other water animals. With fewer animals to eat the algae, they grow even faster.
- When the algae eventually die and are decomposed by bacteria, the remaining oxygen in the water is used up and everything in the lake dies.

The River Nishua in the USA: a dead river, polluted by waste from paper mills.

Oil pollution

Nearly a billion litres of oil are spilled into the sea each year from wrecked tankers, tankers washing their tanks, oil rigs, and factories. Oil poisons sea birds and clogs their feathers so that they cannot fly. Oil, and the chemicals used to disperse oil spills, kill animals and plants in the sea and along the sea-shore. They also contaminate holiday beaches.

A puffin killed by oil from a stricken tanker.

In time, oil is broken down and made harmless by microbes in the sea. But many pollutants, including some pesticides, PCBs (polychlorinated biphenyls), and toxic heavy metals such as lead and mercury, are not broken down in the environment or inside living things. They build up in soil, water, and living tissue to high concentrations.

Pesticide kills harmful pests – and harmless wildlife too.

Pollution of food chains

These pollutants are especially dangerous when they enter living things because they accumulate in fat and other tissues – a process called **bioaccumulation**. This can continue up food chains, from one trophic level to the next, until the top predator has very high, often lethal, amounts in its body. This process is called **biomagnification.**

The diagram opposite shows how pesticide spray can enter water and biomagnify. The same can happen to heavy metals and PCBs. Microscopic plants absorb these pollutants directly from the water. The pollutants pass from one consumer to the next, mainly in the food they eat.

When the small crustaceans eat microscopic plants, they take in only tiny amounts of pollutants. But all of it is retained and builds up in their tissues, so that when small fish eat the crustaceans they take in larger amounts of pollutants. Again, all the pollutants are retained and accumulate in the small fish, so that larger fish get a bigger dose. Fish-eating birds (and of course this also applies to fish-eating humans) get the biggest dose of all.

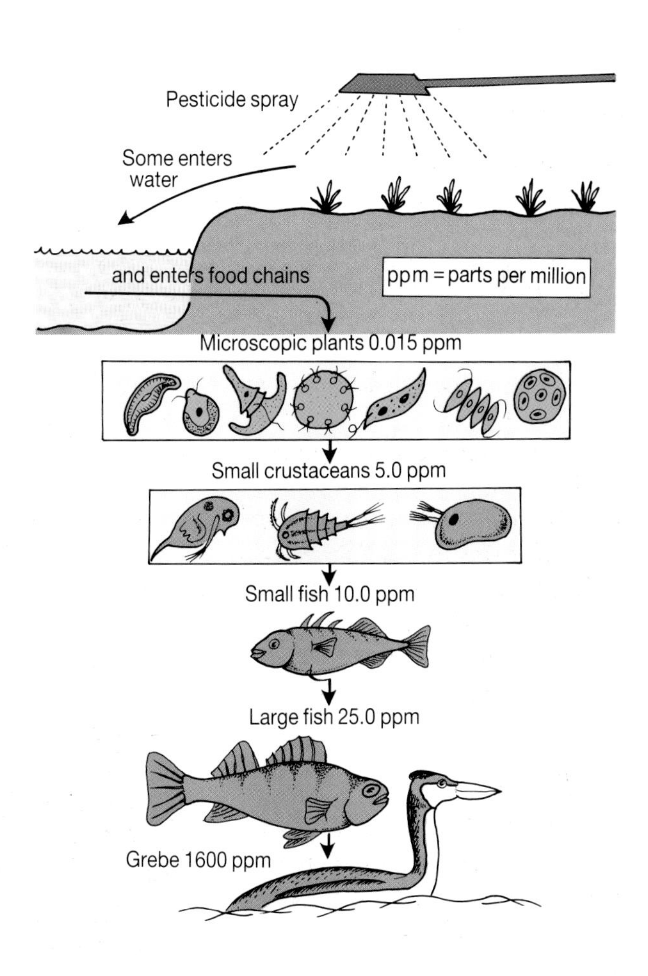

Heavy metals

Humans are in great danger from the biomagnification of heavy metals such as lead and mercury. Lead accumulates in the liver and kidneys, and damages the nervous system. Mercury severely damages the nervous system and causes birth abnormalities.

Questions

1 Some washing powders contain phosphates. How might these help cause the death of living things in a lake?

2 Around 1960, the sparrowhawk population decreased dramatically because pesticide in their bodies caused females to lay eggs with shells so thin, they broke before hatching. How could pesticide have got into their bodies?

3 Why are consumers at the top of a food chain, such as sparrowhawks, in more danger from pollutants than those at the beginning?

9.8 More about pollution

Air pollution

Air is polluted by smoke, dust, and harmful gases. Most of these come from power-stations, factories, cars, buses, and lorries.

Smoke contains tiny particles of carbon (soot). This blackens buildings. It covers plant leaves so that photosynthesis slows down. The smoke from car engines also contains lead, which can cause brain damage in young children.

Dust from quarries, saw mills, and asbestos factories can cause lung disease.

Harmful gases The main ones are **sulphur dioxide** and **nitrogen oxides**. These form when coal, oil, and petrol are burned in power stations, factories, engines, and homes. They damage plant leaves, and aggravate diseases like bronchitis.

When sulphur dioxide and nitrogen oxides rise into the air they can dissolve in clouds and form acids. Then they fall back to Earth as **acid rain**. This rain corrodes metal railings and bridges, and eats away stonework on buildings. See Unit 9.9.

It's difficult for a leaf to photosynthesize when it's completely covered in soot!

Acid rain damage

Acid rain eats away stonework. This was once an angel.

Spruce trees in Germany, killed by acid rain.

Some pollution from cars...

...falls as acid rain.

Destruction of the ozone layer

The ozone layer lies between 20 and 25 km above the Earth. It is very important because it acts as a protective screen which filters out ultraviolet (UV) radiation from the Sun, especially a harmful form of this radiation called UV-B.

Since the 1960s, ozone levels over parts of Antarctica have dropped by 40 per cent, causing an **ozone hole** through which UV-B can pass. The hole appears in spring. In recent years, another has appeared over the Arctic regions as well.

UV-B radiation increases the frequency of types of skin cancer and cataracts. It damages plants, and kills algae in the sea which are the producers of all marine food chains. Already, Australia, New Zealand, and parts of the northern hemisphere are exposed to increased UV-B. Its harmful effects will increase as ozone holes expand.

The main cause of ozone destruction is the presence in the atmosphere of chorofluorocarbons (CFCs), mostly from aerosol cans. CFCs are broken down by UV-B into chlorine atoms which destroy ozone. Many countries have banned CFCs in aerosols. But CFCs are used in other products and remain in the air for 100 years. So ozone destruction will continue for many years to come.

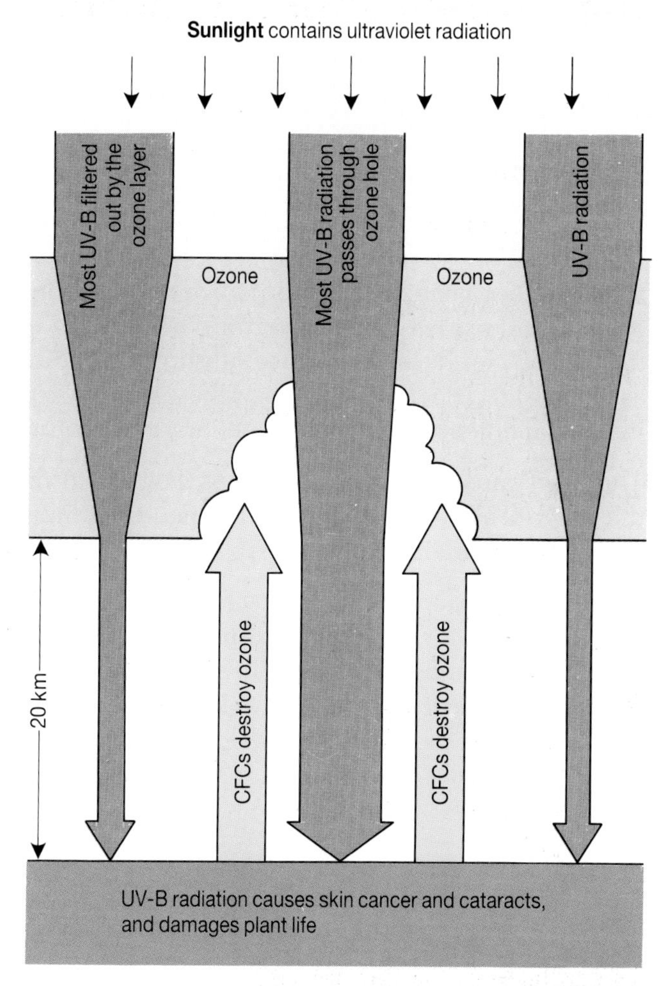

Other kinds of pollution

Radiation There is a risk that radiation will escape from nuclear power-stations and from stored radioactive waste. Radiation can cause cancer and leukaemia.

Noise Noise from cars, motor cycles, aeroplanes, dogs, children, radios, and televisions can be a form of pollution. It can irritate people and cause mental depression. Prolonged loud noise from disco music and factory machinery can make people deaf.

If you live near an airport, you'll know all about noise pollution.

Questions

1 Name two gases which pollute air, describe where they come from and the damage they cause.

2 Describe how acid rain is formed and the damage it can cause.

3 Where does dangerous radiation come from and how can it harm people?

4 What causes ozone holes, and why does their formation harm humans, plants, and food chains?

9.9 More about acid rain

Even unpolluted rain is acid because of a natural process in which carbon dioxide gas dissolves in water vapour in the air to form carbonic acid. This has a pH of 5.6 (pure water has a pH of 7.0).

Acid rain becomes an important form of pollution when sulphur dioxide and oxides of nitrogen from the burning of fossil fuels dissolve in clouds, and produce sulphuric and nitric acids with a pH as low as 4.0. The main sources of these gases are power-stations, industrial boilers, large metal smelters, and motor vehicles.

The gases cause pollution as soon as they are formed and are still dry, as well as after they have combined with moisture in clouds to form liquid acids.

Air pollutants deposited in a dry form are called **dry depositions**, and usually occur close to the source of pollution. Pollutants which form acid in atmospheric moisture are called **wet depositions**, and these can be carried thousands of kilometres from their source by air currents.

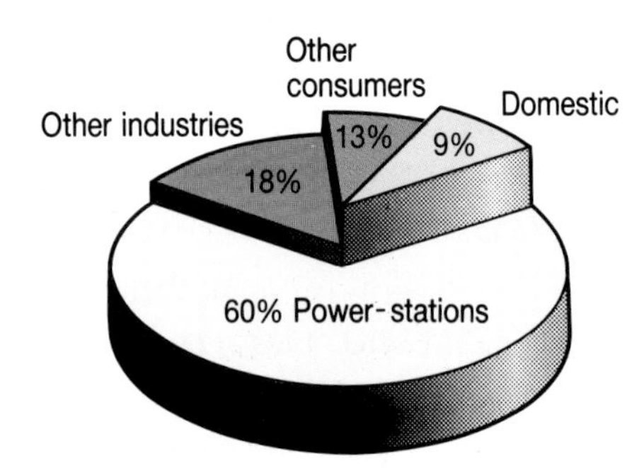

The diagram above shows the main sources of sulphur dioxide in Britain. The diagram below illustrates some of the damage done by dry and wet depositions.

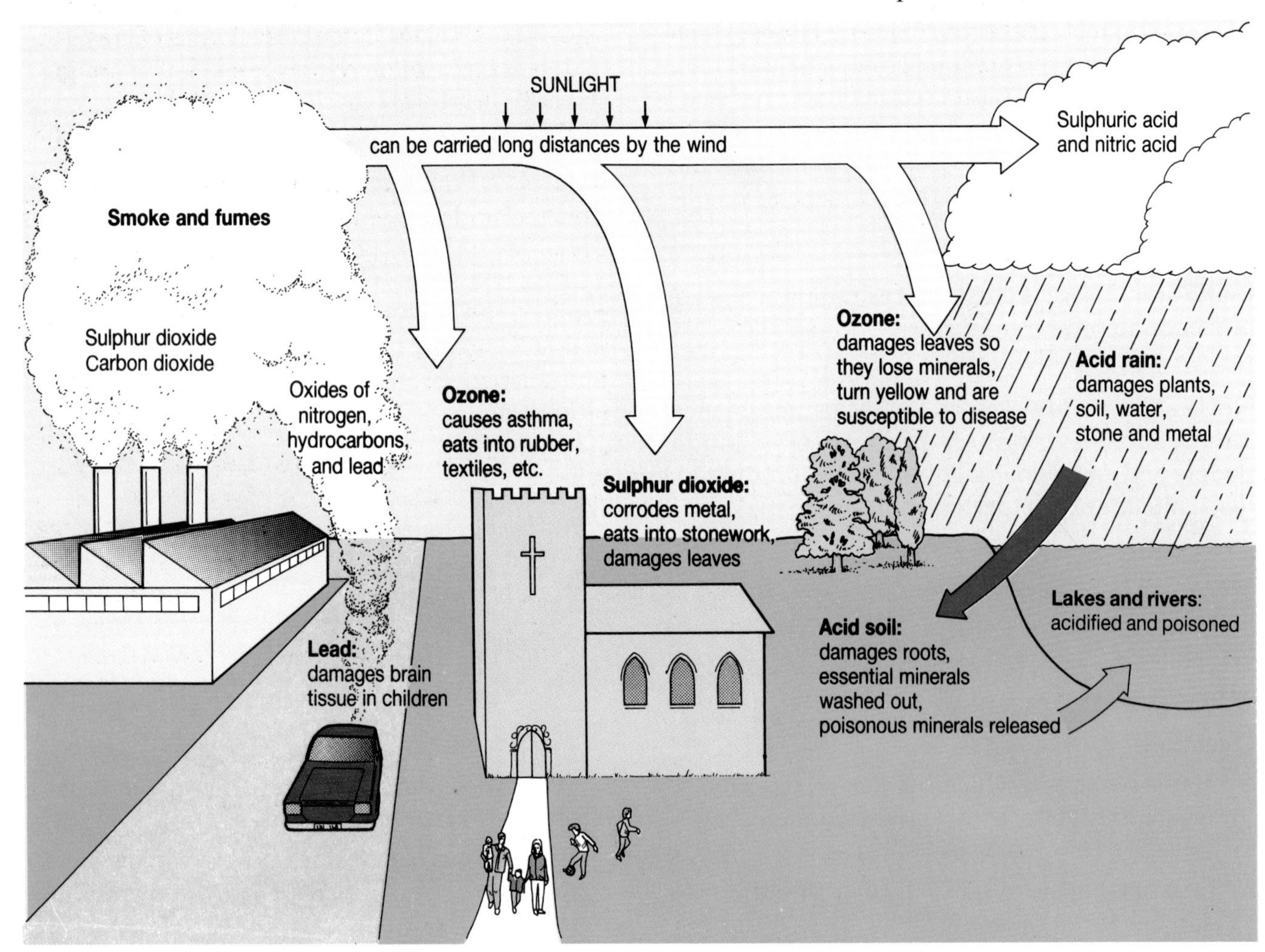

Dry depositions

Sulphur dioxide This gas damages the waxy coating which protects leaves, and it prevents chlorophyll formation. It forms an acid vapour in the lungs which aggravates diseases such as bronchitis. It corrodes exposed stonework causing a skin to form which flakes off. The stone is changed chemically and crumbles away.

Oxides of nitrogen These come mainly from vehicles without catalytic converters. They corrode metal and stone and can have harmful effects on people with lung disease.

Hydrocarbons These come from vehicles without catalytic converters and from burning coal. Some can cause cancer.

Ozone This is formed by the action of sunlight on oxides of nitrogen and hydrocarbons. In the upper atmosphere it shields the Earth from ultraviolet rays. But near the ground, it attacks plant leaves, weakening cell walls. This allows rain to wash away minerals, and stunts plant growth. Ozone eats into rubber, textiles and other materials. It can also cause asthma attacks.

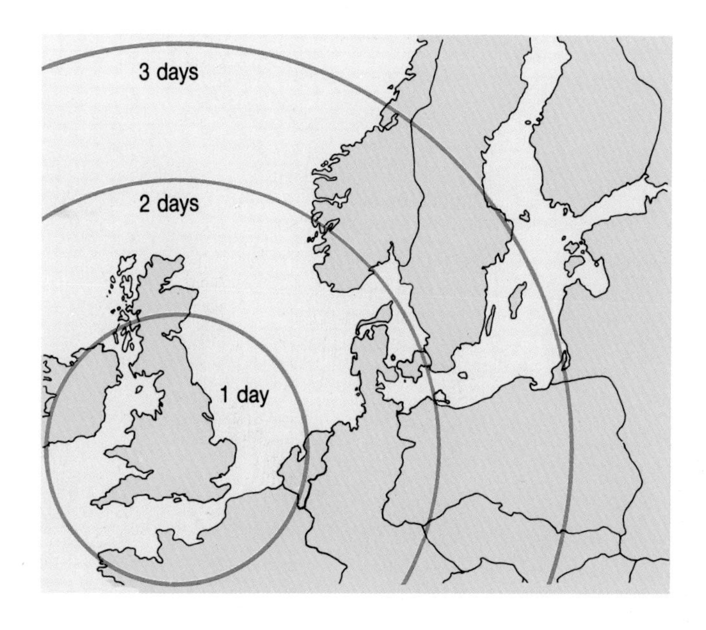

Wet depositions

These include acid rain, acid fog, and melt-water from acid snow.

Damage to plants Acid rain (and dry depositions) are blamed for the death of trees and crop damage in many parts of Europe. Leaves turn yellow and fall off, and roots are damaged so that they cannot absorb minerals. This weakens plants so they are more likely to be killed by drought, severe winters, and attacks from insects and fungi which would have little effect on healthy plants.

Damage to soils Acids cause essential minerals to be washed away by rain. These release poisonous chemicals, including aluminium, which are normally insoluble and harmless.

Damage to water life At around pH 6.0, crustaceans, molluscs, and small fish such as minnows, die. These are near the bottom of many food chains and so their loss endangers aquatic life even if water does not become increasingly acid. At pH 5.6, the external skeletons of crayfish and their eggs soften so they are killed by fungi. In addition salmon, roach and trout die. At pH 5.0, perch and pike die, and at pH 4.5, eels die.

If the pH falls below 4.2, aluminium is released from soil and may enter water. Aluminium upsets the ability of fish to control movement of water in and out of their gills by osmosis. It causes mucus production which suffocates fish by clogging up their gills. And it interferes with skeleton formation in young fish, so that fewer develop into adults.

Nature can provide defences against acid rain. Alkaline soils, like those in parts of North America, neutralize acid rain and so protect plant life. The same happens in lakes over alkaline rocks, such as the limestone of southern England. But lakes in Scotland and Scandinavia, over acidic granite rocks, are hard hit by acid depositions.

Questions

1 Air pollutants can spread from their source at 500 km a day. Norway and Sweden have accused Britain of 'exporting' acid rain to them. Find the direction of prevailing winds in the area of the map above and decide if their claim could be correct.

2 How can power-stations damage churches?

3 How can car exhaust fumes damage a child's brain?

9.10 The greenhouse effect

The world is warming up

There is mounting evidence that the world is getting warmer. Since 1850 world temperatures have risen by about 0.5 °C and could rise another 4.5 °C by the year 2050. Some of the reasons why this apparently small rise is causing alarm are described later. First you should understand why **global warming** is happening.

Global warming is caused by **the greenhouse effect,** illustrated in the diagram opposite. Light from the Sun can pass through the glass of a greenhouse. It is then absorbed by the floor, earth and other contents which warm up and radiate heat, in the form of infrared radiation. Unlike light energy, the infrared cannot escape through the glass. So it is trapped inside the greenhouse, which warms up.

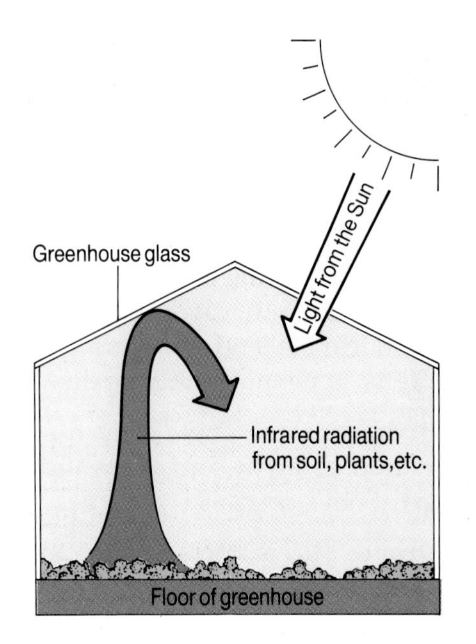

The ways in which a greenhouse warming and global warming are similar, are shown in the diagram below.

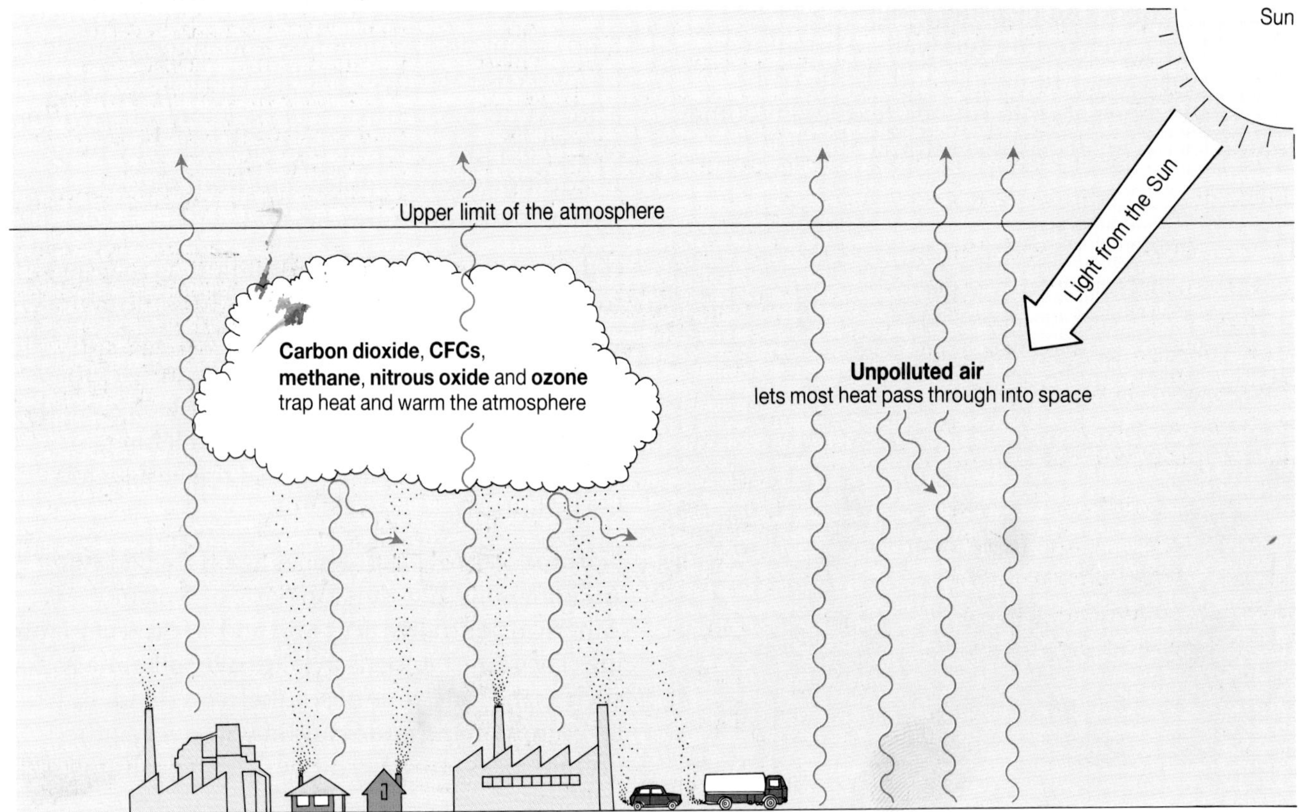

Sunlight warms up the Earth's surface which emits infrared radiation

Sunlight passes through the atmosphere and is absorbed by the Earth's surface, which warms up and radiates infrared (heat) energy. If there were no air, all of this heat would escape straight into space. But carbon dioxide and water vapour in the air act like the glass of a greenhouse, trapping and absorbing some heat, which warms the atmosphere. This natural global warming provides an essential source of heat for living things. Without it, the world would be freezing at −17 °C and life could probably not exist.

The ever-increasing amounts of carbon dioxide, chlorofluorocarbons (CFCs), methane, nitrous oxide, and ozone in the polluted air of our modern world trap more heat than clean air. These are the **greenhouse gases** which could change our climate, landscape, crops, wildlife and other important features of the Earth.

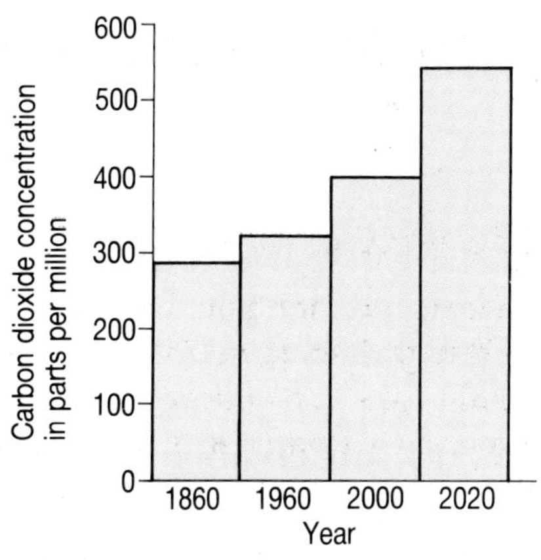

The amount of carbon dioxide in the air should remain constant at about 0.003 per cent, but it is rising all the time, as shown in the graph above. This is happening because our modern industrial society produces this gas much faster than it can be absorbed by plants. We are making this problem worse by destroying vast areas of plant life (such as tropical forests). This is reducing global photosynthesis.

Some effects of global warming

Melting ice-caps Temperature rises could be especially high around the Poles, causing extensive melting of the ice-caps. This, combined with expansion of sea water as it warms up, could raise sea-levels. This seems to be happening already. Glaciers are in retreat, and sea-levels have been rising since 1900. A rise of more than 1.5 metres is predicted by the year 2050, which would cause disastrous flooding of cities and farmland in low-lying areas such as parts of eastern England, Holland, Florida, and the Ganges delta. The cost of new sea defences, increased coastal erosion, and reduced soil fertility due to flooding with salt water, will be enormous. It will strain most nations' finances to breaking point.

Climate change Uneven heating of the world could cause some of our most important food producing areas in North America and Central Europe to become much drier and less fertile. There is also an increasing risk of violent wind storms like those which have hit the Caribbean and Pacific in recent years, and extensive flooding caused by rainstorms like the 'flood of the century'. In 1992 this devastated nine States along the Mississippi river in the USA.

Effects on crops Weeds thrive on extra carbon dioxide. Some crop plants do not: these will be more difficult to cultivate. Leaf tissue becomes less nutritious with increased carbon dioxide, so crop pests will eat far more.

Effects on wildlife Creatures adapted to cool climates will have to spread northwards. Animals can do this quickly, but many plants could be killed by warming.

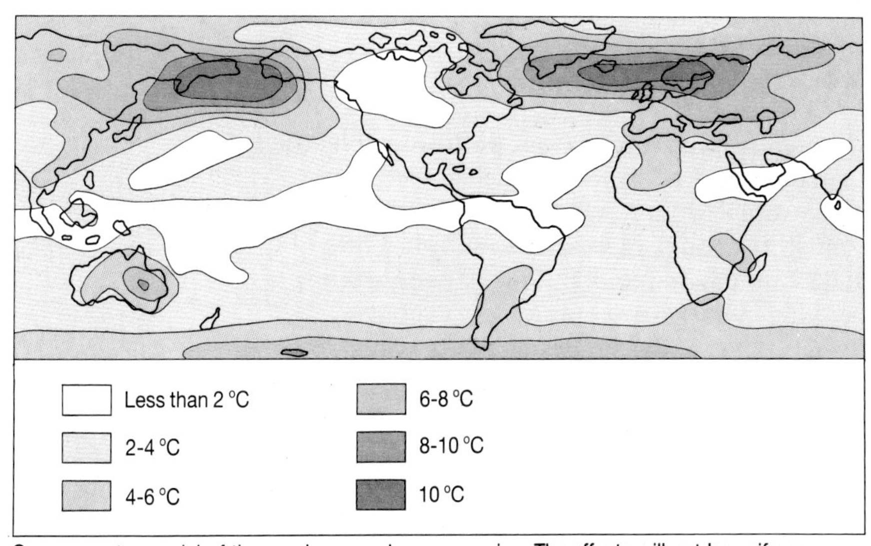

One computer model of the coming greenhouse warming. The effects will not be uniform.

Questions

1 How is the world similar to, and different from, a greenhouse?

2 How could cutting down tropical rain forests increase the rate of global warming?

3 Insects once found only in southern Europe are now found in the south of England. Explain why this is evidence of global warming.

4 How could we reduce carbon dioxide production?

9.11 Controlling pollution

We have the power to destroy our world. We also have the power to save it, and maintain it as a place fit for both humans and wildlife. One of the most important ways of doing this is to control pollution.

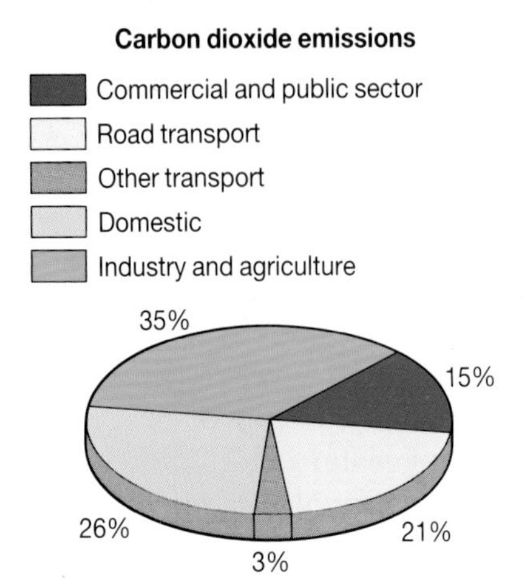

Energy conservation – getting more for less

We waste huge amounts of energy. Up to 60 per cent of the energy we get from fossil fuels goes to waste as heat. Burning fuels also causes acid rain and global warming. Power-stations alone produce more than half of the gases which pollute our air. Therefore saving energy helps prevent pollution.

Energy efficient homes and factories Providing energy for the average home produces 7500 kg of CO_2 a year. Britain could reduce energy demand by 70 per cent, which would help reduce the greenhouse effect. Energy is saved by the following:

- Insulating the roof, cavity walls, hot-water storage tanks and pipes which carry hot water; drawing thick curtains to insulate windows and reduce draughts, and double-glazing windows.
- Draught-proofing windows and doors, and avoiding over-ventilation.
- Turning down heating systems and putting on warmer clothing, and turning off lights when leaving a room.
- Buying the new energy-efficient cookers, refrigerators, boilers and low-energy electric light bulbs.

Low-energy light. A 25 W low-energy bulb gives out as much light as a 100 W filament bulb but lasts 8 times longer, using only a quarter of the electricity.

Energy production without pollution Electricity can be produced in ways which are clean and renewable. Modern **wind power** machines could generate enough electricity to supply a fifth of the UK's annual demand. But many people think wind generators are ugly. A less visible alternative are machines which use **wave power** – the up and down motion of waves – to generate power. Another alternative is **geothermal power** which uses hot rocks deep underground to make steam which is used to work generators and so to heat homes.

Transport

Energy-efficient engines Modern cars can travel around 30 per cent further per litre of fuel than cars built 20 years ago, and research into more efficient engines and lightweight materials could increase this to 100 per cent. Reduce fuel consumption and you reduce pollution as well.

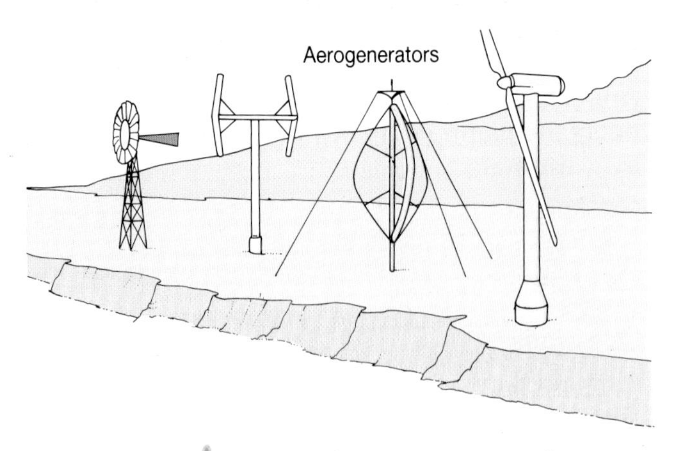

Wind power In areas where strong, steady winds blow, windmills can be used to produce electricity.

Exhaust fumes For every kilometre it travels, an average car produces up to 3 litres of oxides of nitrogen, 2.5 grams of hydrocarbons (both of which form ozone in sunlight), 20 litres of carbon monoxide and lead if leaded petrol is used. By using lead-free petrol and a device called a **three-way catalytic converter**, these pollutants can be greatly reduced (see graphs).

Lead-free petrol reduces some forms of pollution.

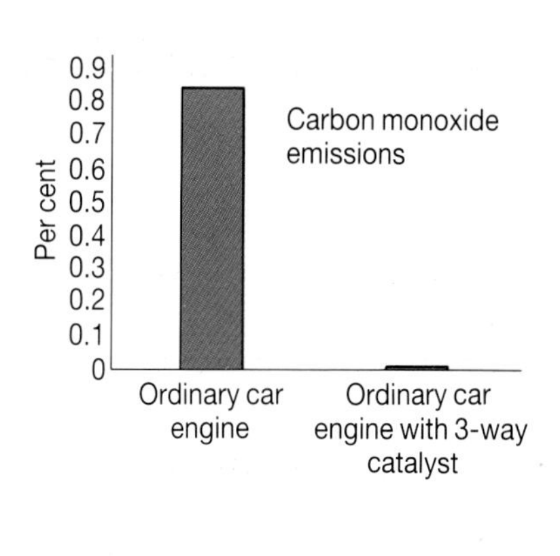

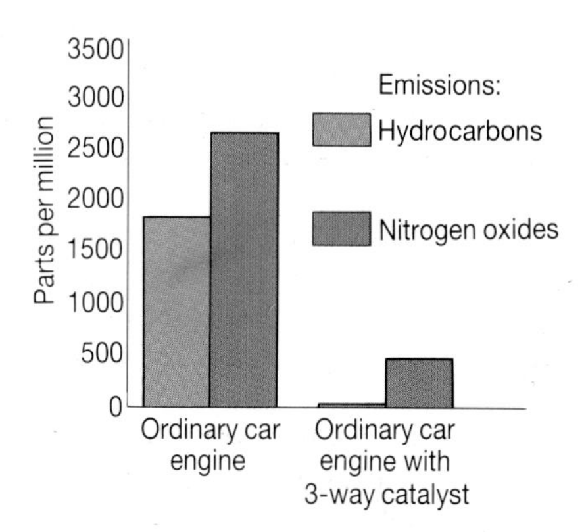

Public transport A train or bus full of passengers carries far more people per litre of fuel than a car. So public transport systems can reduce the number of cars, and therefore reduce pollution.

Agriculture

Fertilizers Genetic engineering could be used to reduce the need for nitrate fertilizers which cause pollution. Non-leguminous plants can be given a gene which allows them to fix their own nitrogen (see Unit 9.4). However, there are risks. No one can be sure of the long-term effects of genetically engineered plants.

If **organic fertilizers** such as manure and bonemeal are used instead of chemicals, they add humus to soil and release plant nutrients slowly. This way more are used by plants and less become pollutants.

Pesticides Plants can be genetically engineered to make their own pesticides. The use of pesticides can be reduced by encouraging natural predators of insect pests, such as spiders and ladybirds. This is called **biological control.** Pest numbers can be reduced by breeding large numbers of sterile male insect pests. When they mate, the female lays infertile eggs, which soon reduces pest numbers.

Cycling 1600 miles uses about the same amount of energy as there is in a gallon of petrol, without causing pollution. What other advantages are there to cycling?

Questions

1 Why would pollution be reduced if all homes were better insulated and used low-energy lights?

2 How do catalytic converters help reduce pollution? Use information in this Unit to calculate:

a) the amounts of oxides of nitrogen, carbon monoxide, and hydrocarbons a car produces between London and Birmingham;

b) the same amounts when the car is fitted with a catalytic converter which removes 80 per cent of these pollutants.

9.12 Don't throw it away!

We live in a wasteful, throw-away world. We buy things in packaging which we tear off and throw away. Bottles and cans are used once and thrown away. Some things are deliberately made so they can't be mended, or to wear out quickly and be thrown away. But nearly three-quarters of what we throw away *could* be used again!

Reuse

Millions of metal, glass and plastic containers are thrown away which could be used again. Old clothes can be sold, given away or used to make something else. Bits of old cars can be used to mend other cars. Think of other things we can reuse.

Recycling waste materials

Recycling means turning materials from old or broken things into new goods. Paper and card can be made into pulp and used to make recycled paper. Rags can be shredded and made into cheap cloth. Glass, metal and certain plastics can be melted down and used to make new goods.

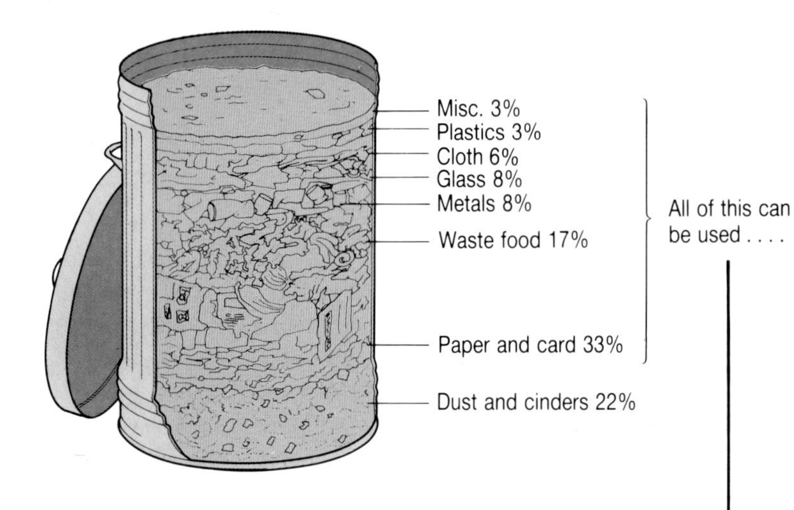

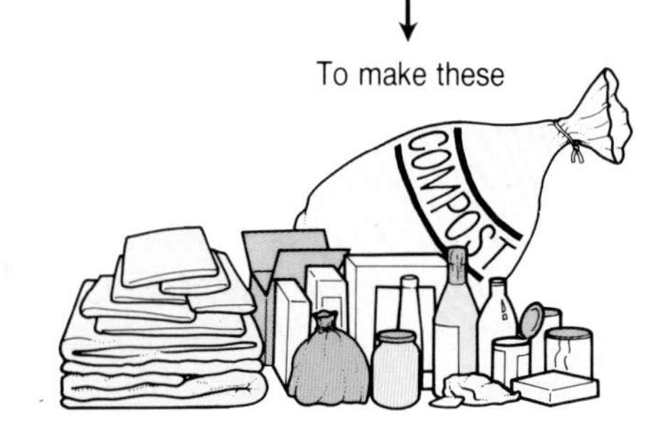

Why recycling is important

1 Recycling means less land is needed for ugly, smelly rubbish tips, and it reduces the risk of pollution from rubbish tips.

2 Recycling slows the rate at which the Earth is ripped up for minerals and forests are cut down for wood and paper.

3 Recycling saves energy. When you throw away metal or glass you also throw away the energy used to make it. It takes far less energy to melt down scrap metal and use it again than to produce it from metal ore.

Composting

In the UK, up to 35 per cent of domestic waste could be turned into compost. Waste which decomposes into compost includes **organic materials** such as weeds, grass mowings, hedge clippings, kitchen waste (apart from meat and fat which attract rats), also twigs, card, and paper if they are shredded first.

Any heap of organic material will produce compost eventually, but the process can be speeded up by turning the waste over regularly in a compost box. The waste must be kept moist, but never waterlogged. Composting produces a valuable soil conditioner, which reduces the need for peat. It also reduces the quantity of material sent to landfill sites.

All of this can be turned into …

Turning waste into energy

Gas from waste Decomposing organic matter produces gas which is 50 per cent methane. This **biogas** is produced in useful amounts by large-scale commercial composting plants, and also by decomposing organic matter in landfill sites. Biogas can be collected in buried, perforated tubes and used as a fuel for heating, or to generate electricity. A landfill site can continue producing useful amounts of biogas for up to 50 years, and perhaps even for as long as 100 years.

Solid fuel from waste Another useful fuel can be obtained by sorting refuse, to separate out combustible material which can be compressed into pellets or briquettes. When burnt, this fuel is able to produce at least half the energy of the same amount of coal.

… odourless compost which makes an excellent soil conditioner.

Landfill mining – the ultimate form of recycling

Recent work in the USA and other parts of the world has shown that there are good reasons for opening up old landfill sites. Useful materials can be reclaimed, such as saleable compost, metals, and glass, in addition to combustible materials for waste-to-energy projects. Landfill reclamation reduces waste volume by up to one half. So the site can be used for fresh refuse disposal, or the site can be reclaimed for alternative use. Sites leaking dangerous chemicals can be cleaned and made safe.

If you must throw away these substances, don't pour them down the drain. They can damage sewage disposal systems.

Questions

1 What is the difference between reusing and recycling? Describe ways in which you could reuse and recycle things.

2 How does recycling help reduce pollution and damage to the environment, and save energy?

3 Give reasons why organic waste should be used to make compost.

4 Describe ways in which waste can make money and energy.

9.13 Clean water and sewage disposal

Supplies of clean water

Every day you use about 160 litres of water, for drinking, cooking, washing and so on. Where does it all come from, and how is it made pure enough to drink?

Some comes from deep wells called **bore holes**, which take water from water-filled rocks, or **aquifers**.

Some comes from **reservoirs**. These are made by building a dam across a river valley so the valley fills with water.

Some is taken from rivers. This is called **river abstraction**.

Screening Metal grids placed where water is taken from rivers and reservoirs keep out floating weeds and rubbish.

Sedimentation The water is pumped into tanks where large particles settle to the bottom. They form a **sediment** which is piped away.

Coagulation Chemicals are added to the water to make small bits of dirt stick together (coagulate) so they can be filtered out.

Chlorination Then chlorine is added to kill germs.

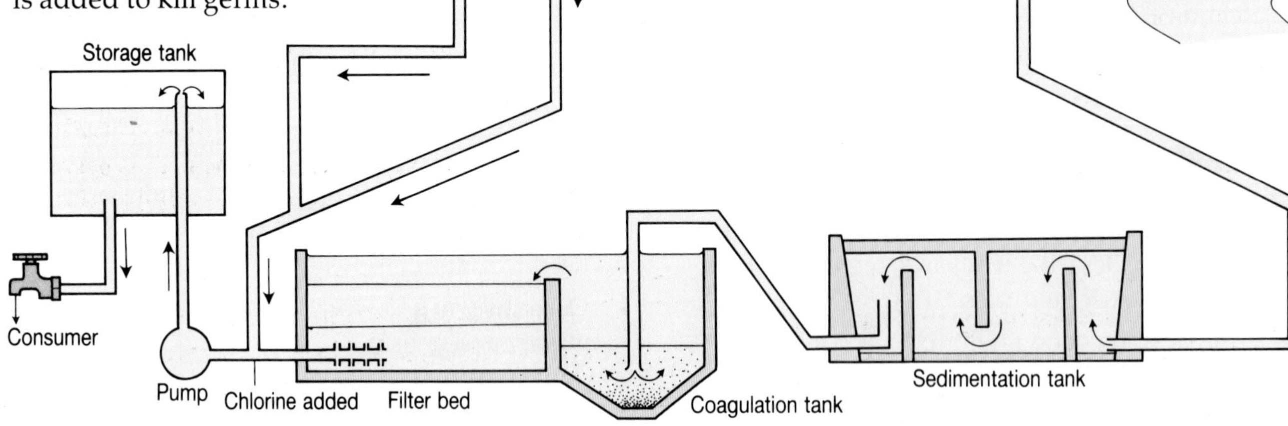

Sewage disposal

Sewage consists of liquid and solid waste from our bodies (faeces and urine), water from washing and some industrial wastes. It contains harmful germs. It must be disposed of so that it does not get into drinking water. This is done in a sewage treatment works.

1 Screening First wire nets strain out sticks, paper, rags and so on. These are burned or buried.

2 Grit tanks The screened sewage is passed into large tanks where grit, stones and other heavy objects sink to the bottom.

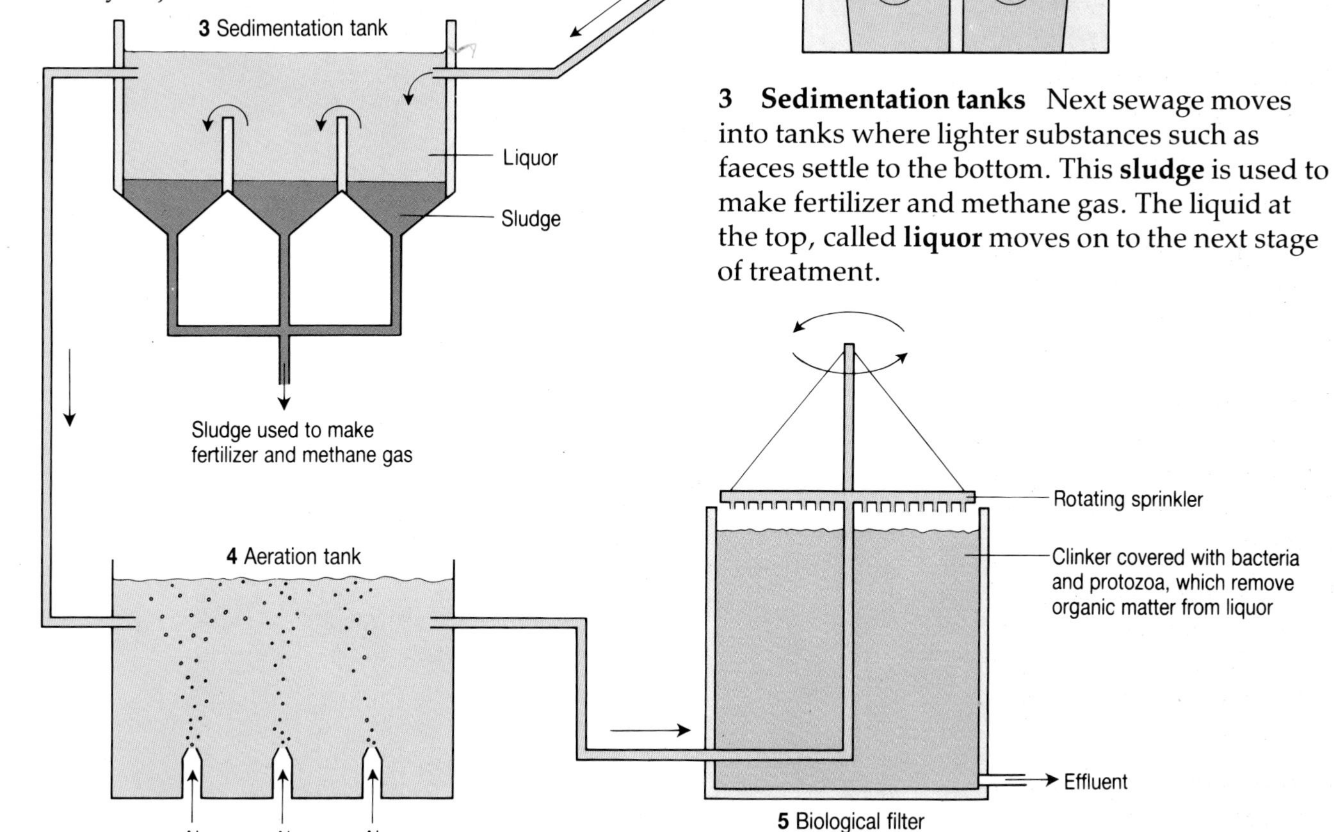

3 Sedimentation tanks Next sewage moves into tanks where lighter substances such as faeces settle to the bottom. This **sludge** is used to make fertilizer and methane gas. The liquid at the top, called **liquor** moves on to the next stage of treatment.

4 Aeration Air is bubbled through the liquor to encourage growth of microbes such as bacteria. These digest wastes and change them into carbon dioxide and other harmless substances.

5 Biological filter Then the liquor is sprinkled on to tanks filled with stones and clinker (the remains of burnt coal). These are covered with microbes which digest more wastes. This leaves a clean liquid called **effluent** which is pumped into rivers or the sea.

Questions

1 **a)** Name the three sources of water for drinking.
b) Which of these do you think has the cleanest and the dirtiest water. Give reasons.

2 Some cities pump untreated sewage into rivers and the sea. Think of ways this could harm people, and other living things.

9.14 Soil

All life on land depends on soil. Land plants need it for support, water, and minerals. Land animals need it because they eat plants, or other animals which eat plants.

Soil is made from small bits of **rock**, and from **humus** which is the decayed remains of dead animals and plants. The rock is broken into pieces by the weather. This is called **weathering**.

How soil is formed by weathering

1 Rock gets broken up when **rainwater** soaks into it and freezes. Rain also contains weak acids which can dissolve rock and carry it away.

2 **Heat and cold** make rocks expand and contract. After a time this can make them crack.

HEAT
COLD
RAIN

First, weathering splits the rock into small pieces.

Then weathering breaks these down into smaller rock particles. Soil is formed when they mix with humus.

Some rock particles are washed into rivers and worn down into sand, clay, and silt.

What soil is made of

If you shake up soil with water it settles into different layers.

Humus floats to the top. It is important because:

1 It is rich in minerals.

2 It contains nitrogen-fixing bacteria.

3 It sticks rock particles together so they don't get blown or washed away.

4 It keeps soil moist because it soaks up water like a sponge.

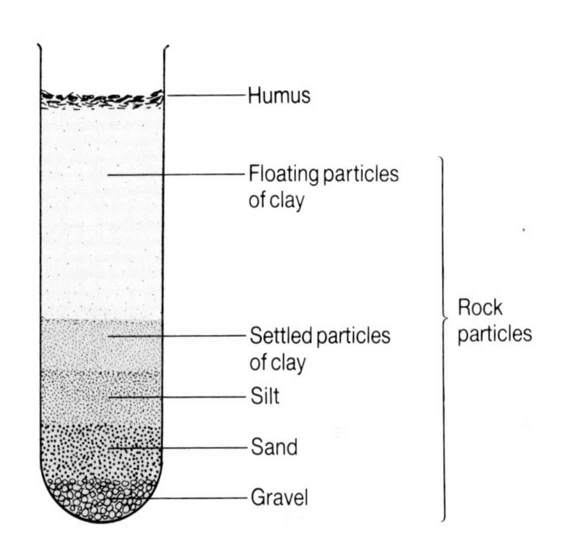

Fertile soil

The most fertile soil is called **loam**. Its rock particles are stuck together by humus into clumps called **soil crumbs**.

Soil crumbs stop minerals being washed from soil. They soak up water, and have air spaces in between them. So roots have minerals, moisture, and air. Soil crumbs also make soil easy to dig.

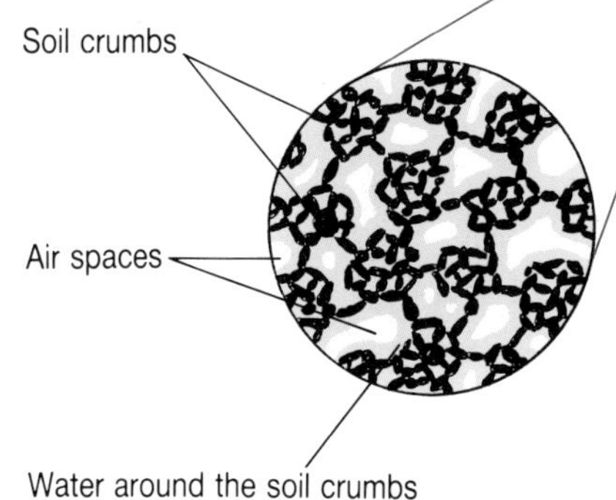

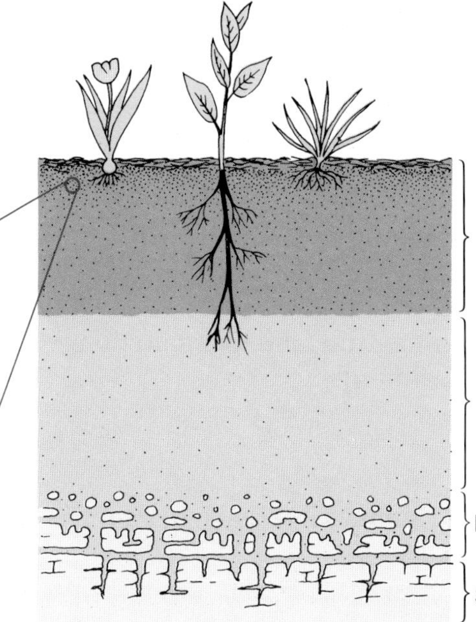

Topsoil is very fertile. Humus gives it a dark colour. It is rich in minerals, and contains many living things.

Subsoil is lighter in colour because it has no humus. It is rich in minerals but has few living things, except plant roots.

Fertilizers

Garden and farm soil must have fertilizers added, to replace the important minerals that are lost when plants are harvested.

Organic fertilizers include manure, well-rotted compost, sewage sludge, and bonemeal. These add humus as well as minerals to the soil. But they work slowly because they must decompose first.

Inorganic fertilizers are chemicals made in factories. They dissolve in soil water, so plants can use them immediately. But if they are used over a long time, they harm soil by destroying soil crumbs.

Soil life

The most important living things in soil are the bacteria, fungi and earthworms. They all help to keep soil fertile. The bacteria and fungi decompose dead plants and animals. When the earthworms burrow through soil, they mix its layers together, and help air and water get into it. They also drag in leaves which rot to form humus.

An earthworm in its burrow. Earthworms eat the soil as they burrow through it. They digest some, and pass the rest out as worm casts.

Questions

1 How does weathering turn solid rock into soil?

2 What is humus?

3 Why is humus an important part of soil?

4 How can chemical fertilizers make soil less fertile?

5 How do earthworms help keep soil fertile?

9.15 Conservation and ecology

We share the Earth with millions of different living things. But we are destroying wildlife so fast that only the toughest species such as rats, mice, and sparrows may survive.

How wildlife is destroyed

1 Wildlife is destroyed when land is cleared for crops, farm animals, housing estates, roads, factories, mines, and quarries.
2 It is destroyed by pollution and waste tipping.
3 It is destroyed by people who hunt for fun, and to obtain things like ivory and furs which we can do without.

Some wildlife habitats

A **habitat** is a place where plants and animals live. Each habitat has its own collection of plants and animals.

Grassland can have twenty or more different grasses, and scores of different flowers. Hundreds of different insects may live there. Most wild grassland has been ploughed for crops, or seeded with fast-growing grasses for cattle.

Heathland is covered with heather, gorse, broom, ferns, bilberry, lichens, and mosses. Snakes, lizards, and many birds live there. It is cleared mainly for commercial forestry, and sand and gravel quarries.

Wild woods contain oak, ash, beech, birch, shrubs, ferns, and many flowers. Most wild woods have been cleared to make way for crops or neat rows of spruce, pine, and other conifers.

Ponds and **marshes** are being drained and filled in so fast for crops and cattle, that many of their plants, amphibia, birds and butterflies have become rare, and may soon disappear altogether.

How wildlife can be conserved

Over a hundred special wildlife habitats are destroyed in Britain every year. We need land to grow food, and for houses, factories, and roads. But we must not destroy all our wildlife to get it. Some places must be left untouched, for ever.

Farmers can help save wildlife by using pesticides only when it is really necessary. They can leave wild habitats untouched, and plant wild species of trees and shrubs in places where farming is difficult.

You too can help save wildlife, by following these rules:

1 Never uproot wild plants, or take birds' eggs.

2 Never leave fish hooks or lead weights where they can be swallowed by animals. The fish hooks will injure them and lead will poison them.

3 Never start fires or leave litter in the countryside.

Warning – toads crossing! Signs like these help preserve our wildlife.

To make up for the habitats we destroy, we must find ways to create new ones. An abandoned quarry makes an ideal wildlife habitat. This one has been turned into a nature park.

A wildlife garden. It contains only wild plants.

Ecology

Ecology is the study of habitats and the wildlife that lives in them. Ecologists find out which plants and animals live in each type of habitat. They study how each is adapted to live where it does. They are concerned about the environment. They advise farmers, industries and governments to consider the environment before making their development plans.

Questions

1 Describe some of the ways that habitats are being destroyed.

2 How can farmers conserve wildlife?

3 What can you do to protect wildlife?

9.16 More about ecology

Living things depend on other living things for food. They also depend on non-living things like sunlight, soil, water and air. They need these for energy, support, shelter and the raw materials of life.

Ecologists study the ways living things depend on each other and on the non-living world. They find the links in food chains and food webs. They look at how organisms are adapted to their way of life and how they are affected by climate, soil, landscape and pollution.

Each type of tree has a different community of animals which feed off it or shelter in it. Nearly a hundred different creatures live on oak trees.

Communities

A **community** is a group of living things which share a habitat such as a rock pool, a pond, a meadow or a forest. Communities are made up of many kinds of animals and plants. A rock pool is the habitat of seaweeds, anemones, winkles, crabs, fish and so on.

A coral reef community is one of the richest on Earth.

Populations

A **population** is a number of creatures of the same kind. A forest may have populations of squirrels, owls, spiders, oak trees and bluebells.

A population of penguins.

A population of periwinkles.

Adaptations

Every living thing is adapted to live in a certain way. Lions have teeth for tearing flesh and crushing bones. Butterflies have long tubular mouths for sucking nectar from flowers. Most flowers have colour and scent to attract insects for pollination. An eagle's beak is adapted for tearing flesh.

Sundew leaves are adapted for catching insects.

An eagle's beak is adapted for tearing flesh.

The environment and living things

An organism's **environment** is all those things around it which affect its way of life, such as climate, landscape, soil, and other living things.

Climate The number and kinds of creatures which live in a region depend on the amount of rain and sunshine there, and the average temperature. The animals and plants in hot dry regions are completely different from those in cool wet ones, or icy cold ones.

Landscape The shape of the land affects living things because it causes local differences in climate. A river valley has a different climate from the top of a nearby mountain. The coast has a different climate than inland areas a few kilometres away.

Soil Many plants are very choosy about the type of soil they will grow in. Sandy and clay soils, for instance, have very different plant populations, because sandy soils have large air spaces and dry out quickly, and clay soils have tiny air spaces and hold the water when it rains.

Other living things Everywhere you look organisms affect the way other organisms live. Plants compete with each other for light and water. Animals feed on plants and on each other. Insects pollinate flowers, and animals disperse seeds. Humans clear land for farming, and pollute the environment.

Only tough plants like rock samphire and lichens can live on this exposed sea cliff.

The fence keeps out grazing animals. Note how this affects plant life.

Bluebells and wild garlic are plants which can live in the shade of woodland trees.

Questions

1 What are the differences between a community and a population? Give an example of each.

2 Explain how a sparrow is adapted for flying, perching, and eating seeds. How is it different from an eagle?

9.18 How to study wildlife II

Random sampling of a habitat

Which are the commonest species? Where do they live in the habitat? It is usually impossible to make a study of all the plants and animals in a habitat in order to be able to answer questions like these. Instead you can use various **sampling techniques**.

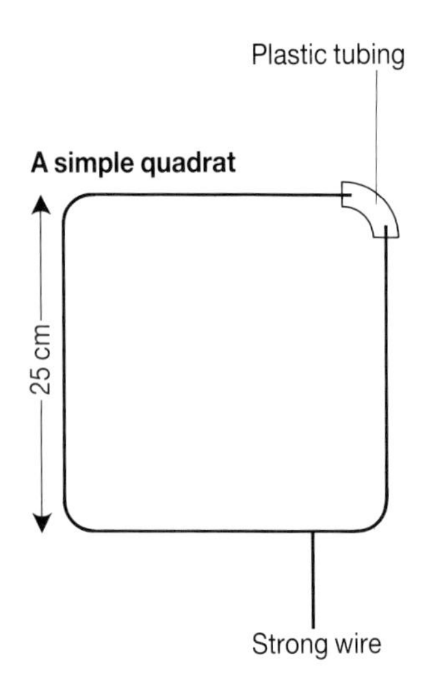

Sampling plant life

To do this you use a square or rectangular frame called a **quadrat** to study small areas (samples) of the habitat, chosen at random. You can place a quadrat at random throughout the habitat, by throwing it over your shoulder. The idea is that it should have an equal chance of landing anywhere, therefore do not deliberately throw it to land on the vegetation which looks the most interesting.

Walk ten paces in any direction, then throw the quadrat without looking. Repeat, after recording the plants inside the quadrat. The records you make depend on the information you need, which could be the **density, frequency,** or **percentage cover** of various species.

Density This is the number of plants or animals in a unit area of habitat (e.g. the number inside a 25 cm^2 quadrat). To discover the density of a plant species in a habitat, you count the number of this species present inside a quadrat each time it lands. Continue doing this until the quadrat has been cast throughout the whole habitat. Then you can calculate the average number of times the species was found.

Frequency This is the number of times that a particular species is found when a quadrat is thrown a certain number of times. To calculate frequency, you count the number of different species within the quadrat each time it lands, and note their names. If, for example, you throw the quadrat 100 times, you have to note the number of times each species was found, and express each result out of a 100. This will tell you the most common (most frequent) species in the habitat, then the next most common, down to the rarest.

Percentage cover This is the percentage of the total area of a habitat which each species covers. It can only be calculated with species which grow in clumps, or large patches. To do this, estimate the percentage cover of the species found inside a quadrat, each time it lands. Then calculate the average cover over a number of throws.

Microscope slides suspended in a pond, or an aquarium, will be colonized by an amazing variety of microscopic plants and animals. Devise experiments in which you study how the numbers of different species change with time, how the populations of these species change with time, and the percentage cover of species such as microscopic algae which become attached to the slides.

Sampling animal life

One aim of studying animal populations is to list species from the most common to the rarest. This is difficult because animals move in and out of an area, especially in open habitats like estuaries and the sea-shore. But the following methods can produce useful results.

Comparative method Carefully search the habitat and compare the numbers of each species you find. Give each species a score: from 5 for the most abundant to 1 for the rarest. If several groups work independently, average scores can be obtained.

Sampling pond life Use a plankton net (see photo on p. 198) to sample the small, swimming creatures in different zones of a large pond. For example, sweep the following zones ten times: the surface, deep layers of open water, and shaded and unshaded shallow water. Collect the samples from each zone in separate white dishes, and record the number of species in each. Try to account for any differences between the samples from each zone: could temperature, or pH differences, be important?

Capture-recapture method You can estimate the total population of a species which can be caught easily and marked in some way, and which disperses quickly when released, e.g. dragonflies, large beetles, snails, crabs, woodlice, and water boatmen. Catch a number of specimens and mark them with a non-toxic paint, then release them where they were caught. After a day or two, when they have thoroughly dispersed, try to capture the same number again. Note the number of marked specimens in the second catch. An estimate of the total population of this species is calculated by this formula:

$$\textbf{Population} = \frac{\textbf{total in first catch} \times \textbf{total in second catch}}{\textbf{number of marked specimens in second catch}}$$

By studying gravestones of different ages encrusted with mosses, algae, and lichens, you can find how the percentage cover of these organisms changes with time. Does the position of the stone in the graveyard affect this growth?

Projects and questions

1. The histograms opposite show the number of plants found in random samples taken in lightly trampled and heavily trampled grassland.
   a) Which species is most affected by trampling and which is least affected?
   b) Study trampled areas and try to find out which plants can live there.
   c) Try to explain how some plants are adapted to withstand trampling.
2. Some students caught 50 crabs in a 25 m^3 area of sea-shore, marked them with paint then released them. Four days later they caught another 50 crabs: 13 of these were marked with paint.
   a) What is the estimated total crab population for this area?
   b) Why is this only a rough estimate of the crab population?

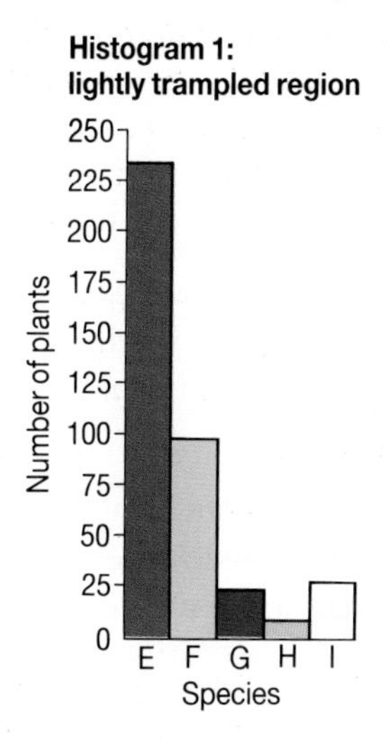

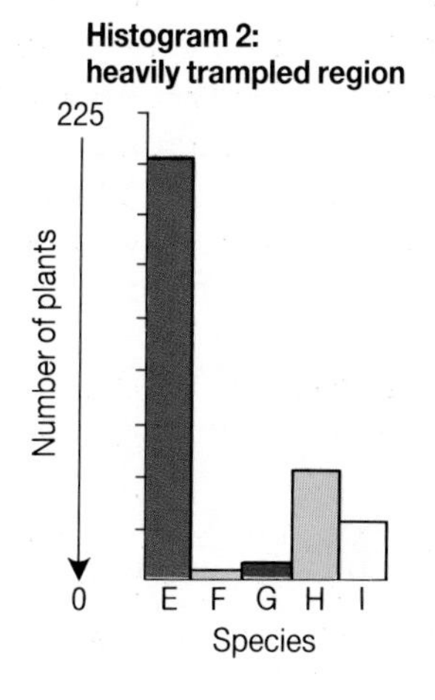

9.19 Managed ecosystems

All creatures which inhabit this planet depend on sunlight energy. This energy is absorbed by plants during photosynthesis, and then transferred along food chains and food webs to us, and other animals throughout the world (Units 3.1 and 9.1).

When humans first evolved, they obtained energy like other animals by gathering grains, fruits, roots, and nuts, and by killing animals. And then, about 10 000 years ago, people began learning how to cultivate nutritious plants, and began to domesticate wild animals.

Over the centuries, farmers have increased food production by improving the efficiency of energy transfer between the Sun, plants and human beings. Modern farming methods have greatly increased yields. In the year 1200, one hectare of soil provided food for only five people. Today it could feed up to fifty. Farm animals have also been improved. Cows produce twice as much milk as in 1945, cattle yield more meat, and chickens lay more eggs than their undomesticated ancestors did.

Clear the land and kill the competition

Vast areas of land have been cleared for cultivation throughout the world. This is only the first step in a never-ending battle against weeds, pests, and diseases – all of which reduce the amount of food produced.

Pests and diseases These destroy at least 30 per cent of world food production. **Insecticides** kill insect pests such as greenfly, weevils, and caterpillars. **Fungicides** kill fungal diseases of crop plants including rusts, smuts, and mildew. They can also be used to control fungi which spoil stored food.

Weeds These are plants which compete with crops for space, water, and minerals. They can be controlled by hoeing, but this is very labour-intensive. An alternative is to kill the weeds by spraying with herbicides.

Protecting crops from the weather

Irrigation Crops can be damaged or killed by drought, but irrigation (artificial watering) can save them. Irrigation can also transform dry, infertile soil into highly productive land. There are various methods of irrigation. Water can be sprayed or sprinkled on to the soil, allowed to run through channels cut between rows of plants, or delivered directly to plant roots through underground pipes.

Insecticides increase yields by destroying insects. Organochlorines (e.g. Dieldrin) paralyze insects and stop gas exchange. Organophosphorus insecticides (e.g. Malathion) can be absorbed by plants and kill sap-sucking insects such as greenfly.

Weeds can be killed with plant hormones which make them grow out of control (compare top and bottom photographs). These are **selective weed killers**: they affect broad-leaved weeds (e.g. plantain) but not grasses and cereal crops.

Greenhouses The ultimate in weather protection is to grow plants under cloches, frames, or in greenhouses. These accelerate growth by providing warmth, humidity, and protection from wind, rain and some pests such as birds. Vegetables, salad crops, and flowers can be germinated under glass up to a month ahead of their normal season. Also plants from warm climates can be grown in cooler regions.

Winter lettuce, tomatoes, cucumbers, melons, aubergines, and peaches are some of the crops which thrive under glass. Flowers such as calceolaria and gloxinia are also grown this way.

Improving crops and livestock

Selective breeding and modern genetic engineering methods continue to increase yields (Unit 2.11) but growth can also be influenced using various **hormones.** These can be used to control fertility (Unit 8.6), growth, and milk yields in animals – though many people are worried about the long-term effects of animal hormones. Plant hormones are used extensively to stimulate fruit production and ripening, and to preserve cut flowers. They are also used to encourage dwarfism in cereals so they are less likely to be blown over by high winds.

More than 90 per cent of hens kept for eggs are housed in battery cages like these.

Improving soil

Agricultural soils rapidly deteriorate into infertile desert unless minerals removed by crops are replaced. This is done by adding fertilizers, which can increase yields by over 50 per cent. **Organic fertilizers** including manure, compost, bonemeal, and dried blood, add humus to soil. They improve its texture and water retention, so that minerals are released slowly. **Inorganic fertilizers** include sulphate of ammonia, sodium nitrate, and sulphate of potash. They release minerals immediately. But they do not contain humus. So their continuous use degrades soil, reduces the vital activities of soil microbes, worms, insects, and other soil life, and can lead to water pollution (Unit 9.7).

Some pigs are kept in 'crates' like these. They have warmth, food and water but cannot move about.

Factory farms

Some modern farms are like factory production units. The majority of pigs and poultry are kept in specially constructed **intensive units** where light, heat, humidity, ventilation, and hygiene are carefully controlled. A large poultry unit has up to 100 000 birds, with mechanized feeding, cleaning, watering and egg collection. The animals are warm and well fed, and production costs are low. But they live in cramped conditions, and cannot behave naturally.

Questions

1. In the past, a field was used for two years then left **fallow** (unused) for a third. Why was this necessary, and why is it unnecessary today?
2. How has machinery increased farm production?
3. What are the advantages and disadvantages of using pesticides?
4. What is a factory farm? List arguments for and against factory farms, and farms on which animals are kept out of doors.

Questions on Section 9

1 *Recall knowledge*
Put each food chain in the correct order:
a) water snail → humans → water weed → fish
b) greenfly → rose leaf → ladybird
c) worms → dead leaves → hawk → thrush
d) slug → grass snake → frog → lettuce leaf

2 *Recall knowledge*
Below is a diagram showing part of the carbon cycle. What do the labels A, B, and C stand for?

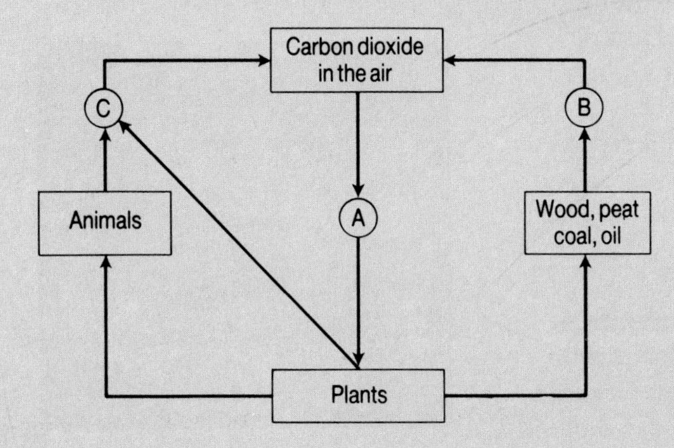

3 *Recall knowledge*
Below is a diagram of the nitrogen cycle.
a) How does clover help to convert nitrogen gas into soil nitrates (label A)?
b) Name the soil bacteria which convert nitrogen gas into nitrates (label B).
c) Describe one other way in which nitrogen gas is converted into nitrates (label C).
d) How are animal proteins converted into nitrates (label D)?
e) How are some soil nitrates converted back into nitrogen gas (label E)?

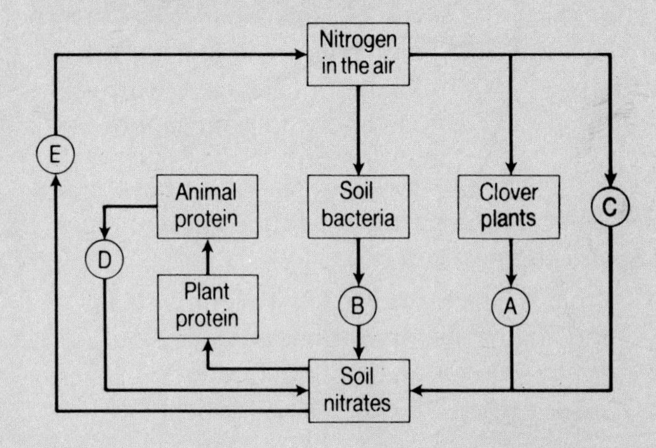

4 *Understanding*
Below is a map of a river.
a) Why do fishermen catch more fish at point A than at point C?
b) Many trees in the conifer plantation are dying. Why do you think this is?
c) One day the sewage treatment works broke down, and untreated sewage entered the river. What happened to the amount of oxygen in the water at point B? Explain your answer. What effect would this have on water animals?
d) When the farmer puts fertilizer on his crops, the water weeds at point C grow bigger. Explain why.
e) Local fish-eating birds get sick when the farmer sprays his crops with pesticide. Explain why.

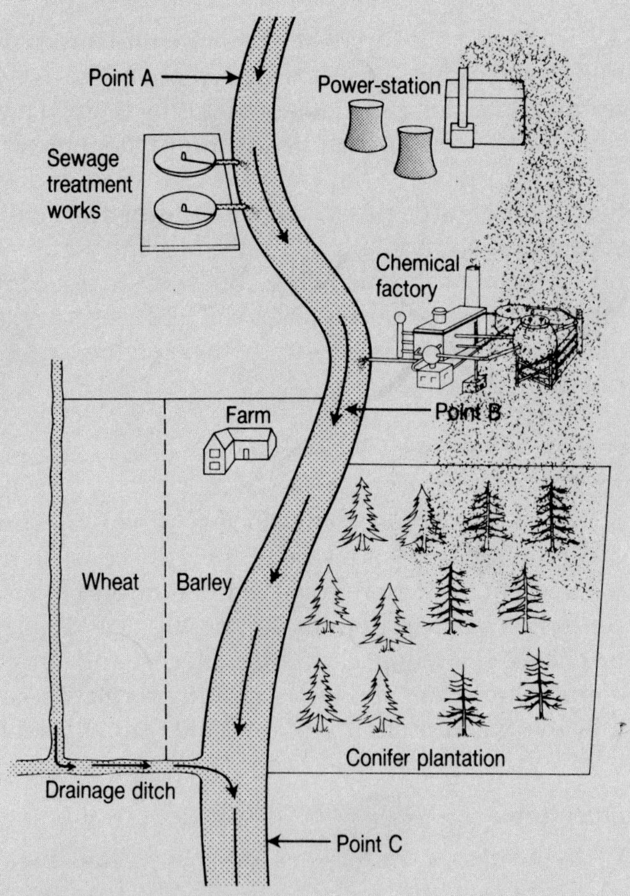

5 *Understanding*
In the autumn, the leaves fall off trees. But chemicals from the leaves may be taken back into a tree through its roots, and used for new leaves. Explain how this happens.

6 *Understanding/interpret data*

Radioactive materials were illegally dumped on a waste tip. Two types of organism began to decompose this material and gave off a radioactive gas. Some time later radioactivity was detected in plants growing nearby, and later still it spread to animals. The amount of radioactivity in two animals is shown in the graph opposite.

a) Name two decomposers.
b) What radioactive gas do you think the decomposers gave off?
c) What process in plants took in this radioactive gas, so that the plants became radioactive?
d) How could grasshoppers and spiders have become radioactive?
e) Why does the curve for spiders lag behind the curve for grasshoppers?
f) Why does the curve for spiders have a higher curve than the curve for grasshoppers?
g) What living processes may have led to the decrease in radioactivity?

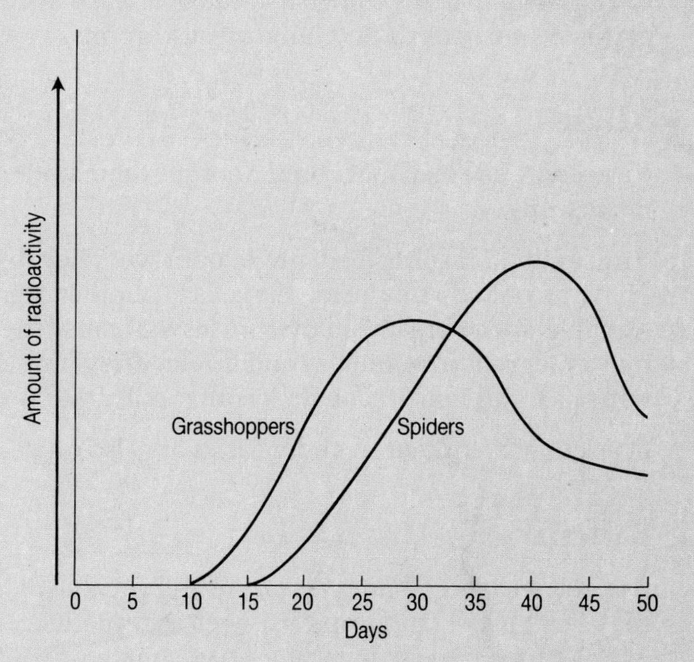

0 = The day radioactive material was dumped

7 *Understanding*

The diagram opposite shows the amount of energy which flows through living things in a grassland habitat each year.

a) What process in grass absorbs energy from the Sun?
b) Calculate the difference between energy absorbed by the grass and the energy passed from grass to other living things. Give reasons for this difference.
c) What would happen to the habitat if decomposers could be removed?

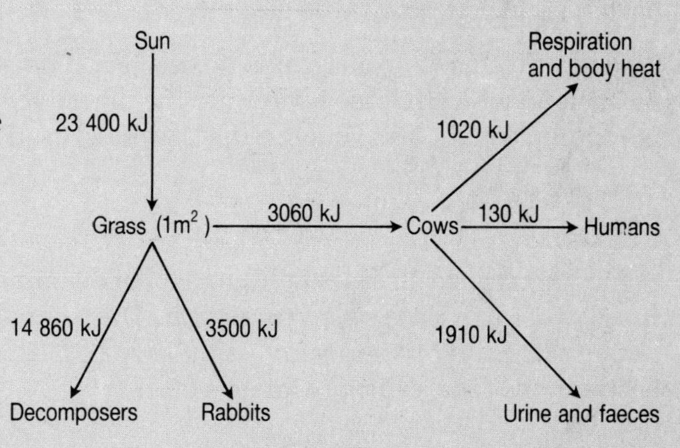

8 *Understanding*

a) Which of the following food chains fits the pyramid of numbers opposite:
- **i)** grass→rabbits→fox
- **ii)** tree→insects→birds
- **iii)** phytoplankton→zooplankton→fish
- **iv)** tree→monkeys→fleas
- **v)** grass→insects→birds

b) What is a pyramid of biomass?
c) Draw the pyramid opposite again so that it is a pyramid of biomass.

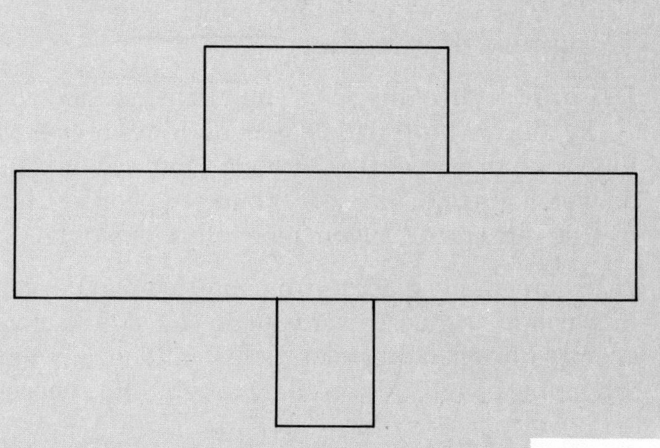

10.1 Germs, disease, and infection

Germs are tiny living things that cause disease. All viruses are germs. Some bacteria and fungi are also germs.

The common cold virus. It is transmitted through the air or by direct contact.

Viruses

Viruses are so small that you cannot see them under an ordinary microscope.

Viruses don't respire, feed, grow, or move. They just reproduce. They can only do this inside the cells of another organism. For example, if you breathe in the viruses that cause the common cold they enter cells in your nose and throat. They turn these cells into virus factories which soon infect other cells.

Chicken pox, influenza, and measles are also caused by viruses.

Bacteria

Bacteria are larger than viruses, but still very small. Unlike viruses they feed, move, and respire, as well as reproduce. Bacteria which live as germs harm living things in two main ways.

1 They destroy living tissue. Tuberculosis is a disease caused by bacteria which destroy lung tissue.

2 They produce poisons, called **toxins**. Food poisoning is caused by bacteria which release toxins into the digestive system. Boils, whooping cough, and venereal disease are caused by bacteria.

Fungi

Some fungi cause diseases in humans. For example, **ringworm fungi** cause a disease called **ringworm**. This shows up as red rings on the scalp and on the groin. A similar fungus attacks the soft skin between the toes, causing **athlete's foot**.

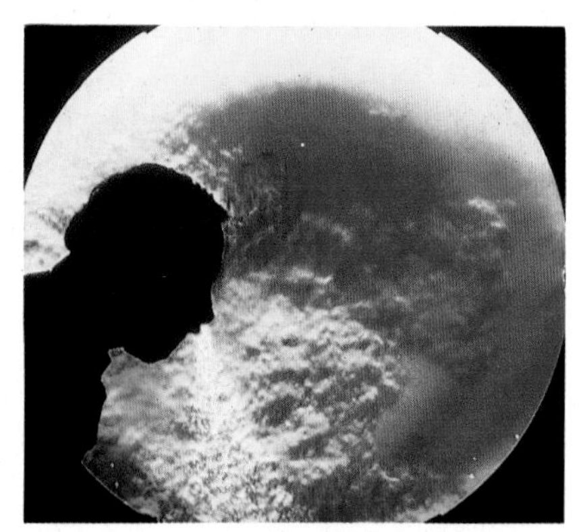

A special camera was used for this photo of a man sneezing. Note the air and droplets of mucus exploding from his nose and mouth. Can you see how coughs and sneezes spread germs?

The spread of infection

These are some of the ways you can be infected with germs:

1 By touching infected people, or things they have used such as towels, combs, and cups. Athlete's foot can be caught by walking on wet floors or mats used by infected people. Chicken pox and measles are spread by touching infected people.

2 By breathing in germs from infected people, especially when they cough or sneeze near you. So you should always cough and sneeze into a handkerchief. Colds, influenza, pneumonia, and whooping cough are spread by coughs and sneezes.

3 From infected food and drink. Food and drink can be infected with germs by coughs and sneezes, dirty hands, flies, mice, and pet animals. Infected food and drink cause food poisoning and dysentery.

Preventing infection

These are some ways you can avoid being infected with germs.

Hair Wash frequently, especially in hot climates, with a medicated shampoo which reduces dandruff and kills insect parasites. Frequent use of a clean comb and brush keeps the scalp healthy.

Eyes Foods rich in vitamin A help maintain healthy eyes. Avoid reading in dim light and exposing the eyes to excessive dust. Use only good-quality sunglasses to reduce glare without distorting your vision.

Food and drink Cook food thoroughly and eat it straight away. Store unused food in a fridge or freezer. Boil water you think may be infected with germs.

Hands Wash thoroughly before meals, before handling food, and after visiting the toilet. Cover any cuts with sterile dressings, which should be changed regularly.

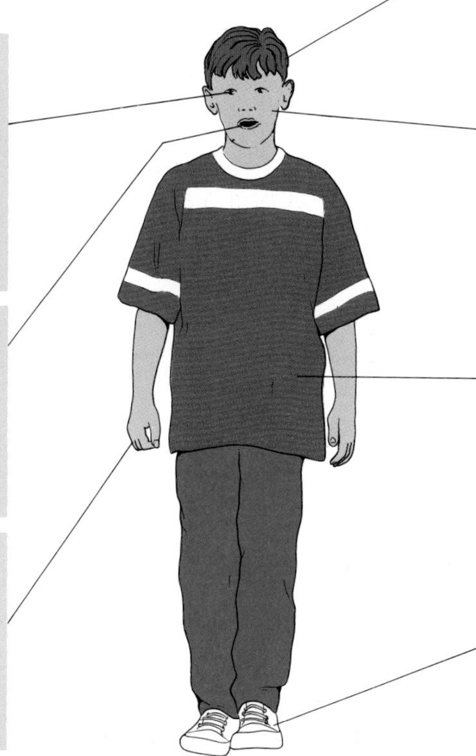

Face Cosmetics must be washed away each night with good-quality soap. Cheap, alkaline soaps can irritate skin.

Body Bathe or shower regularly, especially in hot weather, to avoid skin infections. Excessive use of anti-perspirants in hot climates can cause heatstroke. Change clothing, especially undergarments, regularly.

Feet Wash daily, and dry carefully, especially between the toes. Change socks daily, and avoid using the same pair of shoes every day for long periods, to stop the build-up of sweat and germs.

Questions

1. List all the ways germs are being spread in the drawing above.
2. Why should you never use a stranger's towel?
3. Why should you sneeze into a handkerchief?
4. Why should you regularly wash between your toes?
5. Why must you cook food thoroughly?

10.2 Defences against disease

Your body has many ways of defending itself against viruses, bacteria and other germs. Together these defences give your body **natural immunity** against infection.

Your **nose** and **air passages** are lined with **mucous membrane**. This is a layer of cells which make a sticky fluid called **mucus.** The mucus traps germs and dirt in the air you breathe. Tiny hairs called **cilia** move the mucus to your throat where it is swallowed, passing harmlessly out of your body.

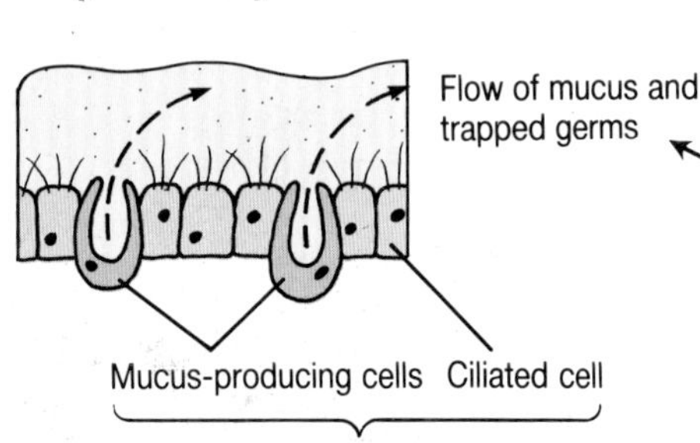

Your **skin** has an outer layer of dead cells which form a barrier against germs. **Sebaceous glands** under the skin make **sebum**, an oily antiseptic liquid which keeps skin supple, waterproof and protected from most germs.

Germs which get into wounds or your **bloodstream** are eaten by white blood cells called **phagocytes**.

The lining of your **stomach** produces acid needed for digestion. This acid also kills germs which enter your body in food and drink.

Your **eyes** are protected against germs by an antiseptic liquid from your **tear glands**. Your eyes are washed with this liquid every time you blink.

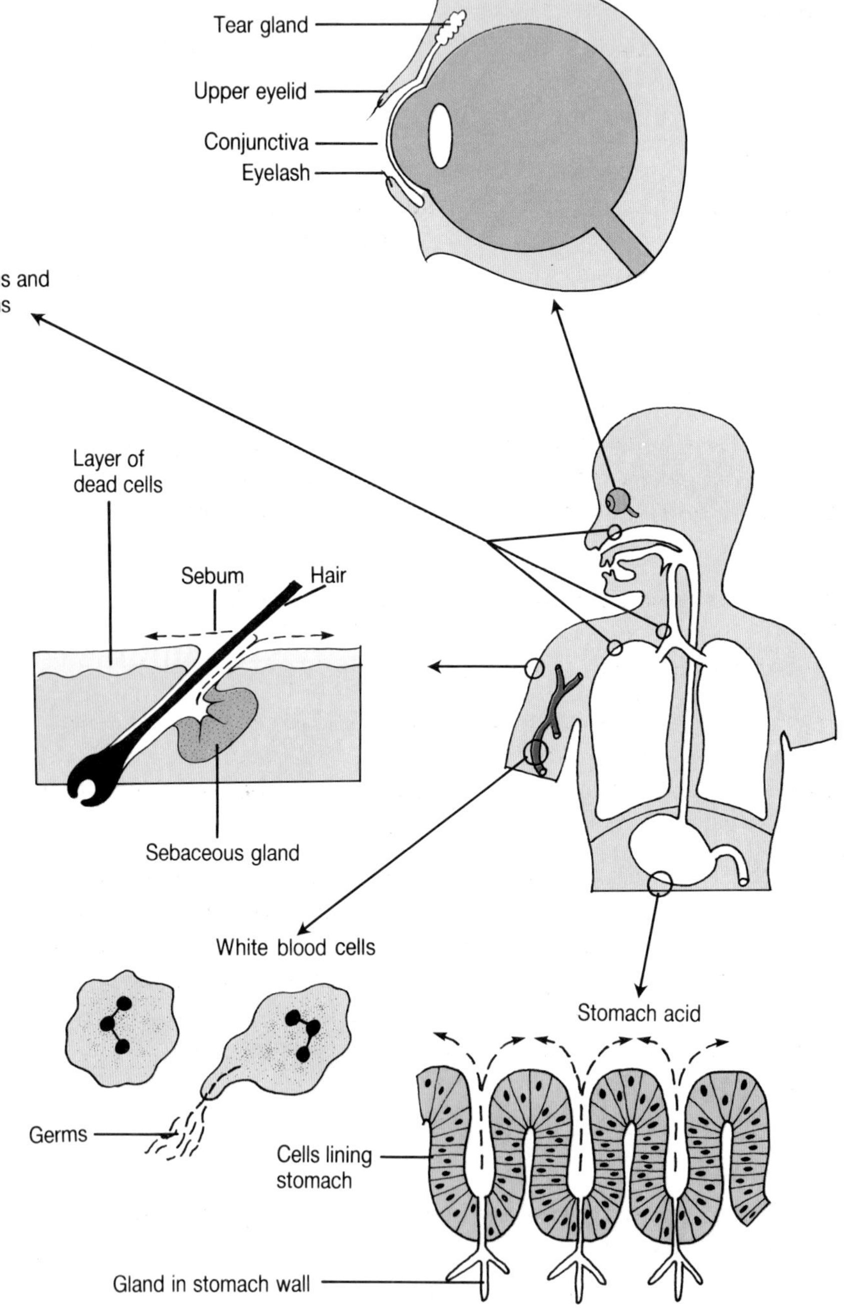

Antibodies – your body's chemical weapons

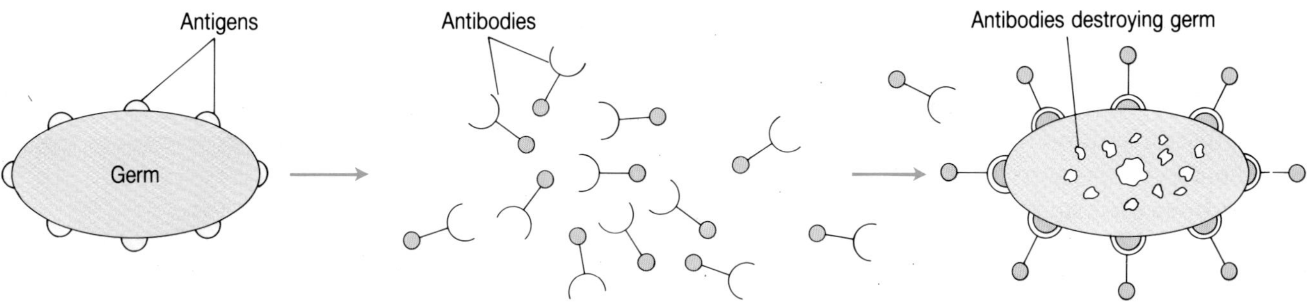

Bacteria, viruses and other germs have substances called **antigens** on their surface.

When your body detects antigens white blood cells called lymphocytes send out chemicals called **antibodies**.

Antibodies combine with the antigens. This kills the germs by making them burst or stick together, or by making it easier for phagocytes to eat them.

Each type of germ needs a different antibody to kill it. Some antibodies called **antitoxins** can destroy the poisons (toxins) which germs produce. Once an antibody is made it can stay in your body for several years making you immune to further attacks from the germs it destroys.

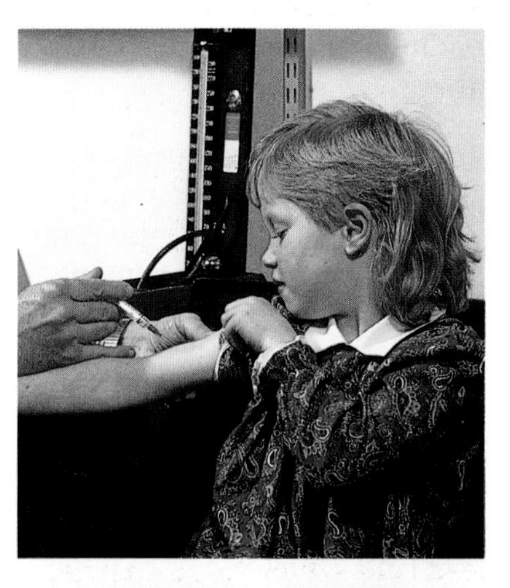

Artificial immunity

You don't have to catch a disease to make antibodies against it. You can be given a **vaccine** which makes your body produce antibodies so you are ready to fight off an infection before the germs arrive.

A vaccine contains either dead germs, harmless germs or a harmless germ toxin. When you are vaccinated your lymphocytes make antibodies just as if real germs had got into your body.

One of the first people to use a vaccine was Edward Jenner. In the early nineteenth century he proved that people who caught a mild disease called cowpox were then immune to a killer disease called smallpox.

He scratched the skin of a healthy boy and rubbed pus from a girl with cowpox into the wound. The boy caught cowpox. When he recovered Jenner inoculated him with pus from a smallpox victim, but the boy did not catch this disease. We now know that cowpox and smallpox germs have the same antigens, so the boy's body already contained the right antibodies to fight off the smallpox infection.

Questions

1 Explain how your eye, skin, air passages and gut fight off germs.

2 Describe two ways your blood fights germs.

3 John caught measles in 1986. His sister caught measles in 1987 but John didn't catch it again. Explain why, using technical words from this Unit.

10.3 Sexually transmitted diseases

Sexually transmitted diseases, or **venereal diseases** (**VD**), pass from person to person during sexual intercourse. **Syphilis** and **gonorrhoea** are examples of these diseases which are caught only by sexual contact. **Acquired immunodeficiency syndrome** (AIDS) is passed on mainly by sexual contact but there are other ways of becoming infected.

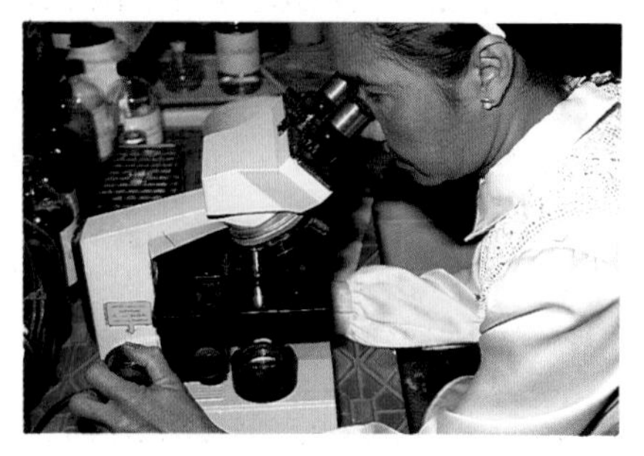

Microscopic examination of human tissue is needed to identify venereal and other diseases. Clinics are available in most towns and cities where confidential advice and help are available for the prevention and treatment of sexually transmitted diseases.

Syphilis

This disease is increasing, but affects more men than women. Early stages can be cured with antibiotics, but syphilis is often difficult to detect. Symptoms usually appear two weeks to one month after contact with an infected person, and the disease has three stages.

Stage one Sores appear around the sex organs, anus, or mouth. The sores look like craters which vary in size between a pin-head and a pea, and have a red centre. They are painless and soon disappear without treatment.

Stage two This is a rash which can appear weeks or months after the first stage. It too is painless and soon disappears.

Stage three This appears years later if the early stages go untreated, and includes blindness, heart disease, and insanity.

Syphilis is dangerous because the early stages are painless and often go unnoticed and untreated. Syphilis can only be treated in its early stages, and it is only during this time that the disease is infectious. It can pass from a mother to her unborn baby, which often dies in the womb or soon after birth.

Gonorrhoea

Gonorrhoea is commoner than syphilis, and its incidence is increasing. It affects both males and females. All cases can be cured. Symptoms appear about ten days after contact with an infected person. These include a burning sensation during urination, followed by a greenish-yellow discharge. Unfortunately, two out of three women have no symptoms at all, and consequently may only know of the disease when their partner develops symptoms.

Later stages of the disease include inflammation of the lining of the heart, inflammation of joints (arthritis), and damage to the iris of the eye. If a pregnant woman is infected with gonorrhoea, she can pass the disease on to her baby as it passes through the vagina during childbirth.

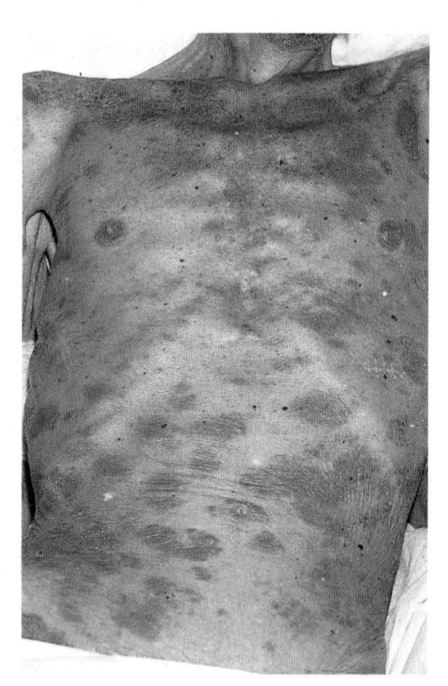

A victim of AIDS. The disease cannot be cured and so must be prevented from spreading. Wearing condoms during sex stops infection. People who use hypodermic needles should use sterilized ones.

AIDS (Acquired immunodeficiency syndrome)

AIDS is caused by the **human immunodeficiency virus (HIV)**. HIV attacks the body's immune system, destroying **T-cells** – a type of white blood cell which produces **antibodies**, (the chemicals which destroy germs). This leaves AIDS patients open to normally harmless diseases which uninfected people fight off without difficulty.

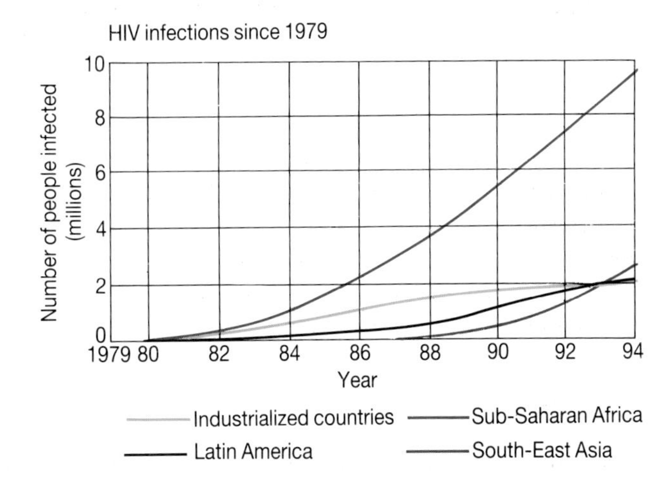

AIDS patients are killed by chronic pneumonia and diarrhoea; fungus infections which block the throat, intestine and lungs; cancer of the skin and bone, and swelling of the brain. The AIDS virus itself can destroy brain tissue causing memory loss, personality changes, and death. At present there is no cure for AIDS.

How you can get AIDS AIDS is passed on by sexual contact, and by contact with body fluids, especially blood.

1 During sexual intercourse, the virus can pass from a man's semen to a woman (or to a man during homosexual contact), or from a woman's vaginal fluid to a man.

2 HIV is in the blood of AIDS patients, so you can be infected if their blood gets into your body.

- This can happen when drug misusers share needles and syringes. The virus can also be passed from person to person on unsterilized tattooing, ear-piercing, and acupuncture equipment.
- You may also be infected by sharing a razor or toothbrush with an infected person (gums can bleed when teeth are brushed vigorously).
- A pregnant woman can pass the virus to her developing baby through the placenta.

How you cannot get AIDS You cannot gets AIDS by sharing a home, school, or workplace with an infected person; by touching them or breathing the same air; or by using the same shower, cutlery, furniture, swimming pool, or toilet as an infected person.

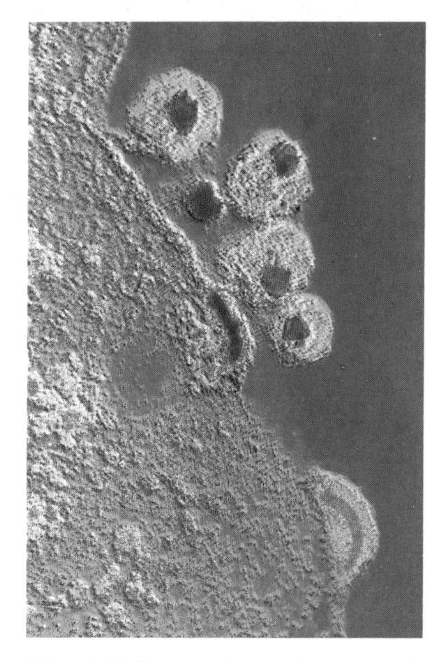

The AIDS virus bursting out of a white blood cell.

Questions

1 **a)** Why is it important that syphilis is detected and treated in its early stages?
b) Why is early detection of syphilis especially important for pregnant women?

2 Why should women with several sexual partners have regular checks for gonorrhoea even if they have no symptoms?

3 How do you think condoms help reduce the risk of infection with VD and AIDS?

4 **a)** Why should first-aid workers wear rubber gloves when dealing with patients who are bleeding?
b) Why should spilt blood be washed away with water containing bleach?
c) Why is it dangerous for drug users to share hypodermic needles?
d) If a friend became an AIDS patient, why would it be unnecessary and wrong to avoid all contact with him or her?

10.4 Cancer

What is cancer?

Cancer is not one disease. There are over 200 different types, each affecting different parts of the body. But they all begin in roughly the same way.

Every cell has genes which control cell division. These genes are 'switched on' when new cells are needed to replace cells that wear out or die. When the right number of cells have been produced, the genes 'switch off' and cell division stops.

Sometimes these genes are damaged or altered. This can be caused by chemicals (called carcinogens), by radiation, or viruses.

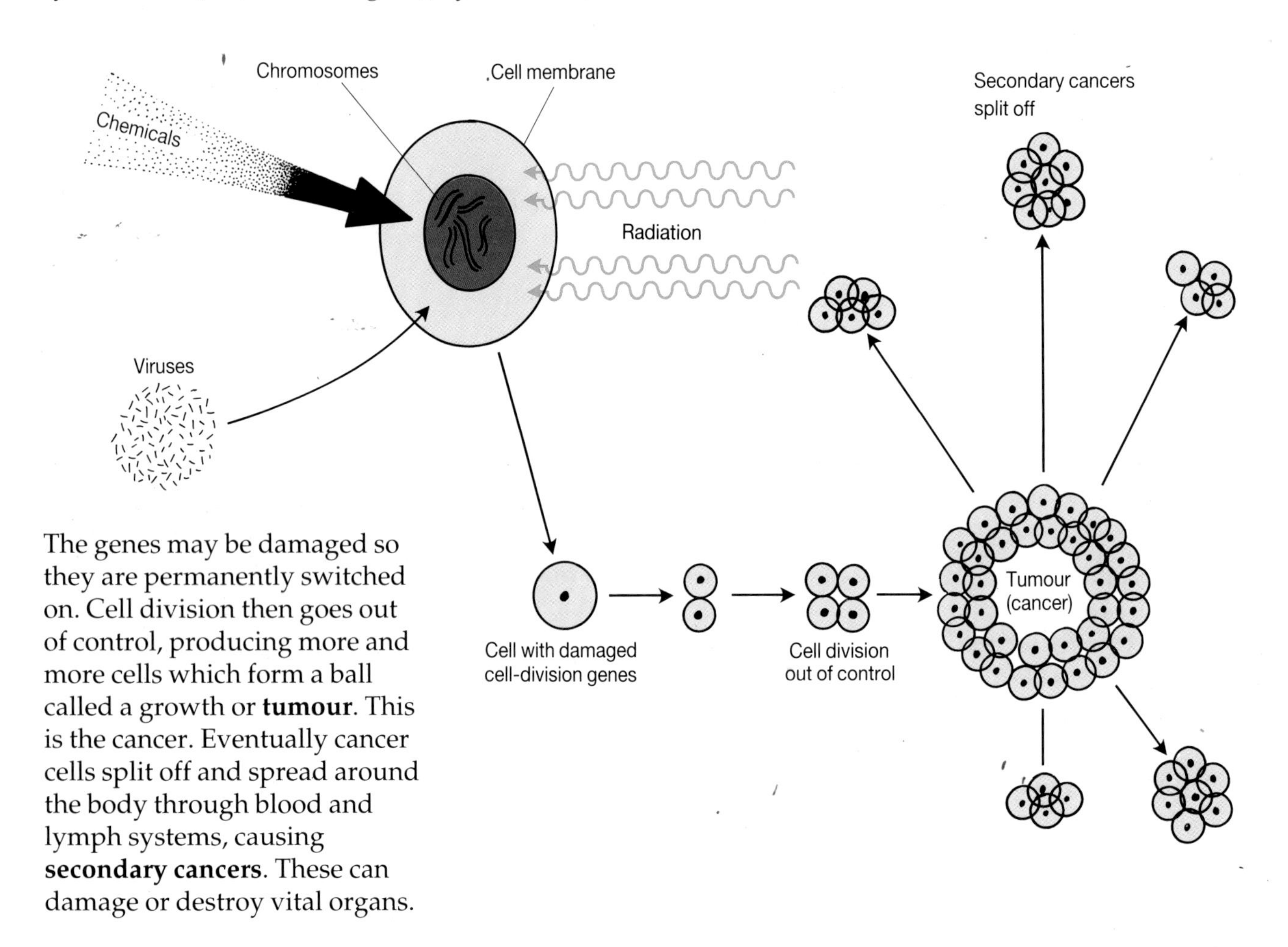

The genes may be damaged so they are permanently switched on. Cell division then goes out of control, producing more and more cells which form a ball called a growth or **tumour**. This is the cancer. Eventually cancer cells split off and spread around the body through blood and lymph systems, causing **secondary cancers**. These can damage or destroy vital organs.

What are the main causes of cancer?

Many people think that radiation and chemical pollutants in the environment are the main causes of cancer. In fact 8 out of 10 cancers could be prevented if people did not smoke (Unit 10.5) and avoided certain foods. Here are some facts.

Fats Fats can cause cancers in two ways. First, the more fat you eat, the more bile acids your liver produces. These may irritate the lining of the bowel and stimulate cancerous growth. Second, fat has an effect on the way breasts and other organs respond to hormones, which may set off cancerous growth. Avoid, or cut down on, fatty foods (Unit 6.3).

Alcohol There is a clear link between excessive drinking of alcohol, and cancer of the mouth, throat, larynx, and gullet. If drinkers also smoke, the risk of all these cancers is greatly increased.

Radiation Some rocks are naturally radioactive, such as the granites found in Cornwall and Scotland. Some release radioactive radon gas which can become trapped in houses. Radiation also comes from medical X-rays. All radiation, no matter how small, can set off cancerous growths. The higher the level of radiation, the greater the risk.

Sexual activity Semen produced by the male during sexual intercourse can cause cancer by damaging cells in the woman's cervix (Unit 8.1). It is essential that all women who have sexual relationships attend a clinic for cervical examination (screening) at least every three years.

Dietary fibre There are fewer cancers of the bowel in areas of the world such as Africa and Asia, where fibre is an important part of the population's diet. A possible reason for this reduced rate of cancers is that fibre absorbs poisonous wastes in food which may cause cancer.

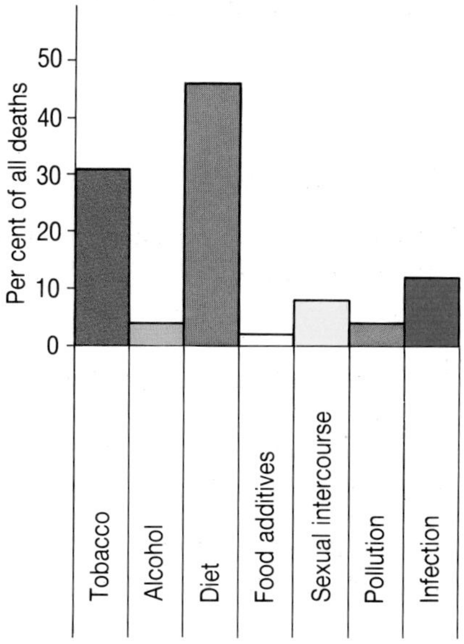

The main causes of cancer.

Study the warning signs

Each year 60 000 cancer sufferers receive treatment allowing them to return to a normal active life. Most cancers can be treated if found early enough, so everyone should look for any of the following warning signs:

- Bad, persistent cough or hoarseness.
- Unusual bleeding from the anus, or when passing water.
- A lump, especially in the breast, neck, armpit, or testicles.
- Any unexplained weight loss.
- A change in a wart or mole.

	Lung cancer	Breast cancer	Other cancer	All cancers
1921–5	0.05	0.26	2.21	2.54
1941–5	0.32	0.26	1.80	2.38
1961–5	0.77	0.27	1.48	2.52
1981	0.68	0.32	1.38	2.38

Deaths per thousand middle-aged people (35–60 yrs) in England and Wales

Questions

1 Study the table above to answer these questions.

a) Which cancer deaths are increasing and which are decreasing?

b) Suggest reasons for these differences.

2 From the graph above list the main causes of cancer in order of importance.

3 How can you reduce the risk of developing cancer?

10.5 Smoking and ill-health

Smoking kills. In Britain nearly 100 000 people die each year from diseases caused by smoking. This is twelve times the number killed in road accidents.

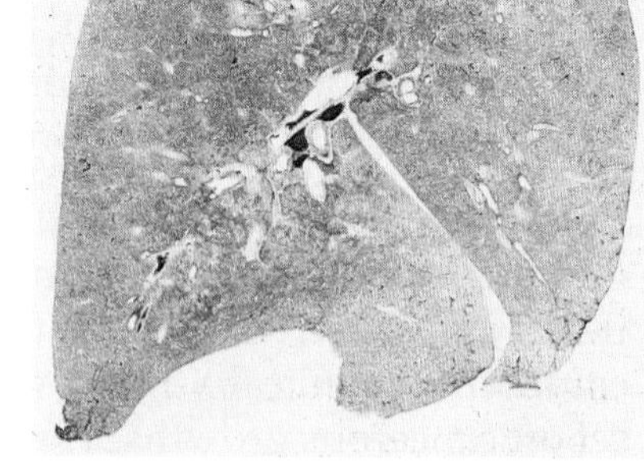

A section through a healthy human lung.

Tobacco smoke is poison

Tobacco smoke contains about 1000 chemicals. Many of them are harmful. Here are some of the poisonous ones.

Nicotine is a poisonous drug. It damages the heart, blood vessels, and nerves. Smokers get addicted to it, which is why they find it hard to give up smoking.

Tar forms in the lungs when tobacco smoke cools. The tar contains 17 chemicals that are known to cause cancer in animals.

Carbon monoxide is a poisonous gas which stops blood carrying oxygen round the body.

Other poisonous gases in tobacco smoke are **hydrogen cyanide, ammonia**, and **butane**. These irritate the lungs and air passages, making smokers cough.

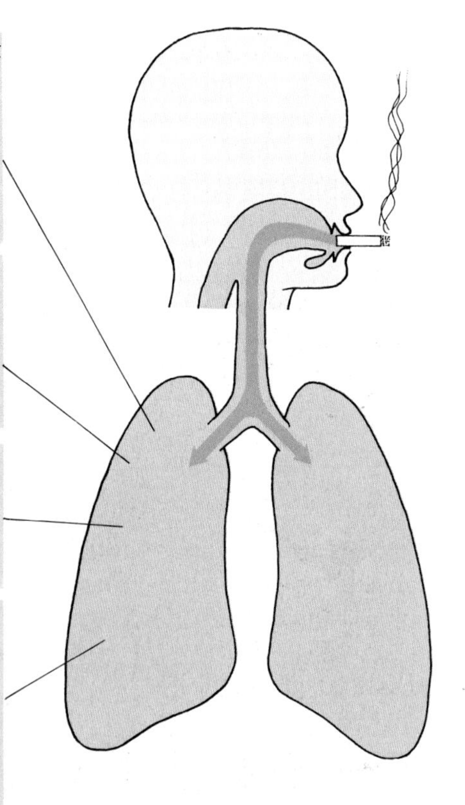

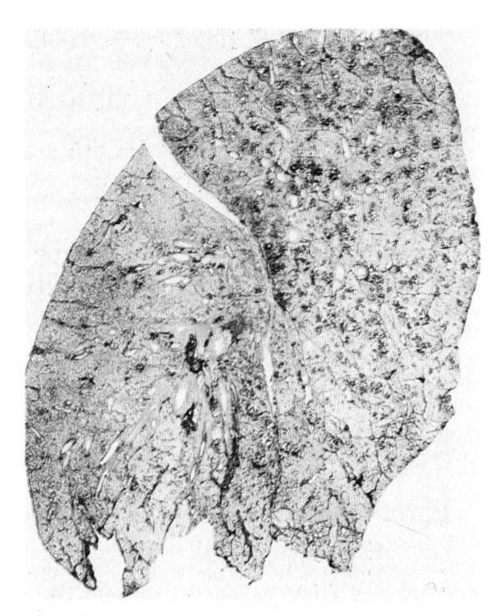

A section through the lung of a smoker. The spots are deposits of tar. They prevent the lungs from doing their job properly.

Diseases caused by smoking

Heart disease is three times more common among smokers than among non-smokers.

Emphysema is a disease in which lung tissue is destroyed by the chemicals in tobacco smoke. The lungs develop large holes which blow up like balloons. Breathing becomes very difficult.

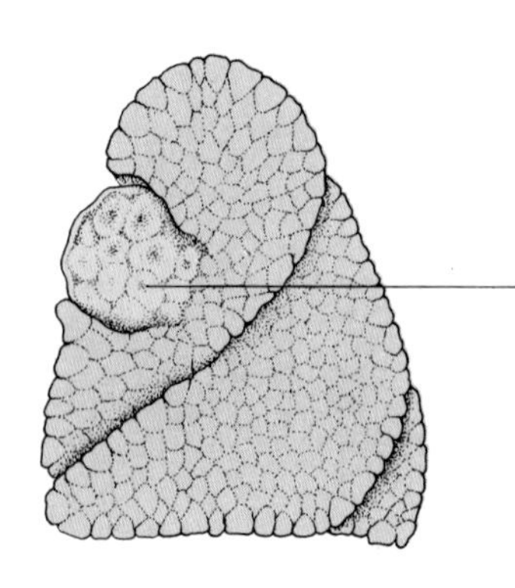

The swelling on this lung is due to emphysema, caused by heavy smoking.

Cancers of the lungs, mouth, gullet, and bladder are caused by chemicals called **carcinogens** in tobacco smoke.

Nine out of ten people who die of lung cancer are smokers.

Bronchitis is mostly a smoker's disease. Smoke irritates the air passages to the lungs, making them swollen and sore. Tiny hairs called cilia usually keep these passages clear. But the cilia stop working, so the lungs fill with mucus.

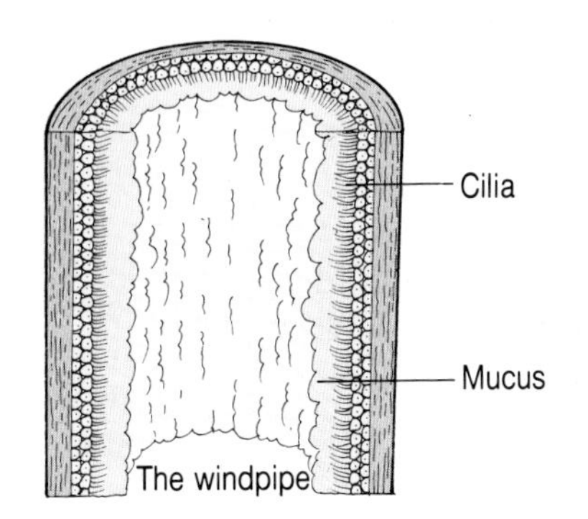

The cilia which keep your lungs and air passages clean stop working when you smoke.

Smokers get up other people's noses

Many non-smokers find tobacco smoke very unpleasant. It makes eyes sting, and can cause sore throats and headaches. It can irritate babies and people with hayfever and asthma.

Non-smokers who work or live with smokers can suffer lung damage and other smokers' diseases.

Pregnant mothers who smoke have smaller babies than non-smoking mothers. They also run the risk of a more difficult birth.

Smokers smell. Their breath, hair, clothes, and homes have a horrible smell caused by the ammonia, tar, and other chemicals in smoke.

Questions

1 Work out how much it costs to buy 20 cigarettes a day for a year. List the things you could buy if you saved this amount by giving up smoking.

2 Which chemical in smoke is addictive?

3 What harm does carbon monoxide do?

4 What is emphysema?

5 List three other diseases linked with smoking.

6 Why is it bad for pregnant mothers to smoke?

10.6 Health warning

Most people don't drink too much, or sniff glue, or take drugs like heroin and cocaine. But everyone should know the harm these things do.

Drugs and their effects

Heroin is made from opium poppies. It gives a feeling of power and contentment. But when this wears off, users become anxious and depressed. So they take more heroin to feel better. Soon they feel they cannot live without it. They are now heroin addicts.

Cocaine is made from the leaves of the coca bush. At first it gives a feeling of energy and strength. But after about 20 minutes users feel confused and anxious. They are unable to sleep for some time.

LSD is made from a fungus. Users may see things around them changing in colour, shape, and size. Or they may see things which are not there (hallucinations). Sometimes the hallucinations are very unpleasant, and are called **bad trips**. They can cause fear, depression, and mental illness.

Cannabis is made from the leaves and resin of the Indian hemp plant. Users seem to see, hear, feel and think more clearly. But they can become confused and do reckless and dangerous things.

Solvent and glue sniffers breathe in fumes from things like glue, paint, and cleaning fluid. They get a floating feeling and hallucinations. But the fumes get into the blood and damage the heart. They also cause sickness, depression, and bad facial acne.

Infections. Some drugs are taken by injection. Dirty needles can cause sores, abscesses, blood poisoning, and jaundice. AIDS can be passed from one person to another if needles are shared.

Addiction. Drugs can cause addiction. People feel they cannot live without the drug. They may turn to crime to get money for it. Once you are addicted to a drug, giving up becomes a difficult and painful experience. People begin taking drugs to feel good, they end up having to take drugs to stop feeling very bad. Drug addicts cause themselves and their families great unhappiness and stress.

Addicts can get help from Family Doctors who sometimes recommend hospital treatment. Support groups exist to help addicts and their families cope, and they may suggest ways of curing addiction.

The risks of taking drugs

Overdoses. A drug overdose can cause unconsciousness, damage to the heart and other organs, and even death. It is easy to take an overdose by accident, because it is hard to tell how strong a drug is, or how much to take. Drugs are far more dangerous when they are mixed. Their effects are multiplied, so a much lower dose can kill.

Mental harm. A drug user may suffer strong feelings of anger and fear, or terrifying hallucinations. This can cause mental damage.

Changes in behaviour. Drug users can become irritable, and lose interest in their friends, hobbies, and work. They may stop looking after themselves.

Accidents. People taking drugs run the risk of accidents, because they get confused. They can fall into water and drown, or walk in front of traffic, or fall out of windows. Glue sniffers can suffocate while their mouths are covered by plastic bags, or choke to death on their own vomit.

Alcohol

All alcoholic drinks contain a chemical called ethanol. This is what makes people drunk.

1 Ethanol is carried to the brain by the blood. From there it affects all the nervous system. A little ethanol makes people feel pleasant and relaxed. But if they keep drinking they feel dizzy and cannot walk straight. Speech becomes slurred and they begin to see double. They may become quarrelsome. Eventually they cannot stand up. If they drink more they become unconscious, and may even die.

2 Even a little alcohol causes people to make mistakes in things like typing and driving. It is very dangerous to drink and drive.

3 The effects of ethanol can last for hours. It takes about an hour for the body to get rid of the ethanol in just half a pint of beer.

4 Heavy drinking over a long period causes the brain to shrink. It can also cause stomach ulcers, cancers of the digestive system, and liver and heart disease.

5 Some people become addicted to alcohol. They turn into alcoholics.

6 If a woman is pregnant, ethanol in her blood is carried to the baby. It can damage the baby's brain and heart, and slow its growth.

Questions

1. Some drugs, and alcohol, are addictive. What does this mean?
2. Write down ten things you could say to convince a friend not to take drugs.
3. List the changes which take place in the body of a heavy drinker.
4. What changes in a person's behaviour can tell you that he or she may be taking drugs?

Questions on Section 10

1 *Understanding/interpret data*
This diagram shows how bacteria are affected by temperature.

a) Between which temperatures do bacteria multiply the fastest?
b) The temperature inside the main part of a refrigerator is about 4°C. Why is it unwise to keep foods, like meat, in this part of a refrigerator for more than three days?
c) The temperature in a domestic freezer is about −20°C. Why is it unwise to keep food in a freezer for more than three months?
d) Bacteria reproduce by dividing in two. If they have food and warmth they do this about every 30 minutes. If just one 'food-poison' bacterium landed on a piece of meat, and reproduced at this rate, how many would there be after 24 hours?
e) It is dangerous to cook a large piece of frozen meat before it has thawed out properly. Explain why.

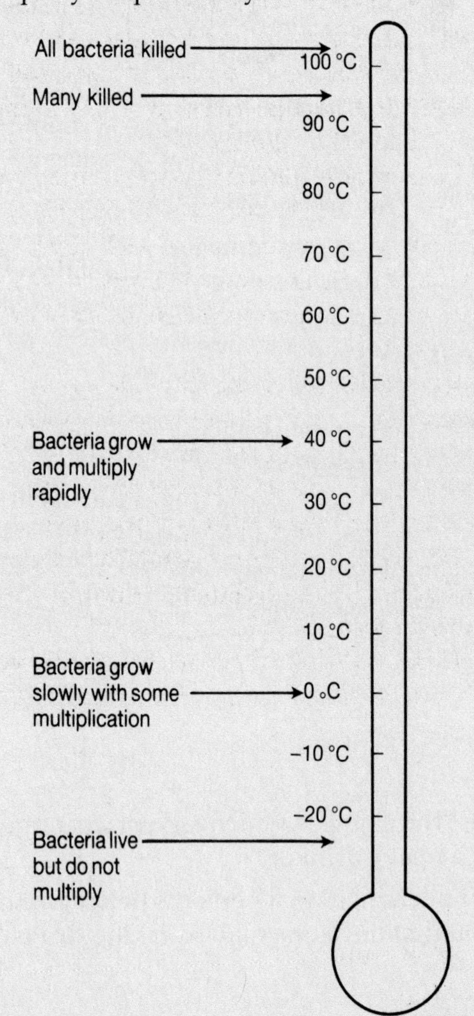

2 *Recall knowledge/understanding*
Why should you:

a) never use someone else's comb or towel?
b) always wash your hands after visiting the lavatory?
c) always wash your hands before handling food for other people?
d) never drink water from a pond or stream?
e) never bathe in the sea near a sewage pipe?

3 *Understanding*
Here is some advice for people who want to give up smoking.

a) Give up smoking at the same time as a friend does.
b) Every day, put the money you would have spent on cigarettes in a money box.
c) Eat fruit or raw carrots, or chew gum whenever you feel the urge for a cigarette.
d) Tell everyone you know that you are giving up smoking.

Take each piece of advice in turn, and explain why it is worth trying.

4 *Interpret data*
What do people know about alcohol? Try this 'true or false' test on as many people as possible, then give them the correct answers. Draw histograms to show the number of correct and incorrect answers.

a) Drinking beer is less harmful than drinking wine or spirits. (False – all alcoholic drinks can be as harmful as each other.)
b) If a man and a woman drink the same amount of alcohol, the woman will be more affected than the man. (True – on average it takes less alcohol to damage a woman's health than a man's.)
c) Alcohol is a stimulant. (False – alcohol dulls the brain.)
d) Alcohol warms you up. (False – alcohol enlarges blood vessels in your skin so you may feel warmer when in fact your body is losing heat more quickly.)
e) Three pints of beer can put someone over the legal blood alcohol limit for driving. (True – an average-sized adult reaches the legal blood alcohol limit for driving after drinking about two-and-a-half pints of beer.)

5 *Interpret data/recall knowledge*

The graph below shows the number of men aged 40–79 who die from lung cancer.

a) How many times is a cigarette smoker more likely to die of lung cancer than a non-smoker?

b) Suggest hypotheses to explain why cigarette smoking is more dangerous than cigar smoking.

c) Why do some non-smokers die of lung cancer?

d) Name other diseases caused by smoking.

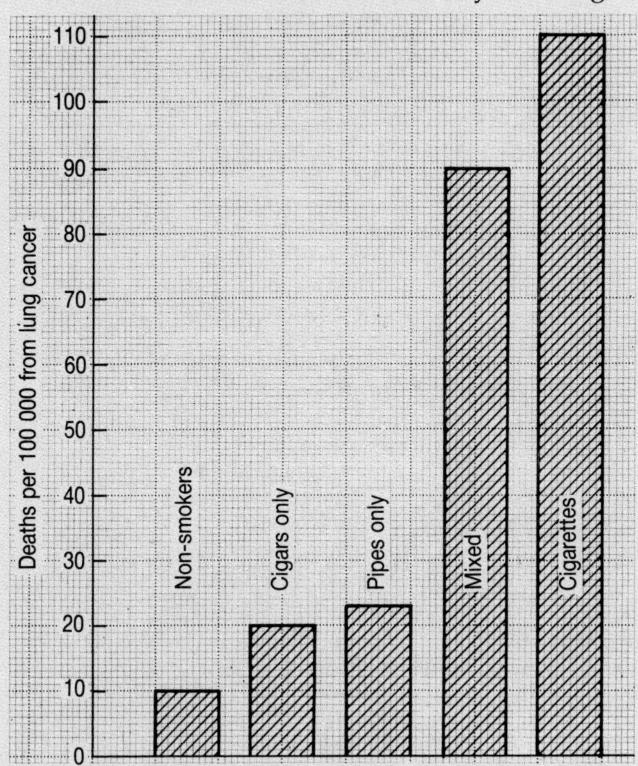

6 *Design/interpret data*

Write a report on drug abuse in your area, or in the UK as a whole. You could get information for your report in the following ways:

a) Newspaper and magazine cuttings. Over a period of several months cut out any articles about drug abuse.

b) Interview people. Ask if you can interview the local police, a doctor, hospital staff, and District Health Authority staff, to obtain information about local drug problems. Some of these people may agree to visit your school and talk to your class.

c) Visit a local hospital (arrange it in advance) to ask what percentage of casualty patients became injured under the influence of drink or drugs.

7 *Understanding*

Read the paragraph below, then answer the questions.

'AIDS is caused by a virus which can only live in blood, semen, and vaginal fluid. People can only catch the AIDS virus through sexual intercourse with an infected person, or by getting infected blood into their bloodstream.'

a) What is semen, and what is the vagina?

b) Why is it impossible to get AIDS by touching someone, or from cups, toilet seats, cutlery, clothes, door knobs or swimming pools?

c) Why does having many sexual partners increase the risk of catching the AIDS virus?

d) Why do drug addicts who inject themselves and share needles, increase their chances of catching the AIDS virus?

e) Why does the use of a condom reduce the risk of catching the AIDS virus?

8 *Design/interpret data*

Investigate the smoking habit in your school. Circulate a questionnaire and use students' answers to make graphs and charts to illustrate your results. Make it clear that nobody will see anyone else's answers. Here are a few of the things you could try to find out.

a) How many smokers, non-smokers, and ex-smokers are there in your school? Calculate the percentage of each type.

b) What is the age of each smoker? Draw a histogram to show 'numbers of smokers' against 'age'.

c) Why do smokers smoke? Design a questionnaire with four or five reasons on it and ask students to choose the one that's true for them. Write a report on your findings. Do your answers vary with age and sex?

d) How many smokers would like to give up the habit? What percentage is this of the total number of smokers?

e) How many smokers are *trying* to give up? What percentage is this of the total number of smokers?

Index

If more than one page number is given you should look up the **bold** ones first.

Acknowledgements

The publisher would like to thank the following for their kind permission to reproduce the following photographs:

p 6 Oxford Scientific Films /S Dalton (left), /G Bernard (bottom), Ardea /P Morris (centre), **p 7** Oxford Scientific Films /J Cooke (bottom left), /D Thompson (centre), Colorsport (left), **p 10** Oxford Scientific Films /J Cooke (all), **p 12** Oxford Scientific Films /E Woods (top), /P Dobson (centre), **p 13** OSF /S Osolinski (left), Bruce Coleman (centre), /H Reinhard (right), **p 16** OSF /P Parks (left, right & bottom centre), /G Bernard (centre right), Science Photo Library /M Abbey (centre), **p 17** OSF /J Cooke (centre top), /B Watts (top right), /R Packwood (bottom centre), **p 18** OSF /J Cooke (left & bottom centre), /G Bernard (centre), /P Parks (right), /G Bernard (bottom left), SPL /CNRI (bottom right), **p 19** OSF /D Thompson (top left), /J Cooke (centre top and right), /G Bernard (centre & right), **p 20** Heather Angel (centre), OSF /S Dalton (top right), /P Parks (right), /N Woods (bottom right), **p 21** OSF /G Bernard (left & centre right), /J Cooke (bottom left), **p 22** OSF /S Dalton (left), /F Amsler (bottom left), /P Parks (centre), /G Thompson (bottom right), **p 23** OSF /A Ramage, /G Bernard (top right, centre & bottom left), /M Fogden (bottom right), **p 24** OSF /D Allan (top right), /G Bernard (left), /B Walker (centre), /G Thompson (right), Bruce Coleman (bottom), **p 25** OSF /M Tibbles (centre), /R Packwood (bottom left), **p 27** OSF /G Thompson (bottom centre), Ardea /W Wagner (right), **p 30** SPL /E Grave (top), /Dr T Brain (left), /CNRI (centre), / M Kage (bottom right), **p 31** SPL /Dr J Burgess, **p 36** SPL /CNRI (top), OSF /G Bernard (centre), S & R Greenhill (bottom right), **p 43** G Cox, **p 44** Poly (top & right), Holt Studios (centre & bottom right), **p 45** Biophoto Associates (top left), SPL /Royal Free Hospital (top right), OSF /G Bernard (left & centre), SPL /R Hutchings (bottom right), **p 49** Bruce Coleman /E Crichton (top), /H Reinhard (centre & bottom), **p 50** OSF /R Toms (right), /S Tupper (centre), **p 55** OSF /P Parks (all), **p 56** SPL /Sinclair Stammers, **p 61** OSF /G Bernard, **p 64** OSF /C Hvidt, **p 65** OSF /H Taylor (right), **p 71** OSF /M Silver (left), **p 72** OSF /Dr Burgess (left), Ardea /I Beames (right), **p 74** Holt Studios (centre), OSF /G Bernard (bottom centre), **p 75** OSF /G Bernard (left), /R Packwood (right), Biophoto Associates (bottom centre), **p 82** OSF /P Parks (top), **p 83** OSF /J Cooke (left), /G Bernard (right), **p 85** OSF /G Bernard (both), **p 86** Colorsport, **p 88** J Hopkins, **p 90** SPL /B Longcore (top), **p 91** Colorsport, **p 97** Science Photo Library, **p 99** OSF /P Parks, **p 100** SPL /L Mulvehill, **p 102** SPL /CNRI (centre), **p 103** SPL /Dr Brody (top), **p 104** SPL /J Stevenson (right), **p 106** Oxford Scientific Films (top), Tropix (bottom), **p 108** SPL /CNRI (top), **p 109** OSF /C Kroeger/Animals Animals, **p 110** SPL /H Morgan (bottom), **p 114** S & R Greenhill (left), Supersport /Eileen Langley (centre), Biophoto Associates (right), **p 117** SPL /P Menzel, **p 123** Science Photo Library, **p 132** OSF /P O'Toole (left), **p 137** OSF /G Bernard (centre), /Z Lesczyski/Animals Animals (right), **p 141** Rex Features, **p 143** A Lawson (both), **p 144** SPL /D Lovegrove, **p 145** SPL /M Kage, /Dr C Chumbley, **p 147** SPL /M Clarke (top), Allsport /B Martin, **p 150** OSF /P Parks (left), /S Morris (left), /G Bernard (bottom right), SPL /E Grave (right), **p 151** OSF /M Silver, **p 152** S & R Greenhill (both), **p 154** SPL /D Scharf, **p 156** N Bromhall, **p 157** SPL /Petit Format/Nestle (all), **p 158** Biophoto Associates, **p 160** Rex Features, **p 162** S & R Greenhill (top), Tropix (bottom), **p 163** Bruce Coleman /G Cubitt (top), Oxford Scientific Films (right), **p 164** International Coffee Organization, **p 165** OSF /R Cousins (top), Bruce Coleman (right), /M Boulton (centre right), OSF /T Davis (bottom right), **p 166** OSF /R Packwood, **p 167** OSF /G Wren (top), /N Rosing (centre), **p 171** OSF /D Cayless, **p 173** Telegraph Colour Library, **p 177** OSF /D Cox, **p 178** S & R Greenhill (top), Ardea /L & T Bomford, /A Lindar (bottom), **p 180** H Angel (centre), Aspect /D Bayes (right), Network /M Goldwater (bottom right), **p 187** Picturepoint, **p 194** OSF /J Cooke (left), /G Maclean (centre), /P Goddard (top right), /D Cayless (right), **p 195** Heather Angel (left), Fauna & Flora Preservation Society (right), **p 202** OSF /J Cooke (top), **p 203** OSF /J McCammon (top), /C Grzimek (centre), /M Chillmaid (bottom right), **p 206** SPL /A Pasieka (top), **p 207** SPL /Dr Settles (bottom), **p 210** Tropix (top), **p 211** SPL /CNRI, **p 215** SPL /J Stevenson, **p 216** Aspect (right), Reflex Pictures /P Gordon (left), **p 217** Aspect (right), Network /M Abrahams (left).

Additional photography by Brian Beckett, Peter Gould, and Chris Honeywell.

The illustrations are by:

Brian Beckett, Jeff Edwards, David Holmes, Peter Joyce, Frank Kennard, Ed McLachlan, and RDH Artists.

Cover illustration by Danny Jenkins.